Microneuroanatomy and Surgery

Feres Chaddad-Neto
Marcos Devanir Silva da Costa

Microneuroanatomy and Surgery

A Practical Anatomical Guide

Springer

Feres Chaddad-Neto
Neurosurgery Division, Department of Neurology and Neurosurgery
Universidade Federal de São Paulo
São Paulo, SP, Brazil

Laboratory of Microneurosurgery Anatomy
Universidade Federal de São Paulo
São Paulo, Brazil

Department of Neurology and Neurosurgery
Hospital Beneficência Portuguesa de São Paulo
São Paulo, Brazil

Marcos Devanir Silva da Costa
Neurosurgery Division, Department of Neurology and Neurosurgery
Universidade Federal de São Paulo
São Paulo, SP, Brazil

Pediatric Neurosurgeon at Instituto de Oncologia Pediátrica - GRAACC - UNIFESP
São Paulo, Brazil

Laboratory of Microneurosurgery Anatomy
Universidade Federal de São Paulo
São Paulo, Brazil

ISBN 978-3-030-82746-5 ISBN 978-3-030-82747-2 (eBook)
https://doi.org/10.1007/978-3-030-82747-2

This Springer imprint is published by the registered company Springer Nature Switzerland AG
The registered company address is: Gewerbestrasse 11, 6330 Cham, Switzerland

For my wife, Patricia, the love and meaning of my life, who has always encouraged me to fly towards my dreams.
For my daughter, Marina, who has been a gift from the beginning and teaches me, every day, the meaning of happiness.
For my parents, Arlindo Chaddad (in memoriam) and Neide Gomes Chaddad, for raising me, with unconditional love, to believe that anything was possible. My sister Andrea who encouraged me during my journey.
For Prof. Emerson Cocco Lanaro (in memoriam) for inspiring and introducing me to the world of neuroanatomy.
For Dr Evandro de Oliveira (in memoriam), with whom I worked for the past 16 years and owe him all my neurosurgical training and microsurgical techniques.
For Dr. Guilherme Carvalhal Ribas, I truly appreciate his unconditional support throughout the course of my career.
For all the colleagues, students, and fellows, who have taught me the extraordinary path of neuroscience, enhancing my life purpose.
I hope this book will serve you well and bring light to your practice as much as other books have meant in my journey.

Feres Chaddad-Neto, MD, PhD

This book is dedicated to
My wife Vanessa my light, my love, my life.
My children Augusto, Julia, and Vitor – my reasons for living.
My mom Eponina and my father Devanir (in memoriam) my examples.
My beloved family which always supported me.
My Professors that generously transferred their precious knowledge, especially my mentors Prof. Sergio Cavalheiro and Prof. Feres Chaddad-Neto.
My readers, a new generation of neurosurgeons.

Marcos Devanir Silva da Costa, MD, PhD

Foreword

Over the 8 years that Dr. Feres Eduardo Chaddad-Neto has been the associate professor of neurosurgery at the Paulista School of Medicine, he has delivered magnificent work. His methodological rigor in anatomical dissections, bold surgical procedures, and organization are strong qualities.

This book updates the reader regarding the state-of-the-art combination of anatomical and surgical approaches, particularly for vascular diseases.

The chapters are well structured, and the author has an excellent command over English, which makes these chapters easy to read and remember. The illustrations and radiological images are of excellent quality and provide the reader with spatial understanding of several anatomical structures.

As a surgeon, I was impressed by the anatomical description of the different areas of the brain and technical details of surgical procedures.

The 18 chapters present a harmonious balance between anatomical knowledge and surgery, making the book an indispensable tool for practitioners of neurosurgery and neuroscience students. Benefits to the patient are unequivocal.

Only a genius, curious, meticulous, inventive, and tireless worker like him could make this project a success.

The author's eminence is apparent not only in the book but also in his conference talks worldwide and at the university, with his students at different undergraduate and graduate levels, and in hospital wards, operating rooms, and the neuroanatomy laboratory.

Our university is proud to present this book of international scope and to have Dr. Feres Chaddad-Neto as a professor.

We are deeply thankful to Dr. Feres Chaddad-Neto for the work presented here as well as for his pedagogical excellence in neuroanatomy and neurosurgery.

Sergio Cavalheiro
Neurosurgery Department, Federal University of São Paulo
São Paulo, SP, Brazil

Foreword

Anatomy is the foundation of surgery. Our neurosurgical procedures are performed on brains concealed within a forbidding cranium, but once breached, we enter an otherworldly splendor. Neuroanatomy has this alluring beauty that even decades of familiarity cannot diminish. I still marvel at its complexity and never tire of my workplace. When one adds an operating microscope to magnify and illuminate neuroanatomy, this realm becomes even more exquisite. Textbooks rarely do justice to microneuranatomy because there is nothing like the real thing. Textbooks of neuroanatomy are commonplace and quite useful for medical students studying the brain for the first time. But textbooks of microneurosurgical anatomy are rare, especially good ones that relate to the procedures neurosurgeons perform and make neurosurgeons perform better.

In this textbook, *Microneuroanatomy and Surgery*, Feres Chaddad-Neto has compiled a definitive review of the microneurosurgical anatomy related to the procedures neurosurgeons perform and needed to perform microsurgery better. The book focuses on microsurgical anatomy, not just anatomy for anatomy's sake. It begins at the gyri and sulci, then moves to the lobes of the cerebrum, the central core, ventricles, brainstem, and cerebellum. It also includes excellent descriptions of the cisterns and parasellar and pineal regions. The textbook is a beautiful blend of cadaveric dissections, radiographic images, case examples, and operative photographs. A thorough read of this material cannot help but make the reader an enlightened and more knowledgeable neurosurgeon.

I congratulate Dr. Chaddad-Neto on this important contribution to the neurosurgical literature. This textbook is destined to become an invaluable resource in every cranial neurosurgeon's library. It clearly reflects Dr. Chaddad-Neto's passion for microneuroanatomy, the disciplined application of his knowledge in the operating room, meticulous surgical technique, and a genuine dedication to teaching neurosurgeons. I continue to be impressed by his work, both in the operating room and on the page, and the leadership role he has assumed as a neurosurgeon founded in anatomy.

Michael T. Lawton
The Robert F. Spetzler Chairman of Neurosurgery
President and CEO
Barrow Neurological Institute
Phoenix, AZ, USA

Foreword

One of the most gratifying aspects of a long career in medicine is enjoying the success of one's colleagues, particularly those involved in charting the future of the discipline. Feres Chaddad-Neto has time and again proven the same, and so it wasn't a surprise when he asked me to write a foreword for his upcoming book on neuroanatomy. Certainly, this book will serve as a gap between the scissors and mind to reach a lesion safely.

Yoko Kato
Fujita Health University
Toyoake, Japan

Preface

Microneuroanatomy is the essential concept required to approach the brain. Most of the time, neuroanatomy knowledge is passed on as a difficult task to accomplish or as a problematic concept to reach. However, often the problem is related to those who convey this knowledge in classes and lectures or by writing books.

In reality, neuroanatomy is simple and needs to be understood as a tool to approach the different areas of the brain, and not as an obstacle because it is difficult; the only way to overcome this problem is to apply the anatomy and correlate it with different diseases (arteriovenous malformations, aneurysms, tumors, cavernomas, hydrocephalus, etc.)

This book will provide a novel approach to the relation between microneuroanatomy and brain diseases. Every single chapter is based on a specific neuroanatomical region, and correlates all the neuroanatomical key points with diseases that affect each neuroanatomical region. This anatomical correlation provides details and tips to perform a brain surgery regarding that anatomical region safely.

Feres Chaddad-Neto
São Paulo, São Paulo, Brazil

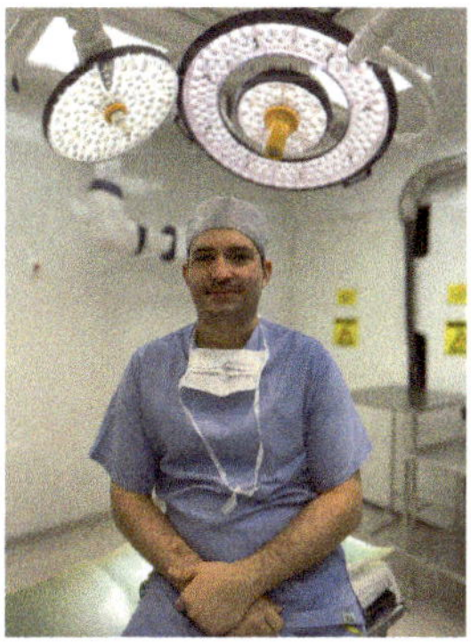

Marcos Devanir Silva da Costa
São Paulo, São Paulo, Brazil

Acknowledgments

This book would never be possible without the commendable work of our valuable collaborators who have helped by writing, dissecting specimens, producing and editing pictures/figures, and editing videos. We are grateful for their collaboration, enthusiasm, and commitment. Their names and institutions are listed below:

Beatriz Souza Renor, MD – Universidade Federal de São Paulo, São Paulo, Brazil

Bruno Augusto Lourenço Costa, MD – Centro Hospitalar Tondela-Viseu EPE, Portugal

Bruno Loof de Amorim, MD – Universidade Federal de São Paulo, São Paulo, Brazil

Daniela de Souza Coelho, Psy, MSc – Universidade Federal de São Paulo, São Paulo, Brazil

Gabriela Miroslava Bustamante Vargas, MD – Hospital Manuel Ascencio Villarroel, Cochabamba-Bolivia

Hugo Leonardo Doria- Netto, MD – Hospital Beneficencia Portuguesa de São Paulo, São Paulo, Brazil

José Maria de Campos Filho, MD – Hospital Beneficencia Portuguesa de São Paulo, São Paulo, Brazil

Joao Vitor Fernandes Lima, MD – Universidade Federal de São Paulo, São Paulo, Brazil

Juan Leonardo Serrato-Avila, MD – Universidade Federal de São Paulo, São Paulo, Brazil

Juan Alberto Paz Archila, MD – Universidade Federal de São Paulo, São Paulo, Brazil

Kaan Yağmurlu, MD – University of Virginia, Health System, Charlottesville, VA, USA

Pedro Góes, MD, MSc – Instituto Estadual do Cérebro Paulo Niemeyer, Rio de Janeiro-RJ, Brazil

Richard Gonzalo Párraga Choque, MD – Neurosurgery Institute of Bolivia (INEB), Cochabamba-Bolivia

Ricardo Chmelnitsky Wainberg, MD – Neurosurgery Division, Hospital São Lucas da PUC-RS - Pontifícia Universidade Católica do Rio Grande do Sul, Porto Alegre, Brazil

Sebastian Anibal Alejandro, MD – Universidade Federal de São Paulo, São Paulo, Brazil

Silvio Sarmento Lessa, MD – Universidade Federal de São Paulo, São Paulo, Brazil

Talita Helena Martins Sarti, MD – Universidade Federal de São Paulo, São Paulo, Brazil

Thomas More Frigeri, MD, PhD – Neurosurgery Division, Hospital São Lucas da PUC-RS – Pontifícia Universidade Católica do Rio Grande do Sul, Porto Alegre, Brazil

Contents

Chapter 1
Surgical Anatomy of the Sulci and Gyri of the Brain

1.1 Definitions

The sulcus is a depression that is delimited by the two neighboring gyri, where the gyri are the folds of the cortical brain. Historically, fissures were defined as a deeper sulcus; however, since 1955, the only recognized fissure is the interhemispheric fissure. The sulci and gyri compose the cortical surface of the brain.

The brain is composed of the telencephalon and diencephalon. The telencephalon is composed of two hemispheres connected by three commissures, which includes corpus callosum, anterior commissure, and hippocampal (fornix) commissure. The diencephalon is formed by the thalamus, hypothalamus, epithalamus, subthalamus, and metathalamus. From our surgical understanding, we considered the thalamus as the center of the encephalon because surrounding the encephalon we have the hemispheres, lateral and basal nuclei (which are also part of the hemispheres), and internal capsule, and to the inferior of the thalamus lies the brain stem.

The brain has three surfaces: superolateral, medial, and inferior; each of which has a group of sulci and gyri. The first important concept is to identify the sulci that are 100% constant in each of them because they will be the landmarks for the surface anatomy recognition.

All the time, as neurosurgeons, we face the challenge of transposing the radiological topographic diagnoses of a certain brain disease through a small opening on the patients' head called as craniotomy, so this chapter and the next chapters were designed for assisting the readers on this hard task.

1.2 Superolateral Surface

The sulci from the superolateral and basal surfaces point in the direction of the ventricles; for this reason, some pathology such as arteriovenous malformation presents a conical shape. There are several strategies for identifying/recognizing the sulci

F. Chaddad-Neto, M. D. Silva da Costa, *Microneuroanatomy and Surgery*,
https://doi.org/10.1007/978-3-030-82747-2_1

and gyri pattern in the lateral surface; however, there is one strategy that can be applied either to anatomical specimens, radiological images, or during the surgery. An end-to-side connection needs to be found between two sulci, one vertical, and another horizontal (Figs. 1.1 and 1.2). Therefore, there are only two options for the vertical sulcus or the sulcus that is connected by the side to be: precentral sulcus or postcentral sulcus (Fig. 1.3).

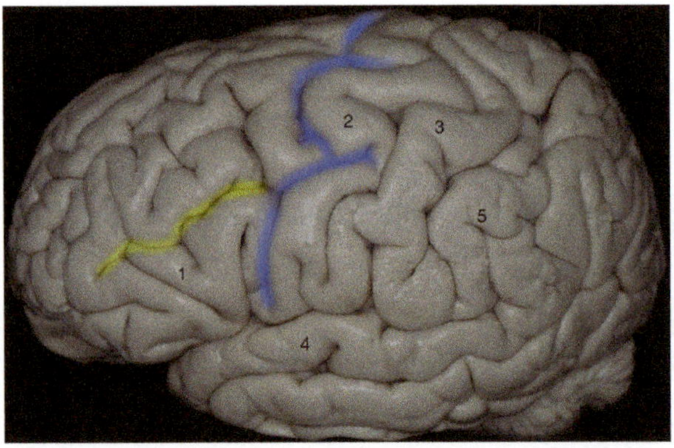

Fig. 1.1 Lateral and superior view showing the end-to-side connection between a vertical sulcus (yellow) and a horizontal sulcus (blue), which is the precentral sulcus. This identification and understanding of this relation allow us to clearly identify the numbered gyri: precentral gyrus (2), postcentral gyrus (3), supramarginal gyrus (5), superior temporal gyrus (4), and inferior frontal gyrus (1)

Fig. 1.2 A real brain magnetic resonance imaging scan in the Cube-FLAIR sequence and its sagittal view. The yellow line (inferior frontal sulcus) is the vertical sulcus that connects with the lateral side of the precentral sulcus (blue). Now it becomes easy for identifying the numbered gyri: precentral gyrus (2), postcentral gyrus (3), supramarginal gyrus (4), inferior frontal gyrus (1), and superior temporal gyrus (5)

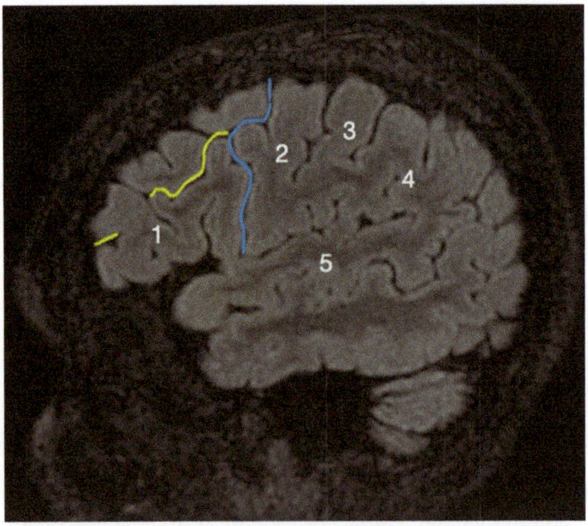

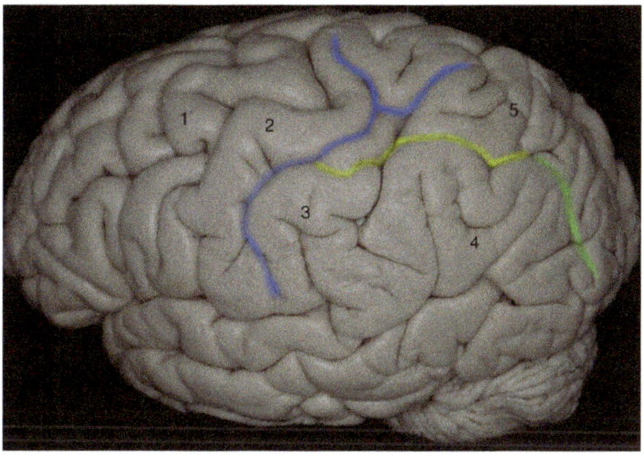

Fig. 1.3 The posterolateral view shows the horizontal sulcus, which is the intraparietal sulcus (yellow) that connects to the lateral side of the vertical sulcus (blue), which must be the postcentral sulcus because the vertical sulcus (yellow) runs from the posterior to the anterior. Here, it becomes simple to identify the numbered structures: precentral gyrus (1), postcentral gyrus (2), supramarginal gyrus (3), angular gyrus (4), and superior parietal lobule (5). The green sulcus is the intra-occipital that is continuous with the intraparietal sulcus (yellow)

1.3 So How to Differentiate the Precentral from the Postcentral Sulcus?

Both the precentral and postcentral sulcus can be identified by the recognition of the relation between one horizontal sulcus connecting to a vertical sulcus (end-to-side connection); the precentral sulcus is always anterior, and the postcentral sulcus is always located posteriorly (Fig.1.3).

It is possible to identify the precentral sulcus, precentral gyrus, central sulcus, postcentral gyrus, and the postcentral sulcus just by identifying one anatomical relation, that is, an end-to-side connection between a horizontal sulcus with a vertical sulcus. If this relation happens posteriorly, the vertical sulcus is the postcentral, and if this happens anteriorly, the vertical sulcus is the precentral.

The central sulcus has an oblique S shape format and rarely connects to a horizontal sulcus. It runs from the posterior to the anterior and from the medial to the lateral; its superior end can be identified 5 cm posteriorly to the bregma, and the inferior end can be identified 7 cm above the preauricular pit.

1.4 Frontal Lobe

The frontal lobe, in the superolateral surface, is formed by three sulci: two are horizontal that delineates three gyri and one vertical, which is the precentral sulcus (Fig. 1.4). The two horizontal sulci, inferior and superior frontal sulcus, delineate from superior-medial to inferior-lateral, superior, middle, and inferior frontal gyrus.

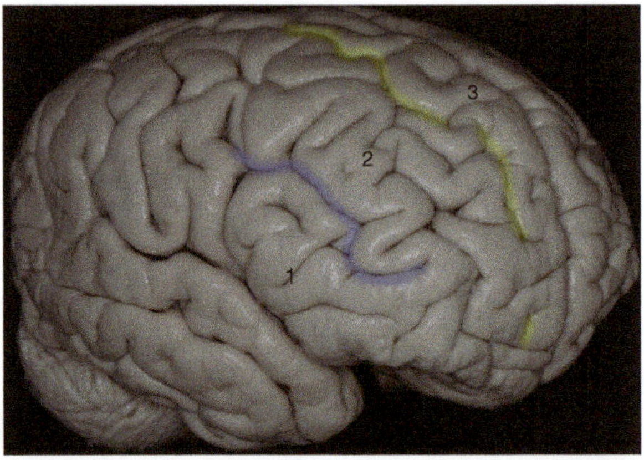

Fig. 1.4 The antero-lateral view of the lateral surface of the frontal lobe, two sulci, delineates three gyri: the inferior frontal gyrus (1), inferior frontal sulcus (blue), middle frontal gyrus (2), superior frontal sulcus (yellow), and superior frontal gyrus (3)

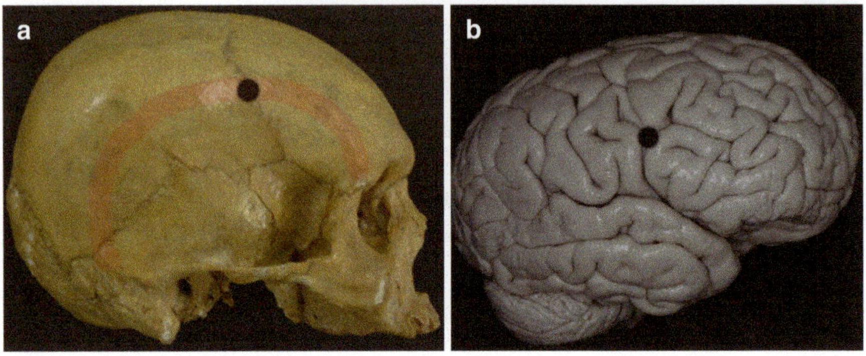

Fig. 1.5 The black dot represents the Stephanium cranial point in (**a**) and the end-to-side connection of the inferior frontal sulcus and precentral sulcus in (**b**). The red line (**a**) is the superior temporal line that meets the coronal suture in the black dot defining the Stephanium cranial point, and representing the end-to-side connection between the precentral sulcus and the inferior frontal sulcus

In 12–16% of the cases, there is an intermediate frontal sulcus, which runs inside the middle frontal gyrus. The central sulcus is the posterior limit of the frontal lobe, while the interhemispheric fissure is the superior-medial limit, and the lateral sulcus (Sylvian fissure) is the inferior-lateral limit.

1.5 Is There a Cranial Landmark for the Inferior Frontal Gyrus?

There is a very important cranial point called Stephanium, characterized by the union between the coronal suture and the superior temporal line; this point is in the same projection of the union between the precentral sulcus and the inferior frontal sulcus (Fig. 1.5). Therefore, below the superior temporal line, we have the projection of the inferior frontal gyrus. Below the inferior frontal gyrus, we have the most important landmark in the lateral surface, the Sylvian fissure.

The lateral sulcus (Sylvian fissure) is the key landmark of the lateral surface, which is divided into a superior-lateral (related to frontal and parietal lobes) and an inferior-lateral (related to temporal lobe) part by the lateral sulcus. It is possible to recognize three rami: anterior horizontal, anterior ascending, and posterior. At the inferior frontal gyrus, the anterior horizontal and anterior ascending ramus, in a V-shaped fashion, delineate the triangular part, with the orbital part located anteriorly and the opercular part posteriorly. The orbital part is located superiorly to the orbital roof and continues as the lateral orbital gyrus in the basal surface, while the opercular part with the triangular part covers the anterior part of the insula as a curtain (Figs. 1.6 and 1.7).

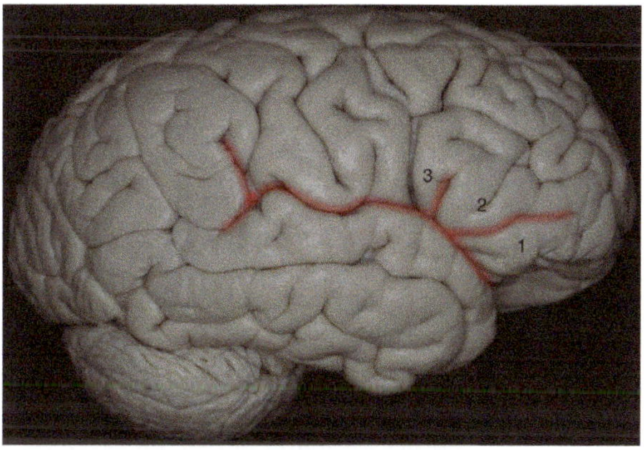

Fig. 1.6 This lateral view of the brain illustrates the Sylvian fissure or lateral sulcus of the brain (red). The sulcus has different rami that separate the inferior frontal gyrus into three parts: orbital part (1), triangular part (2), and opercular part (3)

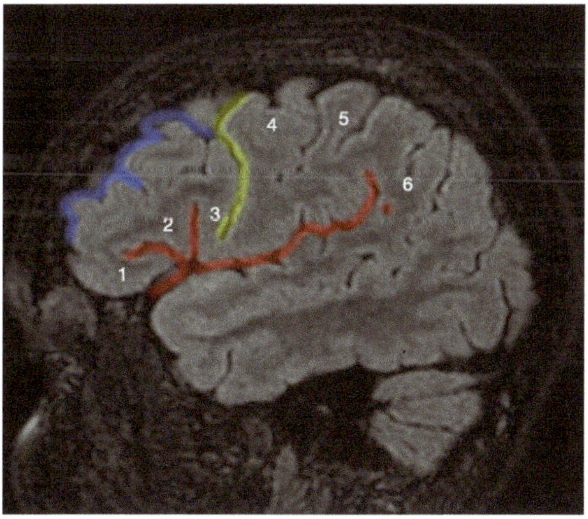

Fig. 1.7 The Cube-FLAIR sequence of a brain magnetic resonance imaging scans reveals some of the anatomic structures studied in the cadaver specimen. The Sylvian fissure (red) and its rami that delineate the orbital part (1), triangular part (2), and opercular part (3). In addition, the meeting of the end of the inferior frontal sulcus (blue) with the side of the precentral sulcus (yellow), precentral gyrus (4), postcentral gyrus (5), and marginal gyrus (6) surrounding the posterior end of the Sylvian fissure is being continuous to the superior temporal gyrus

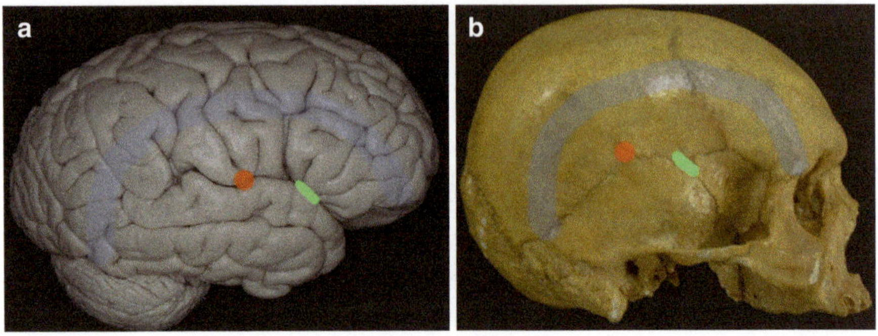

Fig. 1.8 The superior temporal line is represented in B by the blue line, which is also a projection of the inferior frontal sulcus in A, and the projection of the corpus callosum in the brain and cranial surface. The red dot represents the higher point of the squamous suture in B and the point where the central sulcus reaches the Sylvian fissure in A. The green dash represents the Asterion in B and the Anterior Sylvian Point in A

The triangular apex part points out toward the Sylvian fissure. The point of union of these three rami is called as anterior Sylvian point, a cisternal point. The triangular part of the inferior frontal gyrus points out to the anterior Sylvian point and deeply into the lumen of the insula, where the middle cerebral artery changes its axis from M1 to M2 because at this point the middle cerebral artery changes its axes from the medial toward the lateral direction to anterior toward posterior direction.

The lateral sulcus (Sylvian fissure) projects obliquely from the inferior to the superior and from the medial to the lateral, being parallel to the lesser sphenoid wing. This sulcus contains the middle cerebral artery and Sylvian veins.

The Sylvian fissure is represented in the cranial surface by the squamous suture; the highest point of the squamous suture represents the point where the central sulcus reaches the Sylvian fissure. The pterion, which is the region of union of the frontal-sphenoidal suture, sphenotemporal suture, sphenoparietal suture, squamous suture, and coronal suture, also is a reference for the anterior Sylvian point (Fig. 1.8).

1.6 Temporal Lobe

The temporal lobe is located below the lateral sulcus (Sylvian fissure), and it has two horizontal sulci, superior and inferior temporal sulci in the lateral surface, that delineate three horizontal gyri, namely, the superior, middle, and inferior temporal gyri (Fig. 1.9).

The superior temporal gyrus continues posteriorly as the supramarginal gyrus, also known as "temporoparietal operculum." This correlation is of great relevance during the surgery because even with arachnoid membranes and vessels, it is easy to identify the lateral sulcus of the brain. Therefore, the main surgical tip is to identify the Sylvian fissure and infer that inferiorly we can find the superior temporal gyrus that continues by folding posteriorly as the supramarginal gyrus, once the last is the posterior limit of the lateral sulcus (Sylvian fissure).

The superior temporal sulcus divides, as a horizontal "Y," into two rami where one is the posterior limit of the supramarginal gyrus and the other is the occipital ramus that is the inferior limit of the angular gyrus (Fig. 1.10).

The middle temporal gyrus continues posteriorly as the angular gyrus, and deeply it corresponds to the sagittal stratum and to the temporal horn of the lateral ventricle. The inferior temporal gyrus communicates the lateral and the basal surface of the brain and deeply corresponds to the inferior longitudinal fasciculus.

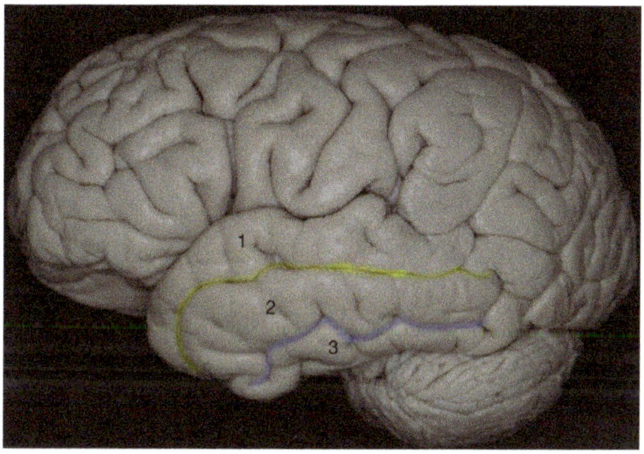

Fig. 1.9 Lateral view of the temporal lobe showing the two sulci separating the lateral surface into three gyri: superior temporal gyrus (1), superior temporal sulcus (yellow), middle temporal gyrus (2), inferior temporal sulcus (blue), and inferior temporal gyrus (3)

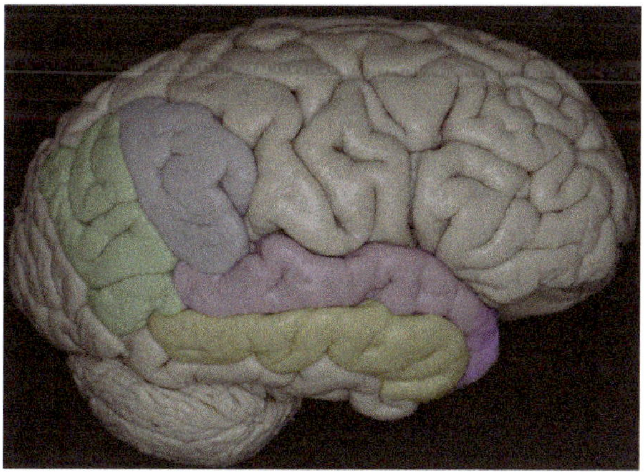

Fig. 1.10 Illustrates the relation between the two neighboring gyri; the superior temporal gyrus (purple) continues to be supramarginal gyrus (blue) by involving the end of the Sylvian fissure. In the same way, the middle temporal gyrus (yellow) continues as the angular gyrus (green)

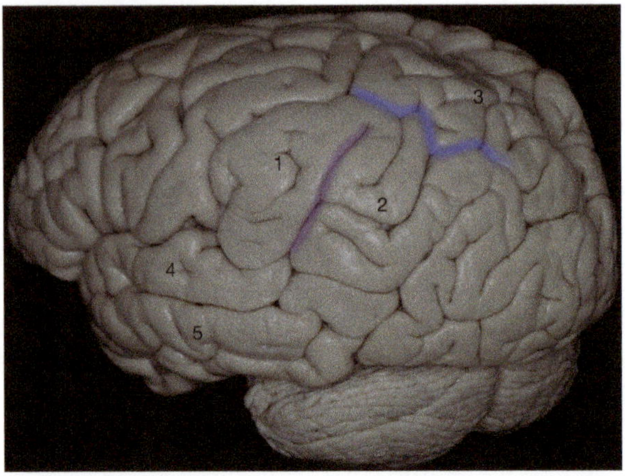

Fig. 1.11 This posterolateral view of the brain emphasizes the parietal lobe; the intraparietal sulcus (blue) is the main sulcus to the lateral surface of the parietal lobe; it divides the superior parietal lobule (3) from the inferior parietal lobule (1 and 2), which is separated by the intermediate sulcus (Jensen – purple) that can come from the superior temporal sulcus or from the intraparietal sulcus. The two gyri numbered 4 and 5 illustrates that the superior temporal gyrus (4) continues as supramarginal gyrus (1) and the middle temporal gyrus (5) continues as the angular gyrus (2). This relation is relevant for observing during the surgical approaches to parenchymal lesions

1.7 Parietal Lobe

The parietal lobe is composed of three principal areas, the post-central gyrus, superior, and inferior parietal lobule, which are separated from each other by the intraparietal sulcus. The inferior parietal lobe has the supramarginal and angular gyrus, which are separated from each other by the intraparietal sulcus (Jensen's sulcus). The connection between intraparietal sulcus with postcentral sulcus is also an example of an end-to-side connection where the horizontal sulcus corresponds to the postcentral and the vertical sulcus to the intraparietal sulcus (Fig. 1.11). The intraparietal sulcus in its depth has an external relation with the atrium of the lateral ventricle. There is a cranial point named intraparietal point that is located 5 cm anterior to the lambda and 4 cm lateral do the sagittal suture. The intraparietal point indicates the point in the intraparietal sulcus that can be used as a pathway for entering the atrium of the lateral ventricle.

1.8 Occipital Lobe

The occipital lobe has no evident separation from the temporal and parietal lobe; thus, it is important to emphasize that the boundaries of the lobe are arbitrary. The anterior limit at the lateral surface is an imaginary line between the pretemporo-occipital notch and the impression of the parieto-occipital sulcus. The superomedial

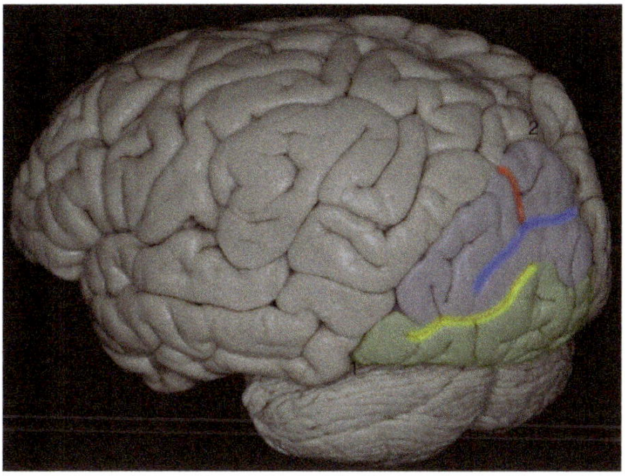

Fig. 1.12 This figure illustrates the posterior view of the lateral surface. The occipital lobe has many possible variations. Here, we show one type of presentation where the lateral occipital sulcus (yellow) separates the occipital lobe in two gyri, the superior occipital gyrus (purple) and inferior occipital gyrus (green). Inside the superior occipital gyrus, we can see the transverse occipital sulcus (blue) establishing a connection with the intra-occipital sulcus (red) that is the direct continuation of the intraparietal sulcus. Number 1 represents the temporo-occipital notch that is the impression of the vein of Labbè at the point where the vein drains to the transverse-sigmoid sinus junction. Number 2 represents the impression of the parieto-occipital sulcus in the lateral surface. By connecting these two numbers as an imaginary line, we delineate the anterior limit of the occipital lobe once there are no clear separations between the occipital lobe and both the temporal and parietal lobe

limit is the interhemispheric fissure, and inferior-lateral limit is the border that is parallel to the transverse sinus. It is important to note that the impression of the parieto-occipital sulcus in the lateral surface is marked in the cranial surface by the union of the lambdoid suture to the sagittal suture; this junction occurs 6–7 cm ahead of the inion, the external occipital protuberance. The intrinsic anatomy exhibits different patterns compared to the previously described lobes. To date, there are at least seven different classifications. Therefore, we will consider the two most prevalent forms, one that considers three gyri and two sulci—the superior, middle, and inferior occipital gyri separated by the interoccipital and lateral occipital sulci—and the other that considers two gyri separated by one sulcus, superior and inferior occipital gyri, and lateral occipital sulcus. However, in this case, the superior occipital gyrus can harbor the interoccipital sulci and transverse occipital gyri (Fig. 1.12).

1.9 Insular Lobe

The insular lobe has three characteristics: it is covered by the frontoparietal and temporal opercula; it lies in the depth of the hemispheric part of the lateral sulcus of the brain (Sylvian fissure). In addition, it is the lateral limit of the central core.

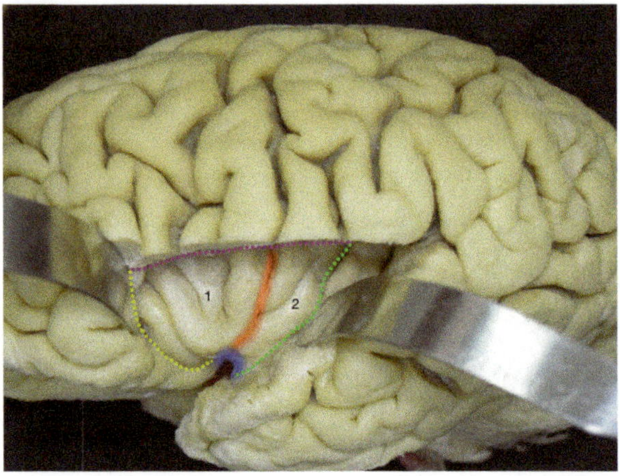

Fig. 1.13 The insular lobe is surrounded by the peri-insular sulcus that is composed of three main parts, the inferior limiting sulcus (green dotted line), superior limiting sulcus (purple dotted line), and anterior limiting sulcus (yellow dotted line). The inferior limiting sulcus (green dotted line) meets the anterior limiting sulcus (yellow dotted line) in the most inferior portion of the insula that is named as the lumen of the insula (blue). The lateral aspect of the insular lobe is separated into two parts by the central sulcus of the insula (red): anteriorly to this sulcus, we found the short gyri of the insula (1), and posteriorly we found the long gyri of the insula (2)

The sulci that encircle the insular lobe are superior, anterior, and inferior limiting sulci that resembles a pyramid with a triangular base. The insular lobe also has two surfaces: one anterior that harbors the anterior transverse gyri and one lateral that has the short and long gyri separated by the central sulcus of the insula, which is in the same projection of the central sulcus of the brain. It is possible to find three to five short insular gyri and one to two long insular gyri (Fig. 1.13).

The limen of the insula is the point where the middle cerebral artery changes its direction and becomes M2 segment (insular); therefore, from a surgical point of view, everything that is medial to the M2 segment is the insular lobe; the triangular part of the inferior frontal gyrus also points to the lumen of the insula that corresponds to the anterior Sylvian point in the lateral sulcus of the brain (Sylvian fissure), which is usually the start point for initiating the Sylvian fissure dissection. The uncinate fibers that connect the fronto-orbital region to the temporal pole are superficially represented by the lumen of the insula.

1.10 Medial Surface

The main landmark of the medial surface is the corpus callosum, which is a telencephalic commissure located on the medial surface of the brain composed of six parts: rostrum, knee, body, splenium, major forceps, and minor forceps (Fig. 1.14). The corpus callosum is encircled by a 100% constant sulcus, the sulcus of the callosum, which contains the pericallosal artery. The corpus callosum is easily

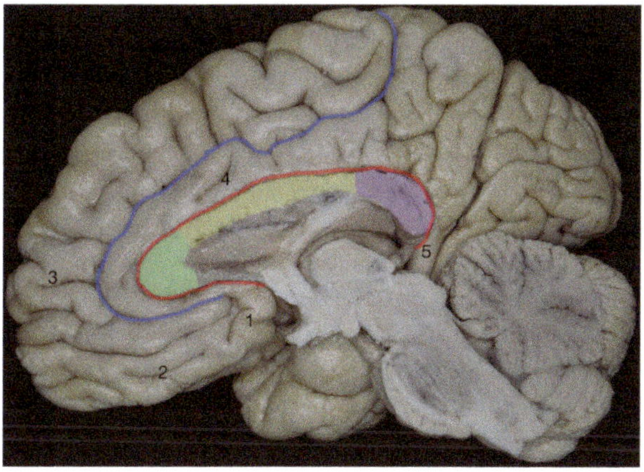

Fig. 1.14 This is a medial view of the brain. The key landmark of this surface is the corpus callosum, which can be separated into four segments at the sagittal view: rostrum (gray), genu (green), body (yellow), and splenium (purple). The corpus callosum is encircled with 100% constant sulcus that is the sulcus of the corpus callosum (red), which separates the corpus callosum from the cingulate gyrus (4) and which is delineated by the sulcus of the cingulate gyrus (blue). We can also identify the subcallosal area (1), rectus gyrus (2), medial frontal gyrus (3), and isthmus of the cingulate gyrus (5)

recognized during the interhemispheric approaches; it has a white surface that is the main tip for differentiating from the cingulate gyrus.

The cingulate gyrus, which is part of the limbic lobe, is delimited by the sulcus of the corpus callosum and cingulate sulcus, and it forms a "belt" around the corpus callosum. When passing below the splenius of the corpus callosum, it has a narrow part, which is called the isthmus of the cingulate gyrus. At this exact point, the isthmus of the cingulate gyrus is a cortical projection of the medial wall of the atrium of the lateral ventricle (Fig. 1.14). Therefore, the isthmus of the cingulate gyrus can be a cortical window for approaching the atrium of the lateral ventricle from the medial surface.

The identification of the corpus callosum sulcus and cingulate gyrus and sulcus is important to differentiate the other structures that belong to the lobes described in the medial surface, for example, the frontal lobe that has the rectus gyrus and the medial frontal gyrus in the medial surface. The precentral and postcentral gyrus are represented in the medial surface as paracentral lobule, which is delineated anteriorly by the paracentral sulcus, posteriorly by the marginal branch of the cingulate sulcus, inferiorly by the cingulate sulcus, and medially by the interhemispheric fissure. The parietal lobe is also represented in the medial surface by the quadrangular lobule or precuneus. The precuneus is also quadrangular, delineated anteriorly by the marginal branch of the cingulate sulcus, posteriorly by the parieto-occipital sulcus, and inferiorly by the subparietal sulcus (Fig. 1.14).

The occipital lobe has its best definition in the medial surface because it is separated from the quadrangular lobule by the parieto-occipital sulcus or fissure that is very deep with 100% constant sulcus. The occipital lobe in the medial surface is composed of two structures separated by the calcarine sulcus, cuneiform, and lingual gyrus; the cuneiform is superior and triangular (Fig. 1.15). The elongated lingual gyrus continues anteriorly as parahippocampal gyrus in the temporal lobe.

The lower edge of the cerebral Foix corresponds to the cingulate gyrus. The cingulate sulcus contains the callosomarginal artery. The cingulate sulcus has a superior prolongation that has denominated marginal ramus of the cingulate sulcus; this sulcus promotes an impression in the postcentral gyrus that is recognized as "pars marginalis" of the cingulate sulcus, which can be easily identified in the brain computed tomography scans and magnetic resonance imaging scans.

The precuneus or quadrangular lobe is the part of the superior parietal lobe between the marginal branch of the cingulate sulcus and the parieto-occipital sulcus. The transition between the parietal and occipital lobes is made by the parieto-occipital sulcus (Fig. 1.15). The cuneiform lobe or cuneus is the portion of the occipital lobe on the medial face between the parieto-occipital and calcarine sulcus.

There is functional importance in recognizing the calcarine sulcus because the neighboring gyri are the primary visual area (Fig. 1.16). However, the calcarine sulcus gives another essential information when performing the occipital transtentorial approach to the pineal region; the best way of finding the cistern of the quadrigeminal lamina is to follow the sulcus anteriorly. It also contains the posterior

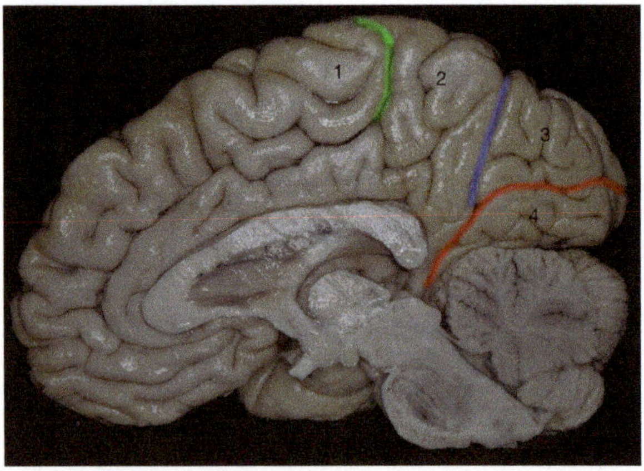

Fig. 1.15 The medial surface of the brain. The green sulcus is the marginal branch of the cingulate sulcus, which causes an impression of the postcentral gyrus named "pars marginalis" and separates the paracentral lobule from the quadrangular lobule (2) that is the representation of the parietal lobe in the medial surface; parieto-occipital sulcus (blue) is one of the major landmarks of the medial surface; it is 100% constant sulcus. Particularly, a deep sulcus is sometimes called as parieto-occipital fissure; its superior end causes an impression in the medial surface, which helps to delineate the parietal lobe from the occipital lobe. The cuneiform lobe (3) represents the occipital lobe together with the lingual gyrus (4); the cuneiform lobe and lingual gyrus as the margins of the calcarine sulcus (red) are also called the calcarine fissure (red), which also has a deep impression in the medial surface, and it is beyond being the representation of the visual primary area. This fissure has an important role in the occipital transtentorial approaches, as demonstrated in the figure if you follow anteriorly this sulcus; its anterior end reaches the pineal region or the quadrigeminal cistern

cerebral artery. The posterior projection of the parahippocampal gyrus is called the lingual gyrus, which makes up the occipital lobe and is an eloquent area.

In addition, in the medial surface, the parahippocampal gyrus which is continuous with the cingulate gyrus represents the lower extremity of the "C" that the limbic lobe performs in the diencephalon. The parahippocampal gyrus has an anterior fold similar to a fish hook called as uncus (Fig. 1.17). The uncus can be divided into

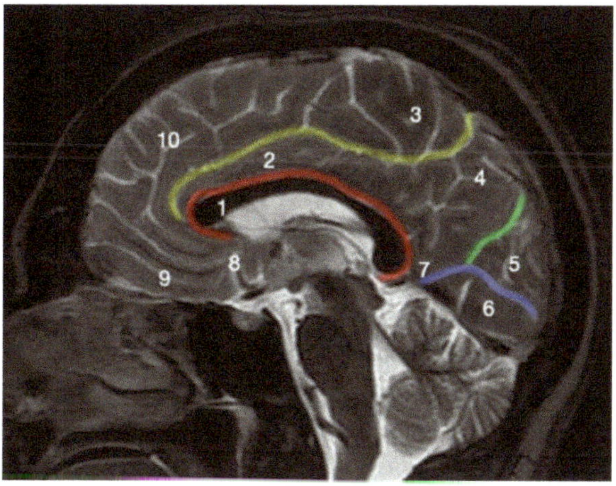

Fig. 1.16 This is a sagittal view of a Cube-FLAIR sequence of the magnetic resonance imaging of the brain that reviews all the major anatomic strictures of the medial surface. The corpus callosum (1) is encircled by the sulcus of the corpus callosum (red). The cingulate gyrus (2) is just above the sulcus of the corpus callosum (red) and, below its own sulcus, the cingulate gyrus sulcus (yellow). This sulcus has a marginal branch that delineates the paracentral lobule (3) from the quadrangular lobe (4). The last is separated by the parieto-occipital sulcus (green) from the cuneiform gyrus (5) that together with the lingual gyrus (6) represents the occipital lobe in the medial surface. They are separated by the calcarine sulcus (blue), which is deep and had an anterior pathway until the quadrigeminal cistern. The other numbered structures are isthmus of the cingulate gyrus (7), subcallosal area (8), rectus gyrus (9), and frontal medial gyrus (10)

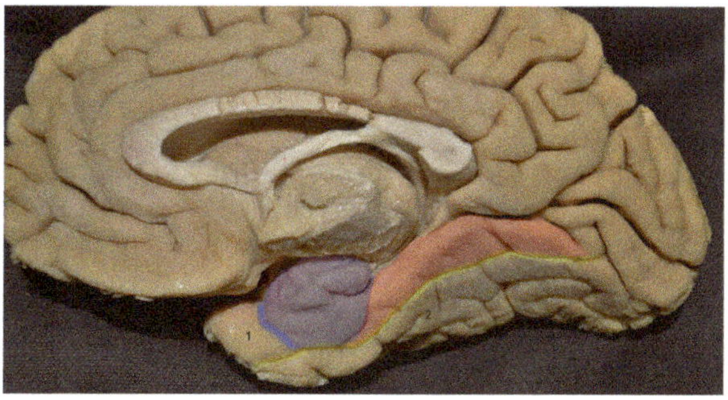

Fig. 1.17 This figure shows a medial view of the temporal lobe; parahippocampal gyrus (red) is the most medial structure in the temporal lobe, and it curves as a fish hook in the anterior end called as the uncus (blue). The parahippocampal gyrus and uncus are separated from the occipitotemporal gyrus by the collateral sulcus (2) and rhinal sulcus (1), respectively

three portions: anterior, apex, and posterior. The anterior internally contains the amygdala; the apex has intimal relation with the third cranial nerve; and the posterior part contains the head of the hippocampus. The parahippocampal gyrus also has a superior surface called as subiculum, which is the "bed" of the thalamus, and it is separated from the dentate gyrus from the hippocampal sulcus. The dentate gyrus is separated from the fimbriae fornix by the fimbriodentate sulcus. It is also interesting to note that the parahippocampal gyrus is present in two surfaces of the temporal lobe: the medial surface and the inferior surface.

1.11 Basal Surface

The frontal lobe lays in the anterior fossa over the cribriform plate, lesser sphenoid wing, and the roof of the orbits; therefore, owing to its direct relation, one of the structures present in the inferior surface of the frontal lobe receives the name of orbital gyri. The orbital gyri are separated from each other by the orbital sulcus that has an "H" shape. Though the four orbital gyri are named lateral, medial, posterior, and anterior orbital gyrus (Fig. 1.18), the lateral orbital gyrus continues in the lateral surface of the brain as the orbital part of the inferior frontal gyrus. The medial orbital gyrus is separated from the rectus gyrus from the olfactory sulcus that rests over the olfactory tract. This relation is a key point during the subfrontal approaches because it is very simple and easy to identify the olfactory tract during the surgery, and once identified, you can determine medially to find the rectus gyrus and laterally to find the orbital gyri (Fig. 1.18). Most of the time when performing the subfrontal retraction, the brain spatulas lie over the orbital gyri.

The temporal and occipital lobes also have representation in the inferior or basal surface. However, there is no clear separation between them since they are

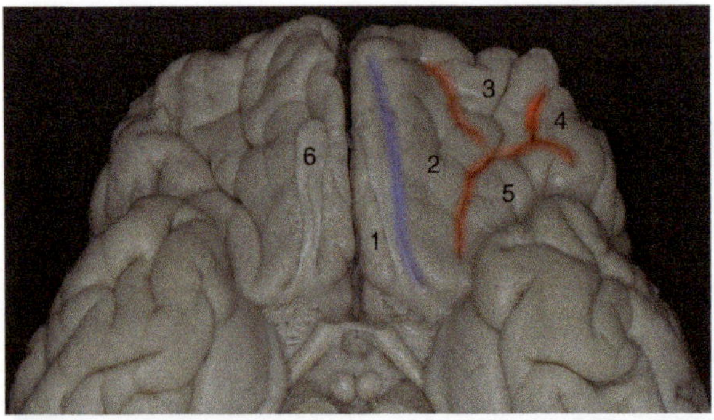

Fig. 1.18 The basal view of the frontal lobe reveals the importance of the olfactory tract and bulb (6) in the identification of the anatomy; once the olfactory tract is identified, it is possible to identify the olfactory sulcus (blue). The olfactory sulcus in its turn separates the rectus gyrus (1) from the orbital gyri. The orbital gyri are separated by the orbital sulcus (red): the medial orbital gyrus (2), anterior orbital gyrus (3), lateral orbital gyrus (4), and posterior orbital gyrus (5)

continuous. The temporal lobe rests over the middle fossa, but the occipital lobe rests over the tentorium. In a lateral view, it is possible to observe that the basal surface has a concave shape owing to the impression of the petrous bone in the inferior surface. The understanding of this disposition of the temporo-occipital inferior surface is crucial for surgical approaches that aim to reach the posterior part of the parahippocampal gyrus; the posterior portion of this gyrus can be also accessed using the supracerebellar transtentorial approach.

The basal surface is composed of the longitudinal sulci and gyri; the lateral occipitotemporal gyrus is present in two surfaces, namely, the lateral and basal surface, so it is the same inferior temporal gyrus that continues posteriorly with inferior occipital gyrus. This gyrus is separated from the medial occipitotemporal gyrus by the occipitotemporal sulcus. The medial occipitotemporal gyrus is separated from the parahippocampal gyrus by the collateral sulcus that is 100% constant (Fig. 1.19). Sometimes depending on the presence of the rhinal sulcus, the medial occipitotemporal can assume a spindle form, which is called as the fusiform gyrus.

This is the end of the first chapter, but it is just the beginning of the book. The next chapters will explore each lobe or anatomical region providing anatomical information and tips to perform surgical approaches. Eventually, we will address functional neuroanatomy, citing cognitive functions with neuroanatomic correlations. However, it is worth mentioning that these are dynamic functions, so we will consider them as areas of predominance of the function, but those are a part of a dynamic system.

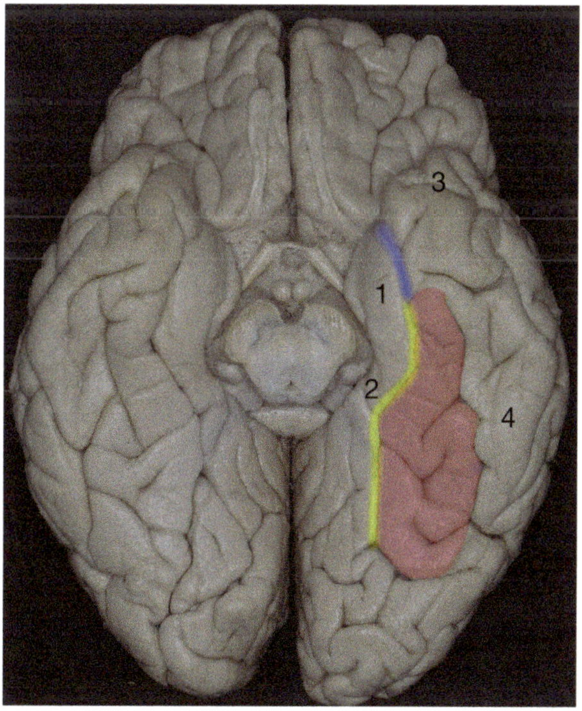

Fig. 1.19 This is the view of the basal occipitotemporal surface. On the right, it is possible to identify the fusiform gyrus (red), collateral sulcus (yellow), and rhinal sulcus (blue). 1- uncus; 2- parahippocampal gyrus; 3- temporal pole; 4- lateral occipitotemporal gyrus

As an example, we can mention the functions related to the lower frontal gyrus that is normally related to the language area. The localization theory and Broca's studies were very relevant for the development of the theory of cognitive functions with an emphasis on language; however, we know that the function of language involves several systems, such as the networks of the lower temporal gyrus than for semantic and phonological processing.

Another example that we can mention is the memory function related to the hippocampus. Recent studies suggest that the memory system starts in the prefrontal cortex and moves on to medial temporal structures for coding and recovery process. In this way, we will deal with the predominant structures throughout the book, as the epicenter of the function, but we understand it as a complex and dynamic system.

Suggested Bibliography

Alves RV, Ribas GC, Parraga RG, et al. The occipital lobe convexity sulci and gyri. J Neurosurg. 2012;116(5):1014–23.

Duffau H. The error of Broca: from the traditional localizationist concept to a connectomal anatomy of human brain. J Chem Neuroanat. 2017;89:73–81.

Herbet G, Duffau H. Revisiting the functional anatomy of the human brain: toward a meta-networking theory of cerebral functions. Physiol Rev. 2020;100(3):1181–228.

Ono M, Kubik S, Aberbathey C. Atlas of the cerebral sulci. Stuttgart: Thieme; 1989.

Ribas GC. The cerebral sulci and gyri. Neurosurg Focus. 2010;28(2):E2.

Ribas GC, Yasuda A, Ribas EC, et al. Surgical anatomy of microneurosurgical sulcal key points. Neurosurgery. 2006;59(4):ONS177-210.

Rodrigues TP, Rodrigues MA, Paz Dde A, Costa MD, Centeno RS, Chaddad Neto FE, Cavalheiro S. Orbitofrontal sulcal and gyrus pattern in human: an anatomical study. Arq Neuropsiquiatr. 2015b;73(5):431–5.

Ruis C. Monitoring cognition during awake brain surgery in adults: a systematic review. J Clin Exp Neuropsychol. 2018;40(10):1081–104.

Rodrigues T, Rodrigues M, Paz D, Costa MD, Santos B, Braga V, Paiva Neto M, Centeno R, Cavalheiro S, Chaddad-Neto F. Is the omega sign a reliable landmark for the neurosurgical team? An anatomical study about the central sulcus region. Arq Neuropsiquiatr. 2015a;73(11):934–8.

Ten Donkelaar HJ, Tzourio-Mazoyer N, Mai JK. Toward a common terminology for the gyri and sulci of the human cerebral cortex. Front Neuroanat. 2018;12:93.

Chapter 2
Surgical Anatomy of the Frontal Lobe

The frontal lobe occupies one-third of the surface of the brain hemisphere. The frontal lobe cortex comprises primary and secondary motor areas and a region anterior to the premotor cortex, called the prefrontal cortex. The prefrontal cortex is related to higher function processing in humans, such as behavior, judgment, personality, responsibility, and social property.

This lobe is composed of three hemispheric surfaces: superolateral, medial, and basal surfaces. The superolateral surface underlies the frontal bone. In this area, the frontal lobe is separated from the parietal lobe by the central sulcus and from the temporal lobe by the lateral sulcus. On the medial surface, the frontal lobe faces the falx cerebri and is separated from the corpus callosum by the sulcus of the corpus callosum and from the parietal lobe by an extension of the superior part of the central sulcus. The basal or inferior surface lies over the cribriform part of the ethmoid bone and the roof of the orbit that is part of the frontal bone, which explains its concave shape.

The medial surface of the frontal lobe was redefined by the International Anatomical Terminology, published in 1998. It standardized the limbic lobe as a distinct cerebral lobe, which medially comprised the cingulate gyrus around the corpus callosum. Therefore, the inferior boundary of the frontal lobe on the medial surface is the cingulate sulcus.

Each frontal surface can be studied as an anatomic unit with its associated characteristics, relations, and pathologies. To better understand this concept, surgical cases will be presented to introduce each surface.

Supplementary Information The online version of this chapter (https://doi.org/10.1007/978-3-030-82747-2_2) contains supplementary material, which is available to authorized users.

F. Chaddad-Neto, M. D. Silva da Costa, *Microneuroanatomy and Surgery*,
https://doi.org/10.1007/978-3-030-82747-2_2

2.1 The Superolateral Surface

After a brief description and study of the frontal sulci and gyri, it is essential to understand their anatomical disposition in relation to deep structures to plan the cranial approach and microsurgery based on the identification of cranial-cerebral relations and landmarks.

The superolateral surface of the frontal lobe is delimited superiorly by the hemispheric fissure and inferiorly by the supraciliary margin that continues posteriorly with the Sylvian fissure. The posterior limit is delimited by the central sulcus, which separates the primary motor and sensory areas located on the frontal and parietal lobes, respectively. The coronal suture crosses the three frontal gyri in front of the precentral sulcus. The central sulcus is located 3.5–4.5 cm posterior to the coronal suture, and its inferior segment is closer than its superior segment.

The superolateral surface in front of the central sulcus is composed of one oblique gyrus (precentral) and three horizontal gyri (superior, middle, and inferior frontal), divided by one oblique sulcus (precentral) and two horizontal sulci (superior and inferior frontal). The precentral gyrus is delimited anteriorly by the precentral sulcus, which is commonly divided into two or more segments connected in different patterns with the frontal sulci, most frequently the superior and inferior sulci in both hemispheres.

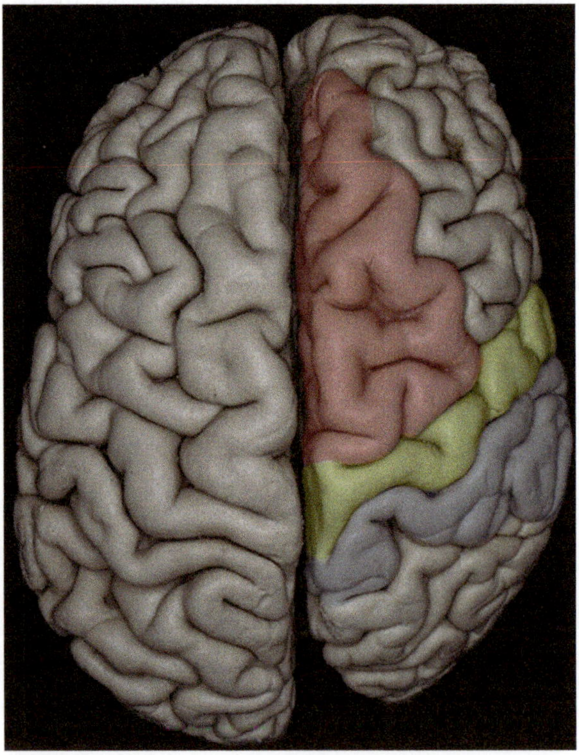

Fig. 2.1 Right hemisphere demonstrating the central sulcus, the posterior limit of the frontal lobe, between the precentral gyrus (yellow) and the postcentral gyrus (blue). Note that this lobe has one vertical gyrus and three horizontal gyri; the one depicted in red is the superior frontal gyrus

The central sulcus (Fig. 2.1) is the second most constant sulcus. It is a continuous sulcus in 92% of both hemispheres and has an S shape with a superior convex turn posteriorly and an inferior turn anteriorly. It ends superiorly at the superior margin of the lateral surface; however, in some cases it extends to the medial surface. Inferiorly, it ends about 2 cm behind the ascending ramus of the lateral sulcus and rarely reaches the lateral sulcus (in 16% of both hemispheres). By closely studying the central sulcus, it is possible to identify two connections between the gyri around it, one superior, named the paracentral lobule (at the medial surface), and one inferior, the subcentral gyrus.

How do we apply this anatomical knowledge to the surgical procedure? Let us consider some real cases and how to interpret the anatomy in clinical practice.

2.1.1 Surgical Case 1

Male, 22 years old, who presented with tonic-clonic seizures at the age of 19 years. MRI (Fig. 2.2) revealed arteriovenous malformation (AVM).

Based on the Fig. 2.2, where is this AVM located? And why? What are the relations with cranial bone and deep brain structures? What are the feed arteries and drainage veins?

The AVM presented here is situated at the depth of the inferior frontal sulcus. The primary reason why the AVM can be considered as present in the inferior frontal sulcus is because the AVM is situated in the first sulcus just above the Sylvian fissure, and this can be confirmed in the coronal view of the MRI. It is also important to note that the inferior frontal sulcus is located at the level of the anterior segment of the corpus callosum. Another practical suggestion on how to recognize this type of relation is to observe the image (computed tomography [CT] or MRI); when you see lateral ventricles in the axial plane, you are below the corpus callosum and consequently you are at the level of the interior frontal gyrus. The inferior

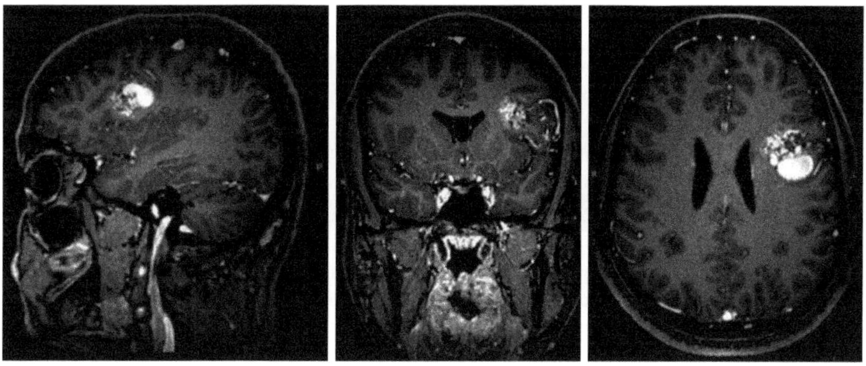

Fig. 2.2 Magnetic resonance imaging showing arteriovenous malformation

frontal sulcus is at the level of the corpus callosum (Fig. 2.3); below this sulcus there is the inferior frontal gyrus, which is divided into three segments according to the lateral sulcus: orbital part (in front of the anterior horizontal ramus), triangular part (between the anterior horizontal and ascending rami), and opercular part (behind the anterior ascending rami and limited posteriorly by the precentral sulcus) (Fig. 2.4). Together, the triangular and opercular parts of the inferior frontal gyri of the dominant hemisphere correspond to the language expression area known as Broca's region. The orbital part is continuous in the basal surface of the frontal

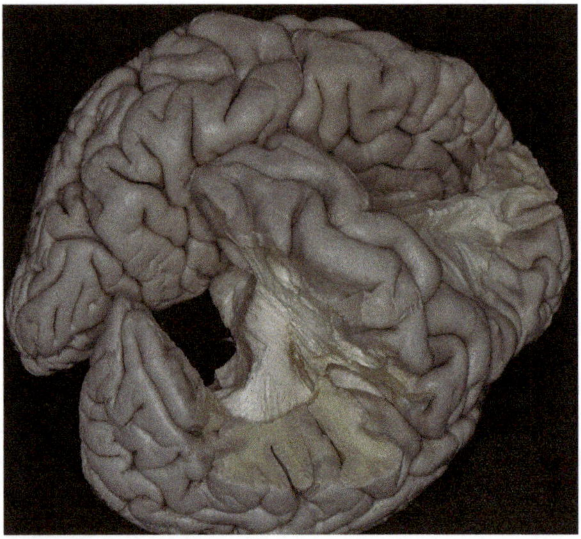

Fig. 2.3 Gyri above the inferior frontal sulcus excluded from the right hemisphere. The inferior frontal sulcus is at the same level as the corpus callosum is seen in this dissection

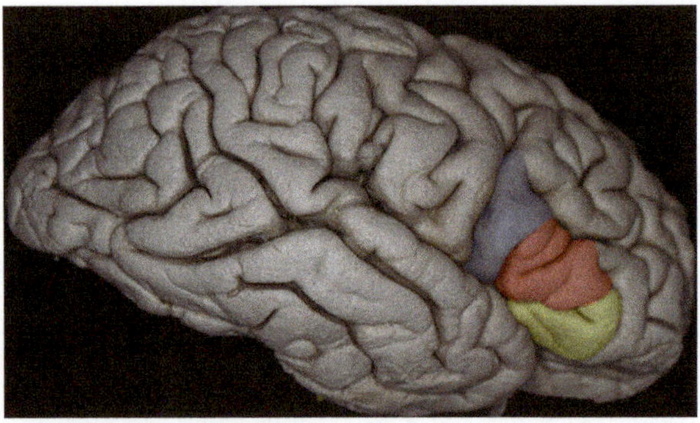

Fig. 2.4 Below the inferior frontal sulcus, the inferior frontal gyrus is divided into three parts, namely, orbital (yellow), triangular (red), and opercular (blue), by the rami of the lateral sulcus: anterior horizontal and anterior ascending. The superior frontal sulcus end points toward the area corresponding to the motor representation of the hand

lobe with the orbital gyri, and the opercular part is continuous with the precentral gyrus (Fig. 2.3).

The triangular and opercular parts of the inferior frontal gyrus are located at the lateral projection of the following structures from the surface to the depth: anterior part of the insula, head of the caudate nucleus, and frontal horn of the lateral ventricle. The apex of the triangular part of the inferior frontal gyrus points toward the trifurcation of the lateral sulcus and to the limen insulae, marking the anterior limit of the basal ganglia and the anterior horn of the lateral ventricle (Fig. 2.5).

The inferior frontal sulcus in almost all cases connects to the precentral sulcus at a point anterior to the motor area representation of the face in the precentral gyrus. At the cranial surface, it corresponds to the stephanion, where the coronal suture crosses the superior temporal line.

The superior limit of this AVM is the middle frontal gyrus, composed of multiple segments and commonly connected to the precentral gyrus. The posterior end of the AVM in the vicinity of the precentral sulcus represents the frontal eye field.

The superior frontal sulcus separates the superior and the middle frontal gyri and in most cases connects with the precentral sulcus posteriorly. The posterior end of the superior frontal sulcus points toward the area of the precentral gyrus corresponding to the motor representation of the hand.

The anterior ascending ramus of lateral sulcus corresponds at the surface to the anterior limiting sulcus of the insula and, additionally, represents the anterior limit of the caudate head and the anterior portion of the frontal horn. Just below the triangular part of the inferior frontal gyrus, there is an enlargement of the lateral sulcus,

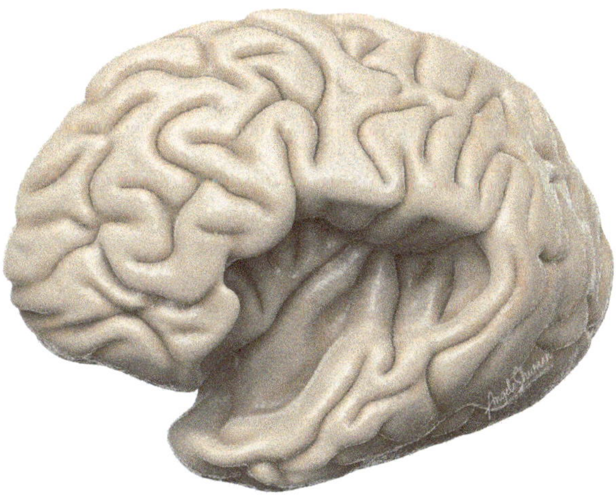

Fig. 2.5 Opening the Sylvian fissure and exposure of the insula. (Illustration created by Angelo Shuman)

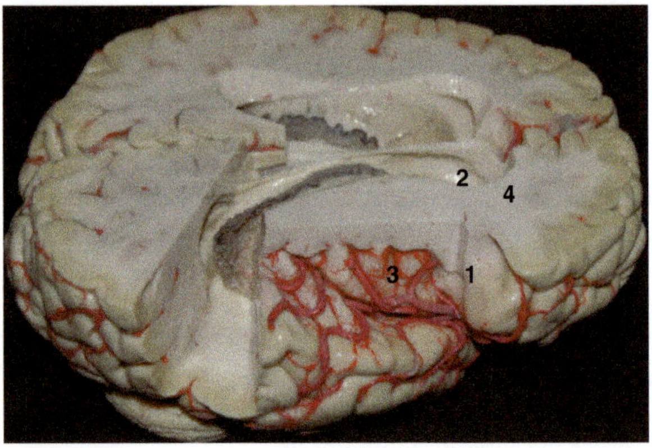

Fig. 2.6 The anterior ascending ramus of the lateral sulcus corresponds at the surface to the anterior limiting sulcus of the insula (1). The relation between this ramus and the frontal horn (2) of the lateral ventricle, (3) insula, (4) forceps minor of the corpus callosum

called the anterior Sylvian point, which corresponds to the point where the middle cerebral artery changes its direction at the limen insulae (Fig. 2.6).

The concepts above are essential to plan surgeries like ventricular catheterization to avoid injury to the basal ganglia, capsula interna, fornix, thalamus, and vessels. Based on these concepts, which would be a safe landmark on the cranial surface for external ventricular drainage and endoscopic surgery?

The tip of the catheter or the endoscope can enter the superior frontal gyrus and pass through the corona radiata, the fibers from the corpus callosum, and ependyma, reaching the frontal horn of the lateral ventricle. To avoid important structures and to reach the correct target, the entry point should be carefully planned. Kocher described this point as "2.5–3 cm from the median line and 3 cm forward of the precentral fissure". Regarding the veins and the superior sagittal sinus, a distance of about 3 cm from the midline is a reasonable position. Entry points just ahead of the coronal suture are sufficient to avoid contacting the precentral gyrus at the coronal plane, considering that the foramen of Monro corresponds to the bregma at the cranial level and the precentral sulcus meets the superior frontal sulcus at a point 3.5 cm behind the bregma.

The white matters of the hemispheres are categorized as association, commissural, or projection fibers. The cortical gray matter of the superolateral surface of the frontal lobe covers the corona radiata, which is composed of projections fibers, and the corpus callosum that contains the commissural fibers. Superficially to the corona radiata, a bundle of association fibers named superior longitudinal fasciculus is present.

The superior longitudinal fasciculus (SLF) surrounds the insula and passes under the frontal, parietal, and temporal cortical gray matter. It is divided in three groups. The first is the frontoparietal part, formed by fibers in a horizontal disposition. The temporoparietal segment is composed of fibers in a vertical orientation and interconnects the superior and middle temporal gyri to the inferior parietal lobule. The frontotemporal fibers, known as the arcuate segment, connect the prefrontal region to the posterior temporal region. The SLF consists of an integrating center of multimodal functions. In the nondominant hemisphere, it is related to spatial awareness and, in the dominant hemisphere, assists different aspects of language (Fig. 2.7).

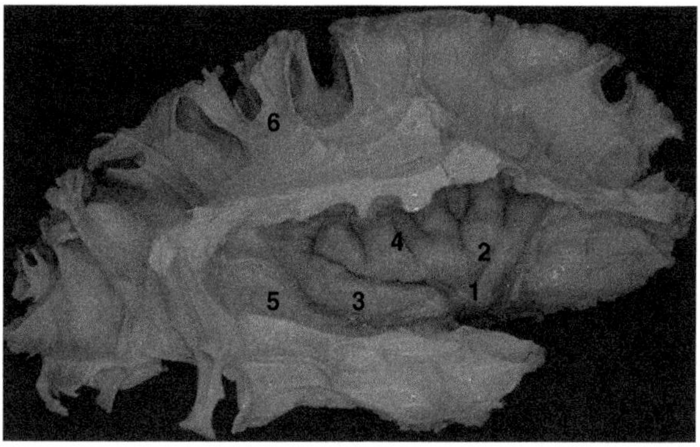

Fig. 2.7 The superior longitudinal fascicle is exposed and is divided into a horizontal segment (yellow) and a vertical segment (blue). (1) Limen of the insula. (2) Insular apex. (3) Long gyrus of insula. (5) Short gyri of insula. (5) Sagittal stratum. (6) Subcortical U-fibers

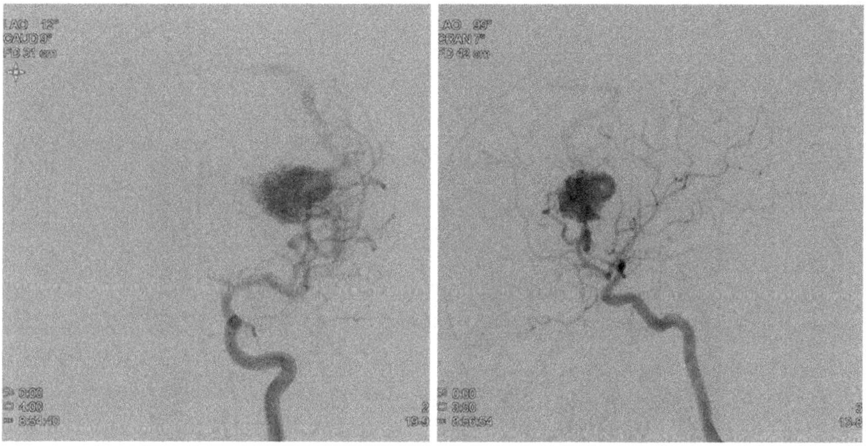

Fig. 2.8 Digital angiography: anteroposterior (on the left) and lateral (on the right) view of the left carotid artery

The projection fibers above the upper edge of the thalamus are known as corona radiata, which continues as the internal capsule. At the level of the basal ganglia, the internal capsule is located lateral to the thalamus and medial to the lentiform nucleus. From medial to lateral, the following structures can be identified: globus pallidus, putamen, external capsule, claustrum, extreme capsule, and insula.

There is one concluding question about surgical case 1: What are the feed arteries and drainage veins? To answer the question, the patient was subjected to digital angiography, as shown in Fig. 2.8.

In summary, the cerebral arteries are divided into cortical, lenticulostriate, thalamoperforating, and choroidal. Note in the anteroposterior view of the angiograph that the AVM is supplied by cortical and perforating/lenticulostriate rami of the left middle cerebral artery (MCA). The segments of the MCA are characterized according to the change in the direction of the artery (Fig. 2.9).

The M1 segment, also referred to as the sphenoidal segment, extends medially to laterally, beginning at the carotid bifurcation and reaching the limen insulae. Its perforating rami, denominated lenticulostriate arteries, pass through the anterior perforated substance to supply the basal ganglia and the internal capsule. This

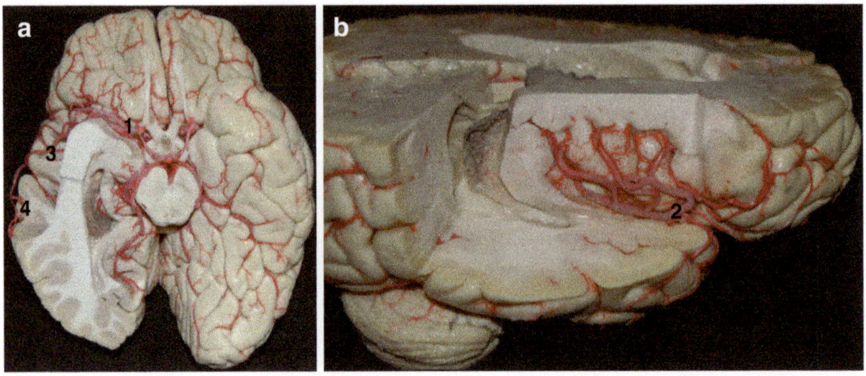

Fig. 2.9 (**a**) The middle cerebral artery is divided into four segments. The M1 segment (1) is identified as the one that passes medial to lateral in the sphenoid portion of the Sylvian fissure. The M3 segment (3) is around the opercula. Finally, the M4 segment (4) gives rise to the cortical branches. (**b**). The M2 segment (2) travels anterior to posterior in the Sylvian fissure and is related medially to the insula

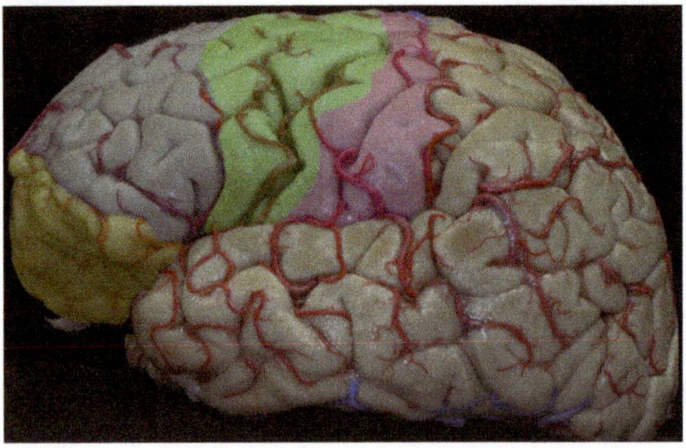

Fig. 2.10 The superolateral surface of the brain is supplied by the branches of the M4 segment. There are 12 areas of irrigation, named after the main artery of each one. The frontal lobe contains four areas: orbitofrontal (yellow), prefrontal (blue), precentral (green), and central (purple), exemplified in the illustration based on the most common pattern. The limits of each region cannot be precisely depicted and varies in each hemisphere. The superior strip of the superolateral surface is a watershed area between the MCA and ACA

segment bifurcates in 78% of cases into a superior and an inferior trunk. The M2 segment, also designated as the insular segment, originates at the limen insulae and courses posteriorly toward the circular sulcus of insula, where it assumes a medial to lateral direction and changes to M3 or the opercular segment. Finally, the M4 or cortical segment arises on the surface of the lateral sulcus, courses superiorly and inferiorly, and terminates at the cortical territory.

Although the MCA courses through most of the superolateral surface of the frontal lobe, it does not extend to the frontal pole (Fig. 2.10). The frontal lobe is composed of orbitofrontal, prefrontal, precentral, and central areas of irrigation, each supplied by a cortical branch of the same name. They are rami of the superior trunk of the MCA in most cases with a bifurcation pattern, although a trifurcation pattern may also be observed. The precentral and central arteries arise either from the superior or the middle trunk.

Looking at the lateral view of the angiography, it can be concluded that the AVM is fed by the prefrontal arteries, specifically by rami of the superior trunk of the MCA. An angiography of the right internal carotid artery demonstrates that the right internal cerebral artery contributes to the irrigation of this AVM (Fig. 2.11).

The anterior cerebral artery (ACA) irrigates a narrow strip of the superolateral surface bordering the interhemispheric fissure. Its leading contribution is to the basal and medial surfaces.

Based on the anatomical relations demonstrated till now, which type of craniotomy should be performed in this case 1? What would be the cranial landmarks?

All these anatomical concepts were assessed to plan the surgery of case 1. Since the malformation was close to the eloquent area, a neuropsychological professional evaluated his cortical function, and no significant change was found, except for below-average verbal fluency based on age and education. A frontotemporoparietal craniotomy was performed (Fig. 2.12) for AVM resection. Postoperative

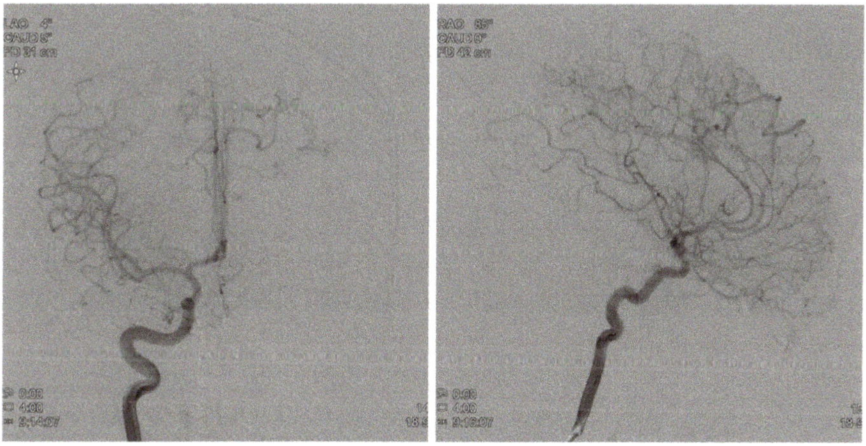

Fig. 2.11 The right internal carotid artery angiography demonstrating that the arteriovenous malformation from surgical case 1 is also supplied by the right cerebral artery

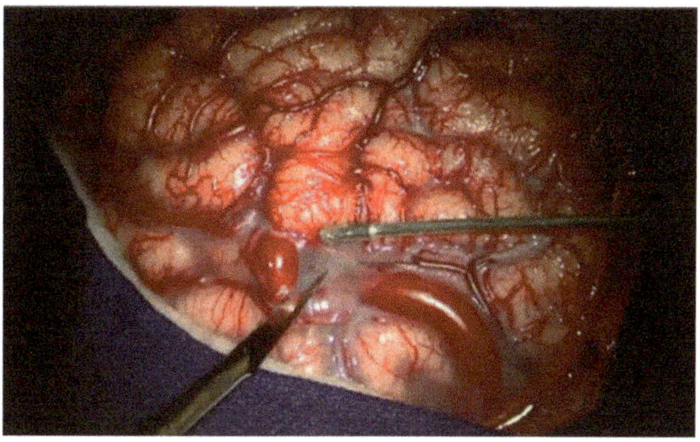

Fig. 2.12 After craniotomy, the arteriovenous malformation is exposed and starts to become squeletized

angiography demonstrated complete exclusion of the AVM. The patient did not have neurological or cognitive deficits, and there was no change in verbal fluency (*another case example is also demonstrated in the* Video 2.1).

2.2 The Medial Surface

2.2.1 Surgical Case 2

A 39-year-old man was investigated after an onset of epileptic crisis. He visited the outpatient clinic with the following MRI (Fig. 2.13).

The MRI demonstrated a cavernous malformation anterior to the lamina terminalis, a part of the anterior wall of the third ventricle. This cortical region is denominated as the subcallosal area and is composed of two parolfactory gyri anteriorly and one paraterminal gyrus posteriorly (Fig. 2.14). They are divided by the anterior

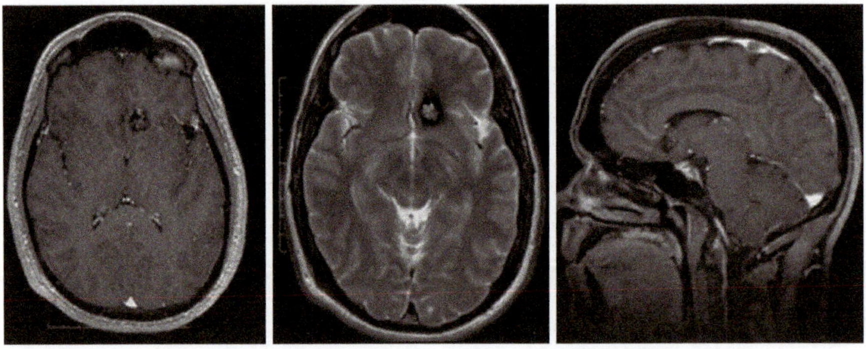

Fig. 2.13 Magnetic resonance imaging showing a cavernous malformation

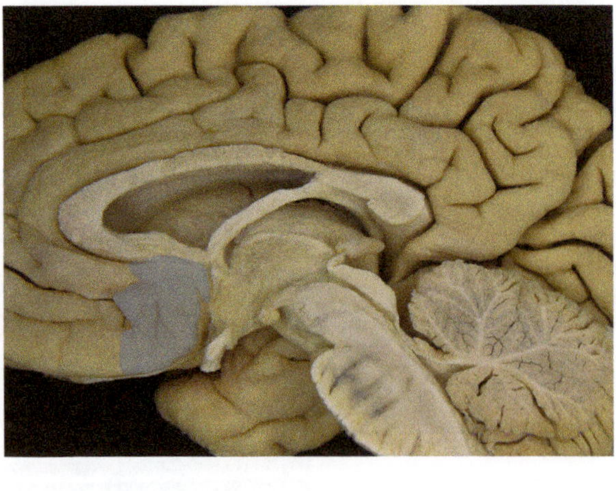

Fig. 2.14 Subcallosal area demarcated in blue

and posterior parolfactory sulci, respectively. The paraterminal gyrus encases the septal nuclei and is part of the limbic lobe. The subcallosal area is interconnected to the mesial temporal lobe by the uncinate fasciculus, located deep in the limen insulae.

Considering the anatomical landmarks, what surgical approach would you apply to this case?

Some gyri connect two surfaces of the hemisphere, such as the superior frontal gyrus and the gyrus rectus in the frontal lobe. At the medial surface, it is limited superiorly by the superior rostral sulcus and connects to the cingulate gyrus anterior to the parolfactory gyri, by the cingulate pole. Therefore, anterior and inferior approaches are reasonable for lesions in the subcallosal area, as are transventricular variations.

The superior frontal gyrus is one of the most prominent cortical structures on the medial surface of the frontal lobe and is also called the medial frontal gyrus (this is not the same as the middle frontal gyrus, which is located on the lateral surface of the frontal lobe). The cingulate sulcus separates the superior frontal gyrus and the paracentral lobule from the cingulate gyrus and gives rise to a branch named the paracentral sulcus, which posteriorly limits the superior frontal gyrus. The paracentral lobule is bounded anteriorly by the paracentral sulcus and posteriorly by the ramus marginalis of the cingulate sulcus. The paracentral lobule is the connection between the precentral and postcentral gyri at the medial surface of the hemisphere, encircling the central sulcus. The central sulcus ends superiorly at the superior margin of the lateral surface; however, in some cases, it extends to the medial surface (Fig. 2.15).

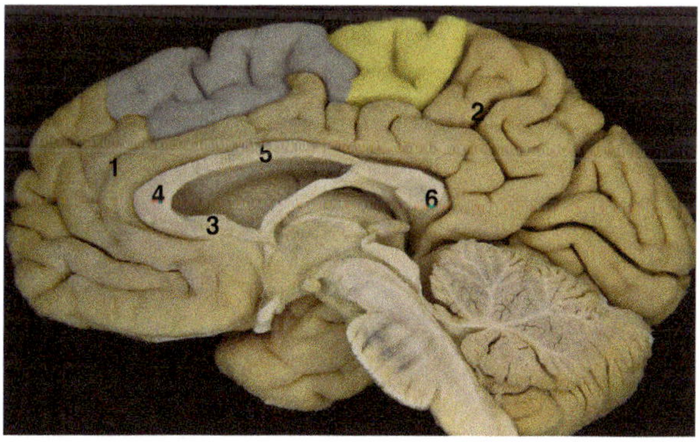

Fig. 2.15 Medial surface of the frontal lobe. The inferior limit is the cingulate sulcus (1). The paracentral lobule (yellow) is composed of the precentral and the postcentral gyri and is limited posteriorly by the marginal ramus of the cingulate sulcus (2). The posterior area of the superior frontal gyrus contains the supplementary motor area (blue region in this figure). (3) Rostrum of the corpus callosum. (4) Genu of the corpus callosum. (5) Body of the corpus callosum. (6) Splenium of the corpus callosum

The posterior part of the medial frontal gyrus contains the supplementary motor area (SMA), a secondary association cortex, responsible for planning and coordination of movement directly controlled by the primary motor area (precentral gyrus).

As can be observed, the corpus callosum is a noticeable landmark that surrounds and constitutes parts of the boundary of the lateral ventricle. From lateral to medial dissection, it is exposed after the removal of the cingulum. As previously mentioned, this structure is composed of commissural fibers that join both hemispheres. At the frontal lobe, the orbitofrontal and the prefrontal areas, one from the basal and the other from the lateral surface, are connected by a segment of the corpus callosum denominated as forceps minor.

The inferior fronto-occipital fasciculus is another bundle of fibers that is composed of white matter structures related to the frontal lobe. It extends from the dorsal and medial part of the prefrontal and premotor regions of the frontal lobe to the dorsal and medial parts of the occipital lobe, passing through the dorsal and medial parietal lobules (Fig. 2.16).

The distal segment of the ACA is divided according to the corpus callosum. The ACA originates at the bifurcation of the internal carotid artery just below the anterior perforated substance and travels above the optic nerve and the optic chiasm, toward the interhemispheric fissure. Above the optic chiasm and adjacent to the lamina terminalis, the anterior communicating artery (ACoA) communicates to the right and the left ACAs. The portion of the ACA distal to the ACoA encircles the corpus callosum and is further divided into four segments, A2 to A5, which supply the medial surface of the hemisphere (Fig. 2.17).

The ACoA, as well as the proximal subdivision of the ACA, the A1 segment, does not outline the corpus callosum. The ACoA complex exhibits several different anatomical variations, including orientation of the arteries, duplication, hypoplasia, diameter, and asymmetry. The neurosurgeon must be aware of these details.

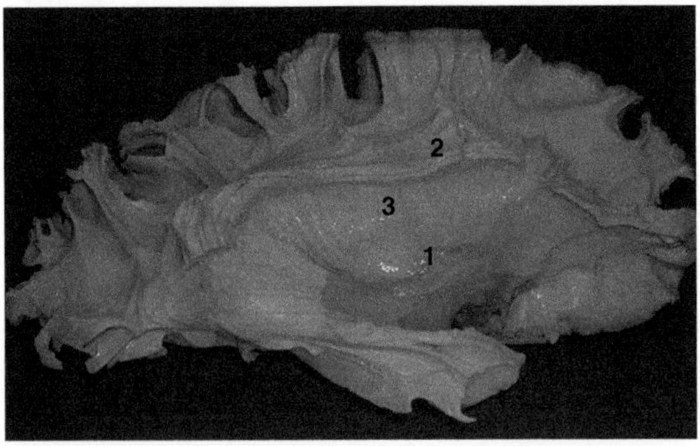

Fig. 2.16 The inferior fronto-occipital (blue) and the uncinate (red) fasciculi are part of the external capsule at the level of the limen insulae. The claustrum (1) becomes evident at the level of the insular apex. The inferior fronto-occipital fasciculus is the main bundle of the stratum sagittal (yellow). (2) Superior longitudinal fasciculus. (3) External capsule

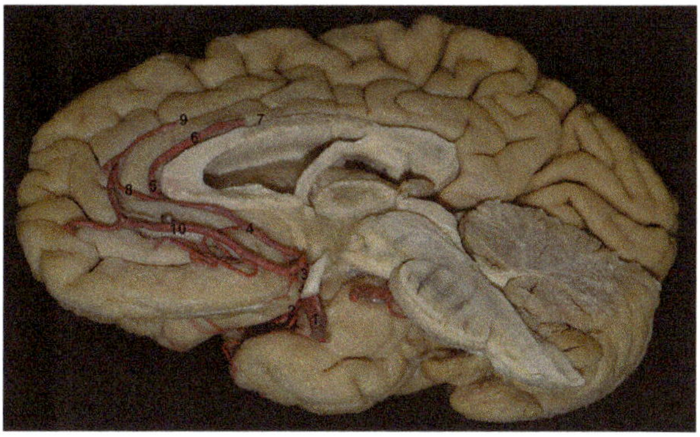

Fig. 2.17 The anterior cerebral artery (ACA) surrounds the corpus callosum into the callosal sulcus (7), and it is divided into five segments. Starting from the internal carotid artery (1), which bifurcates into middle (2) and anterior cerebral arteries (3), the A1 segment (3) is the precommunicating segment. The A2 (34) goes from the communication between the A1s until the rostrum of the corpus callosum; the A3 (5) surrounds the genu of the corpus callosum and more commonly gives rise to the callosomarginal artery (8), which runs into the cingulate sulcus (9). The A4 (6) segment comes from the end of the genu of the corpus callosum and goes horizontally backward into the callosal sulcus (7). The A5 segment is defined in two different ways, one is that it corresponds to the segment that runs into the callosal sulcus posteriorly to point corresponding the foramen of Monro, the other definition is the cortical branches of the ACA. In the figure, it is also possible to see the contralateral pericallosal artery (10)

The vein from the medial surface of the frontal lobe courses toward the superior sagittal sinus, the inferior sagittal sinus, and the basal vein through tributaries from the corpus callosum.

2.2.2 *Surgical Case 3*

A 57-year-old man, presented to the emergency service because of sudden onset of headache, vomiting, and presyncope 3 days earlier. Initial CT scan demonstrated subarachnoid hemorrhage predominantly at the pericallosal cistern and hematoma at the genu and the body of the corpus callosum. He was subjected to digital angiography (Fig. 2.18).

By comparing surgical cases 3 and 4, it becomes apparent that the aneurysm in the latter one is from the distal segment of the ACA, anterior to the genu of the corpus callosum. In front of the ACoA, the ACA is named the pericallosal artery and is subdivided according to its relation with the corpus callosum: A2 (infracallosal), A3 (precallosal), A4 (supracallosal), and A5 (posterocallosal). The A4 segment extends to the limit corresponding to the foramen of Monro in depth. Each subdivision gives rise to eight cortical arteries, a majority leading to the frontal lobe. A2 is the origin of the medial frontobasal and frontopolar arteries. A3 derivatives include callosomarginal, anterior, and middle internal frontal arteries. A4 is the source of the

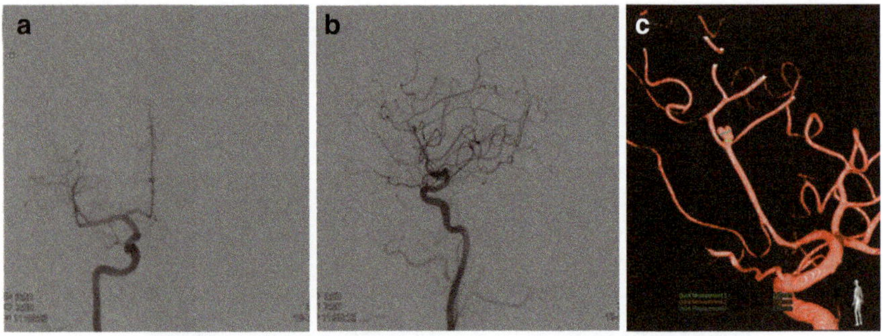

Fig. 2.18 Digital angiography with anteroposterior (**a**) and lateral (**b**) views of the right internal carotid artery. The aneurysm is best detailed in the 3D reconstruction (**c**)

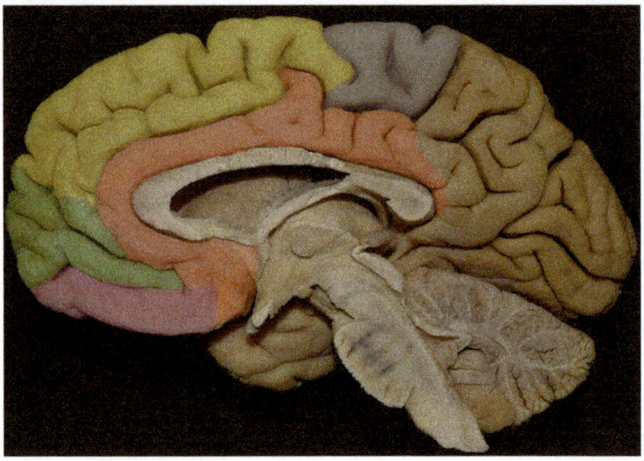

Fig. 2.19 The medial surface of the frontal lobe is irrigated by cortical rami of the anterior cere bral artery. The precise territory of each vessel cannot be established, as it is overlapped by other arteries. The orbitofrontal artery supplies the rectus gyrus (both medial and basal surfaces) and the medial portion of the orbital gyri (pink). The frontopolar artery irrigates the frontal pole (green). The anterior, middle, and posterior internal frontal arteries arise from the pericallosal or the cal losomarginal arteries and supply the superior frontal gyrus with multiple different patterns (yellow). The paracentral artery supplies the paracentral lobule (grey). The cingulate gyrus (red) receives vascular supply from the pericallosal artery

paracentral artery. Many variations can be found, including the absence of the callosomarginal artery and cortical branches arising from callosomarginal artery (Fig. 2.19).

Based on the digital angiography, from which segment is the aneurysm formed from?

The aneurysm arises from the bifurcation of the ACA in front of the genu of the corpus callosum (A3 segment) where, in this case, the pericallosal artery originates from the callosomarginal artery. This point is the most common origin of the pericallosal aneurysms. The callosomarginal artery courses adjacent to the

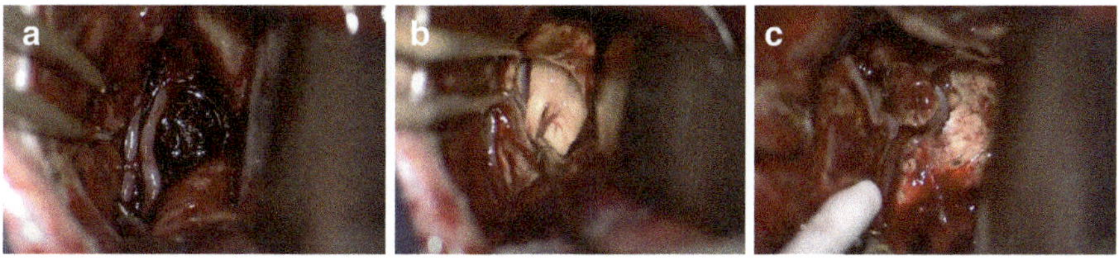

Fig. 2.20 Operative view of the surgical case 4. (**a**) Hematoma near the pericallosal arteries. (**b**) To enlarge the space for surgery, the corpus callosum was aspirated and the lateral ventricle was opened to drain the cerebrospinal fluid. (**c**) Demonstrates the aneurysm at the pericallosal artery

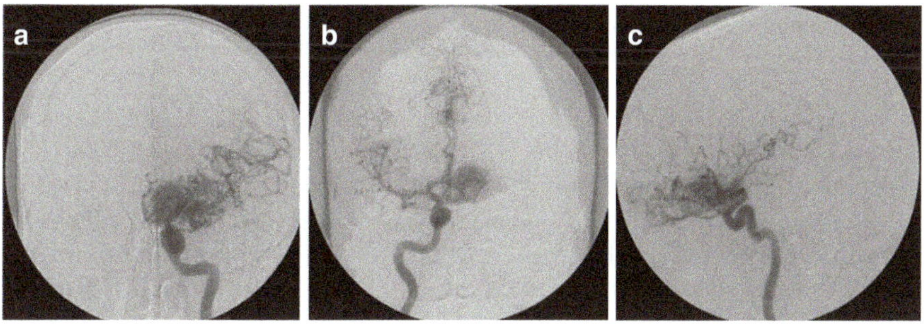

Fig. 2.21 Digital angiography: anteroposterior view of the left (**a**) and right (**b**) internal carotid arteries and lateral (**c**) view of the left internal carotid artery

cingulate gyrus and in parallel to the pericallosal artery, which is in the callosal sulcus.

The patient was subjected to unilateral parasagittal craniotomy to access the interhemispheric space. The falx cerebri is narrower anteriorly, which explains the attachment of both cingulate gyri. Similarly, a major part of the pericallosal artery is below the inferior margin of the falx. In this particular case, beyond the parasagittal vessel drainage and gyri adherence, the intracerebral hematoma associated with the subarachnoid blood reduces the innate brain pathway (Fig. 2.20). Despite the challenge, the aneurysm was clipped and the patient recovered with no neurological deficits.

2.3 The Basal Surface

2.3.1 Surgical Case 4

Male, 54 years old, 4 years ago, presented with a single episode of tonic-clonic epileptic crisis. The onset of crisis was characterized by behavior arrest and mental confusion 3 years previously. His neuropsychological exam revealed that verbal fluency and working memory were below the normal range. He presented for outpatient consultation with the following angiograph (Fig. 2.21).

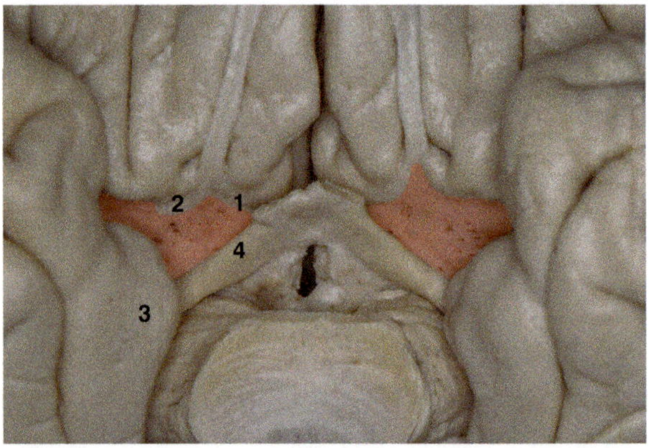

Fig. 2.22 The perforating aspect of the basal surface (red) serves as the point of entrance to the perforating arteries. Its limits are lateral (1) and medial (2) olfactory stria anteriorly, interhemispheric fissure medially, limen insular laterally, anterior portion of the uncus posterolaterally (3), and the optic tract posteromedially (4)

Is it possible to determine the location of this arteriovenous malformation and its feeding arteries?

It can be observed that the feeding arteries are perforating branches from the left ACM and the ACoA complex, as well as cortical branches from the left A2 segment. The perforating rami reach the deeper structures passing through the anterior perforating substance, located at the basal surface of the hemisphere. The anterior perforating substance is localized during surgery by tracing the posterior trajectory of the olfactory tract, which ends in a bifurcation into the medial and lateral olfactory stria at the anterior limit of the anterior perforating substance. It is limited medially by the interhemispheric fissure and laterally by the limen insulae; the posterior limits are the optic tract and the temporal lobe (Fig. 2.22). During surgery, this region is located by tracing the olfactory tract in association with the ACM and ACA. Rami from the choroid part of the internal carotid artery and the choroid artery itself penetrate the anterior perforating substance organized in a specific pattern to supply different territories.

The perforating arteries arise most commonly from the posterior side of the main artery to irrigate the structures above the anterior perforating substance in addition to the adjoining area (Fig. 2.23).

The patient was submitted to an MRI to analyze the relation of parenchymal structures (Fig. 2.24), from which it is possible to understand these areas: caudate nucleus, putamen, internal capsule, globus pallidus, and thalamus.

The perforating arteries from the A1 segment also supply parts of the hypothalamus, including the optic chiasm, and the anterior commissure. The recurrent artery, known as the Heubner artery, is a branch from the distal part of the A1 segment or proximal portion of the A2 segment. It roams in the opposite direction from its origin and passes above the bifurcation of the internal carotid artery to issue branches to the anterior perforating substance (Fig. 2.25). A major part of the frontobasal surface is irrigated by the medial orbitofrontal artery, the first branch of the distal A2 subdivision.

The base of the inferior surface of the frontal lobe is divided by the olfactory sulcus, concealed by the olfactory bulbs and tracts. Medially, the rectus gyrus, which is continuous to the superior frontal gyrus, lies over the cribriform lamina of

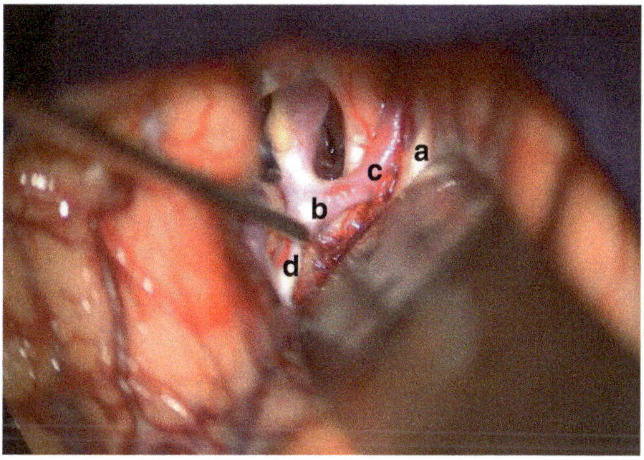

Fig. 2.23 Operative view of the perforating arteries during posterior communicant aneurysm clipping surgery. The dissector is pointing to the lenticulostriate arteries. The spatula is placed in the orbital gyri and moves the frontal lobe away. The medial limit of the orbital gyri covered by the olfactory tract (A) and the carotid bifurcation (B) into the A1 segment (C) and the M1 segment (D)

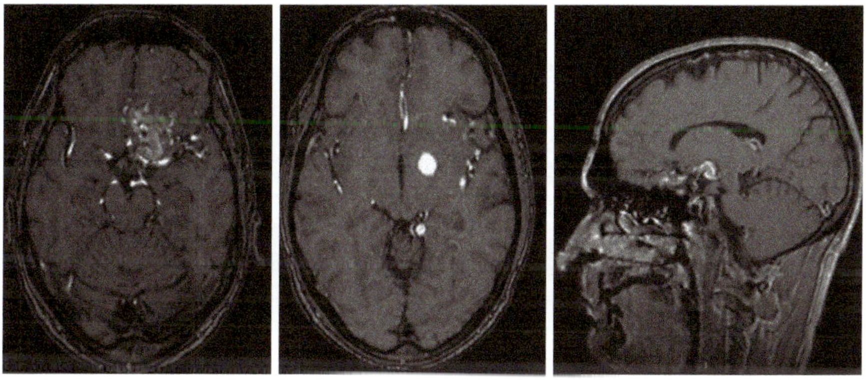

Fig. 2.24 Magnetic resonance imaging demonstrating MAV of the left frontobasal region

Fig. 2.25 Intraoperative view of a rupture anterior communicating artery aneurysm clipping surgery. In this case, the Heubner artery (1) is ramus of the proximal part of the A2 segment (2). (3) Carotid bifurcation. (4) A1 segment. (5) M1 segment

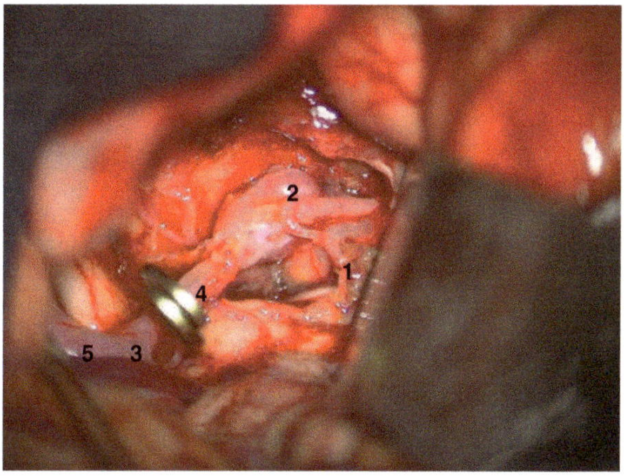

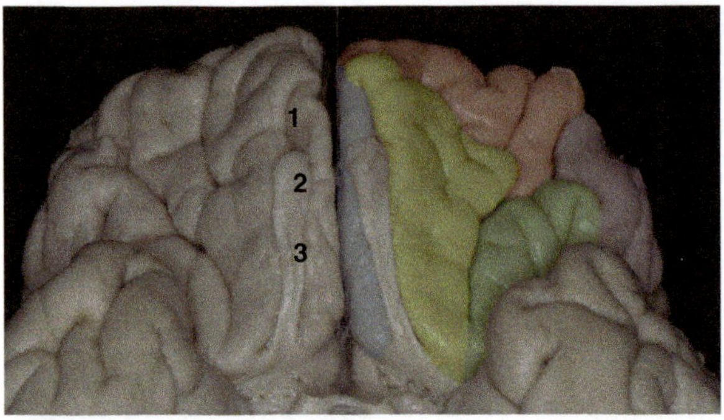

Fig. 2.26 The basal surface of the frontal lobe is divided by the olfactory sulcus (1), covered by the olfactory bulbs (2) and tract (3). Medially, there is the rectus gyrus (blue) and, laterally, the orbital gyri separated by the orbital sulcus into medial (yellow), lateral (purple), anterior (red), and posterior (green) gyri

the ethmoid bone. Laterally, the orbital gyri are divided by an H shape sulcus named the orbital sulcus branches into lateral, medial, anterior, and posterior orbital gyri. The orbital and superolateral surfaces are separated by the frontomarginal sulcus (Fig. 2.26).

Lastly, what are the drainage veins of the ACM? It is necessary to consider the inferior surface of the frontal lobe and the pattern of drainage; anteriorly it has veins that drain into the superior sagittal sinus and, posteriorly, a tributary to the basal vein.

The basal vein (Rosenthal's vein) originates inferior to the anterior perforated substance, where the anterior cerebral vein meets the deep middle cerebral vein. It receives the olfactory and the orbitofrontal veins, besides others from other. The neurovascular bundle between the uncus and the cerebral peduncle corresponds to the optic tract, the basal vein, and the choroidal artery.

Suggested Bibliography

Chaddad-Neto F, Campos Filho JM, Dória-Netto HL, Faria MH, Ribas GC, Oliveira E. The pterional craniotomy: tips and tricks. Arq Neuropsiquiatr. 2012;70(9):727–32.

Chaddad-Neto F, Silva D, da Costa M, Bozkurt B, et al. Contralateral anterior interhemispheric-transcallosal-transrostral approach to the subcallosal region: a novel surgical technique. J Neurosurg. 2018;129:508–14.

Chang Mulato JE, Alejandro SA, Paganelli SL, Vela Rojas EJ, Viera Neves AP, Silva da Costa MD, Dória-Netto HL, Campos Filho JM, Chaddad-Neto F. Microsurgical treatment for a post hemorrhagic cavernoma of the frontal lobe coexisting with an unruptured ipsilateral MCA aneurysm. World Neurosurg. 2021;S1878-8750(21):01331-0.

Kakou M, Destrieux C, Velut S. Microanatomy of the pericallosal arterial complex. J Neurosurg. 2000;93:667–75.

Kocher TE. Chirurgische Operationslehreed. 1st ed. Jena: Gustav Fischer; 1892. p. 38.

Kocher TE. Operative surgery. New York: W. Wood & Company; 1894. p. 50.

Ono M, Kubik S, Abernathey CD. Atlas of the cerebral sulci. Stuttgart: Georg Thieme Verlag; 1990.

Rhoton. Cranial anatomy and surgical approaches. Philadelphia: Lippincott Williams & Wilkins; 2003.

Ribas GC, Yasuda A, Ribas EC, Nishikuni K, Rodrigues AJ Jr. Surgical anatomy of microneurosurgical sulcal key points. Neurosurgery. 2006;59:ONS177–211.

Rodrigues TP, Rodrigues MA, Paz Dde A, Costa MD, Centeno RS, Chaddad Neto FE, Cavalheiro S. Orbitofrontal sulcal and gyrus pattern in human: an anatomical study. Arq Neuropsiquiatr. 2015;73(5):431–5.

Chapter 3
Surgical Anatomy of the Parietal and Occipital Lobes

3.1 Introduction

3.1.1 The Superolateral Surface

In the lateral aspect, the parietal lobe is bounded anteriorly by the central sulcus, which separates the frontal and parietal lobes, superiorly by the interhemispheric cistern, inferiorly by the posterior portion of the lateral sulcus and the anterior portion of the occipitotemporal line, and posteriorly by the superior half of the imaginary lateral parietotemporal line (a connecting line between the parieto-occipital sulcus and preoccipital notch) (Fig. 3.1). The parietal lobe can present gyral connections to the occipital, temporal, and frontal lobes.

The lateral surface of the parietal lobe comprises the oblique postcentral gyrus and two horizontal lobules: the superior parietal lobule and the inferior parietal lobule, which is itself composed of the angular and supramarginal gyri. These three sections are separated by two main sulci: the oblique postcentral sulcus and horizontal intraparietal sulcus (which divides the superior and inferior lobules). They maintain a highly constant perpendicular relation to each other.

Smaller and highly variable sulci can be identified on the lateral surface: the transverse parietal sulcus of Brissaud, which may subdivide the superior parietal lobule; the first intermediate sulcus of Jensen may subdivide the inferior parietal lobule into the angular and supramarginal gyri, and the angular sulcus may divide the angular gyrus.

Supplementary Information The online version of this chapter (https://doi.org/10.1007/978-3-030-82747-2_3) contains supplementary material, which is available to authorized users.

F. Chaddad-Neto, M. D. Silva da Costa, *Microneuroanatomy and Surgery*, https://doi.org/10.1007/978-3-030-82747-2_3

3.1.2 Case 1 Lateral Surface

A 21-year-old, right-handed female patient with complaints of slowness and memory loss. The physical examination results were unremarkable. Her neuropsychological evaluation altered perception and moderate to severe verbal fluency deficits (Fig. 3.2).

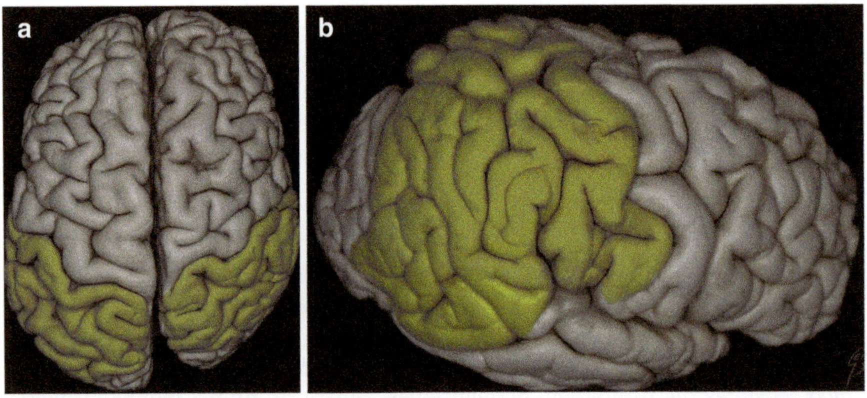

Fig. 3.1 Parietal lobe shaded in yellow. (**a**) Superior view of both hemispheres evidencing the oblique of the parietal lobe. (**b**) Lateral view of the right hemisphere

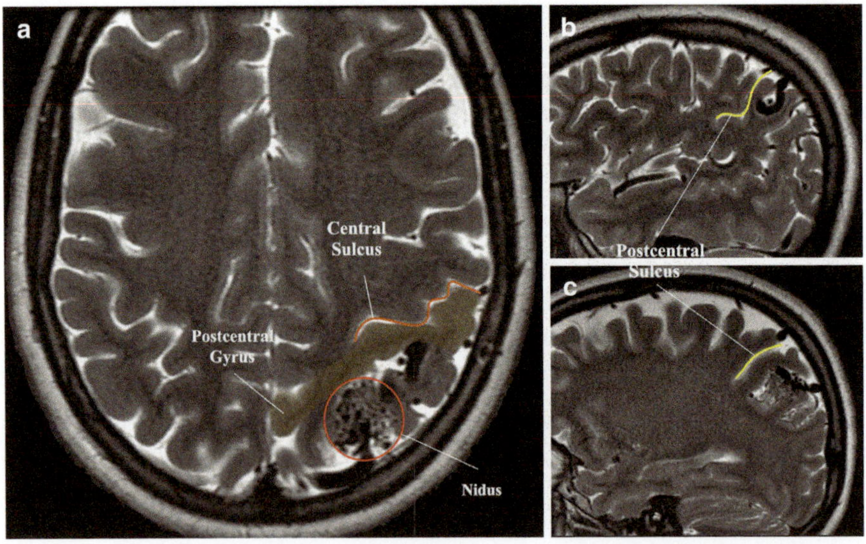

Fig. 3.2 Brain MRI T2 acquisitions obtained from the axial (**a**), sagittal (**b**), and coronal (**c**) planes, showing a left parietal arteriovenous malformation (AVM) to be described below

3.2 Key Questions Regarding Case 1

3.2.1 *What Is the Location of the Arteriovenous Malformation AVM? Which Landmarks Should Be Intraoperatively Identified to Guide Resection?*

The AVM nidus boundaries are the postcentral sulcus and gyrus anteriorly, the parieto-occipital sulcus and occipital lobe posteriorly, and the intraparietal sulcus and inferior parietal lobule inferiorly. Therefore, they correspond to the limits of the superior parietal lobule.

The superior parietal lobule has a quadrangular shape and is usually connected to the postcentral gyrus through a gyral bridge at the end of the postcentral sulcus and to the occipital lobe posteriorly through the parieto-occipital arc. It can be subdivided by a superior vertical branch of the intraparietal sulcus, termed the transverse parietal sulcus of Brissaud. On the medial aspect, it continues as the precuneus.

The postcentral gyrus is oblique, posterior to the central sulcus, which runs parallel to the precentral gyrus. A useful feature to differentiate the pre- and postcentral gyri in the surgical field is that the postcentral gyrus is usually narrower and has a constant relationship with the intraparietal sulcus, coursing perpendicular to it. Often, the postcentral gyrus is crossed by gyral bridges connecting it to the superior parietal lobule and/or the supramarginal gyrus. Its cortex corresponds to the primary somatosensory area.

3.2.2 *What Is the Appropriate Craniotomy to Approach Such Pathology? Which Craniometric Landmarks Can Be Used to Plan It?*

As the AVM comprises the superior parietal lobule, the appropriate approach would be a parietal craniotomy.

The union between a few craniometric and sulcal points can provide imaginary lines resulting in a square surface corresponding to the parietal lobe, which can be used to plan the craniotomy if neuronavigation is not available.

The superior Rolandic point represents the intersection between the central sulcus and the interhemispheric fissure, located 5 cm posterior to the bregma (Fig. 3.3). The inferior Rolandic point represents the intersection between the inferior portion of the central sulcus and the Sylvian fissure corresponding to the connection between the squamous suture and a vertical imaginary line 4 cm above the tragus. A line drawn between these two points corresponds to the anterior limit of the parietal lobe. The superior limit is the sagittal suture, the inferior limit is the squamous suture that represents the Sylvian fissure, and the posterior limit is a line between the lambda and halfway to the asterion.

The intraparietal sulcus is usually at the level of the superior temporal line. Therefore, on the surface of the cranial vault, the superior parietal lobule is between the sagittal suture and the superior temporal line. The inferior parietal lobule is between the superior temporal line and the squamous suture (Figs. 3.3 and 3.4).

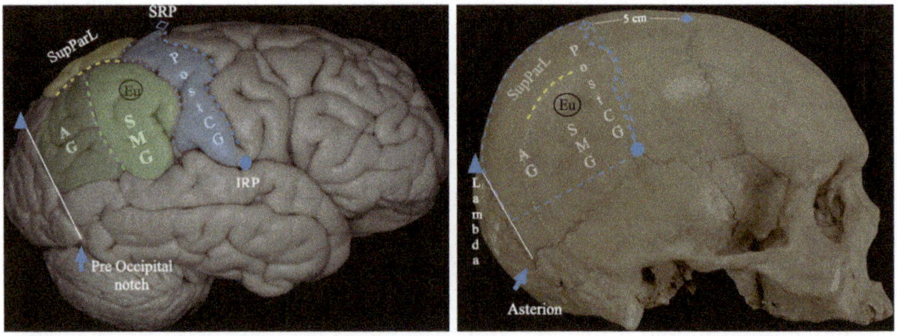

Fig. 3.3 Surface anatomy and correlation to craniometric points of the lateral surface of the parietal lobe. SRP superior Rolandic point, SupParL superior parietal line, SMG supramarginal gyrus, AG angular gyrus, IRP inferior Rolandic point, Eu eurion

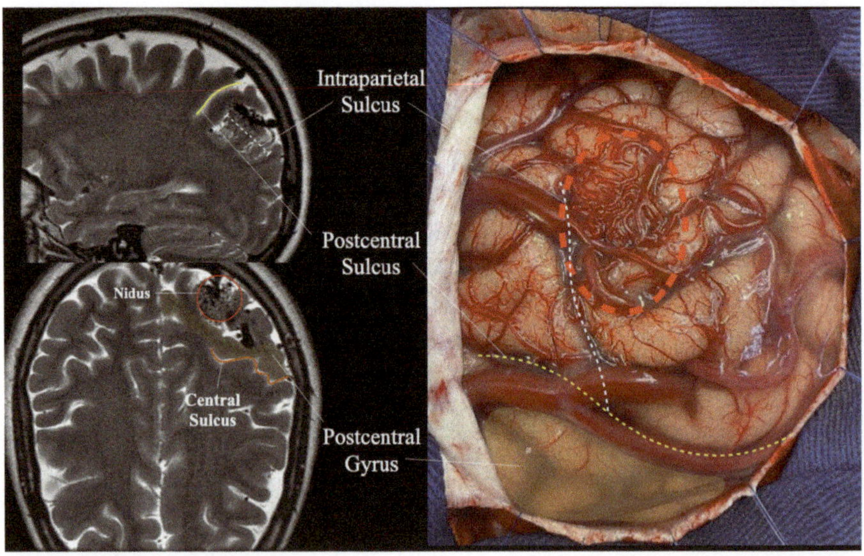

Fig. 3.4 Magnetic resonance imaging findings and surgical field correlation

3.2.3 *Which Deep Structures Are Involved in This Lesion? What Are the Functions Related to It? Is It Compatible with the Neuropsychological Evaluation?*

The parietal cortex is subdivided into the anterior parietal cortex, represented by the postcentral gyrus, which contains the primary somatosensory cortex, and posterior parietal cortex, which is formed by the superior and inferior lobules divided by the horizontal intraparietal sulcus. The posterior area contains the multimodal association cortex.

The superior parietal lobule covers the dorsal part of the posterior parietal cortex and is involved in several neurocognitive functions, such as visuospatial attention, working memory, and spatial orientation, and in multimodal sensory integrative connections.

Subcortical projections are abundant in the parietal lobe, represented by several white matter pathway connections. There are three types of white matter bundles: (1) commissural fibers, which connect both brain hemispheres and are mainly represented by the posterior portion of the corpus callosum; (2) projection fibers that connect the cortex with the deep encephalic structures and the spinal cord, represented by the sagittal stratum and internal capsule, and several others; and (3) association fibers that interconnect different cortical regions of the ipsilateral hemisphere composed of the superior longitudinal, arcuate, middle longitudinal, and inferior longitudinal fasciculi and the cingulum bundle.

The superior longitudinal fasciculus is located lateral to the centrum semiovale and is composed of association bundles that connect cortical areas of the frontal, occipital, and temporal lobes with the parietal lobe. It can be divided into three components: SLF I (superior longitudinal fasciculus) in the superior parietal lobule and superior frontal gyrus, connecting the premotor supplementary motor areas; SLF II, which links the inferior parietal lobule and the frontal cortex and is related to the visual, attention, and perception pathways; and SLF III, which is located in the frontoparietal operculum and is related to language cognition.

The arcuate fasciculus is an association curved bundle connecting Wernicke's and Broca's language areas and is therefore related to auditory and language functions. The middle longitudinal fasciculus connects the inferior parietal lobule cortex with the white matter of the superior temporal gyrus linking multiple paralimbic cortical areas and being responsible for language, organization, and memory. The inferior longitudinal fasciculus connects the parieto-occipital and temporal cortices, which comprise part of the lateral wall of the lateral ventricle; they are related to functions concerning vision. The cingulum bundle is within the cingulate gyrus, links it with the parietal and other cortices, and is involved in the emotional, memory, motivation, and nociception pathways.

On the superior parietal lobule, the association pathways are mainly represented by the superior longitudinal and arcuate fasciculi, which interconnect the superior parietal lobule and the precuneus with the supplementary motor areas in the frontal cortex, and the middle longitudinal fasciculus, which establishes connections between the superior parietal lobule and the temporal cortex. The projection fibers converge to the corona radiata and the commissural fibers to the corpus callosum.

3.2.4 *Regarding Case 1 (Figs. 3.2, 3.4, 3.5, and 3.6), on the Lateral Surface, Considering the Angiographic Images Shown Below, What Is the Related Vascularization?*

In case 1, the AVM is irrigated by the posterior parietal, last middle cerebral artery superior group ascending cortical branch and angular artery (M4 segment), and by the parieto-occipital, posterior cerebral artery cortical branch (P4 segment). The venous drainage pattern is through the superficial system: (1) toward the superior sagittal sinus through the superior parietal veins and (2) toward the transverse sinus through the posterior temporal vein. There is also retrograde venous flow through the vein of Trolard toward the superficial Sylvian vein.

The parietal lobe is irrigated on its lateral surface by the cortical branches of the middle cerebral artery (M4) (Fig. 3.7). The cortical branches dispose radially starting from the Sylvian fissure. There are mainly three ascending branches: the central, anterior parietal, and posterior parietal arteries and a horizontal angular artery. This orientation is useful for guiding angiographic identification.

The Rolandic or central artery is localized within the central sulcus and irrigates the posterior half of the postcentral gyrus. The anterior parietal artery travels through the postcentral sulcus and partially vascularizes the postcentral sulcus and superior and inferior parietal lobules. The posterior parietal artery partially supplies the superior and inferior parietal lobules. The angular branch is directed horizontally following the orientation of the Sylvian fissure and partially supplies the inferior parietal lobule, superior temporal lobe, and part of the occipital lobe (Fig. 3.7).

On the lateral and medial aspect of the parietal lobe, the superior cortical veins drain the superolateral and superomedial cortices into the superficial system through the superior sagittal sinus. They are highly variable, travel centrifugally, and have no valves. Angiographically, on the parietal lobe, they tend to course obliquely to the superior sagittal sinus.

3.3 Case 2 Lateral Surface

A 39-year-old, right-handed male patient with a 13-year medical history of epilepsy, treated with two antiepileptic drugs, was referred for investigation. The neurological examination was unremarkable, but the neuropsychological evaluation alerted

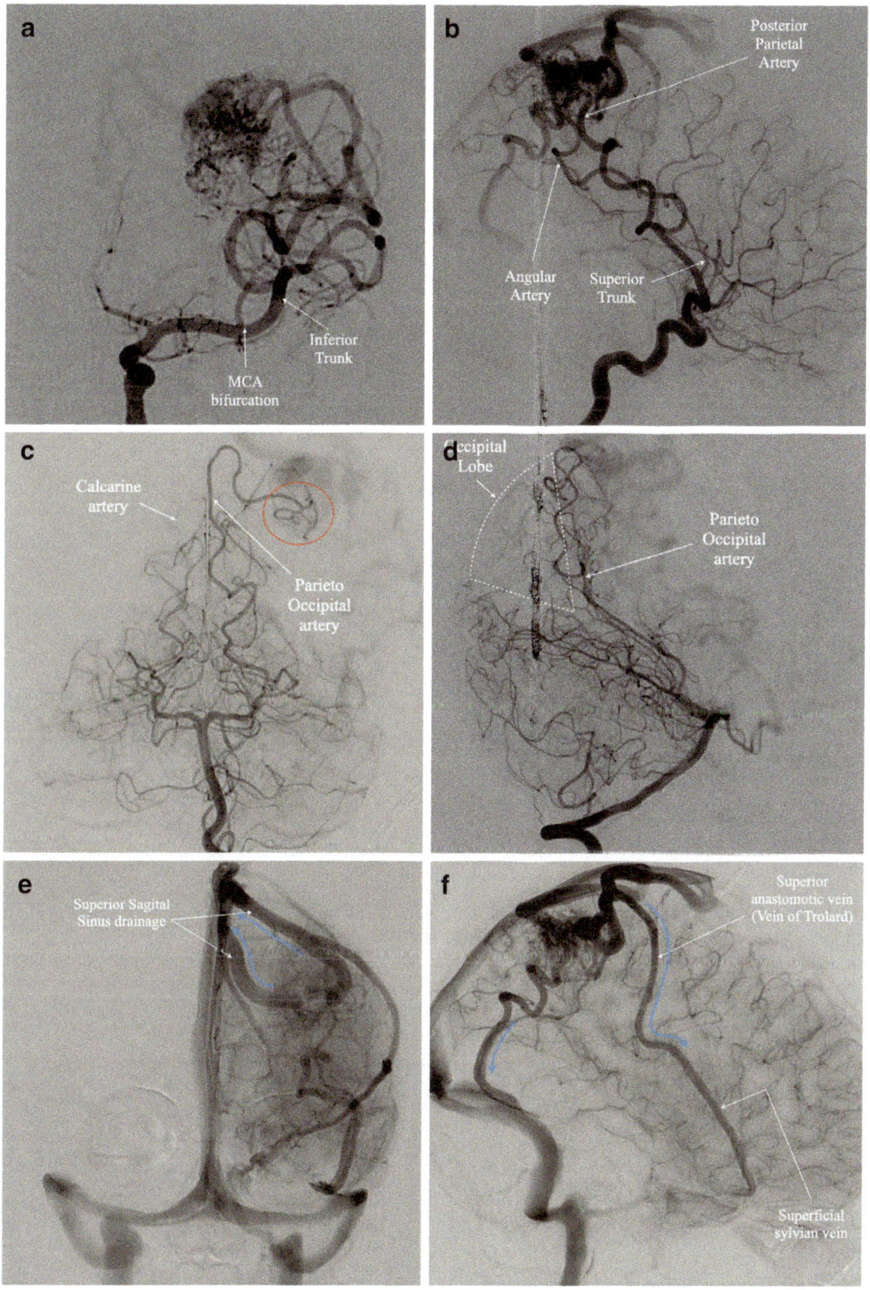

Fig. 3.5 Digital angiography, left carotid (**a**, **b**) injection. Left vertebral injection (**c**, **d**). Venous drainage (**e**, **f**)

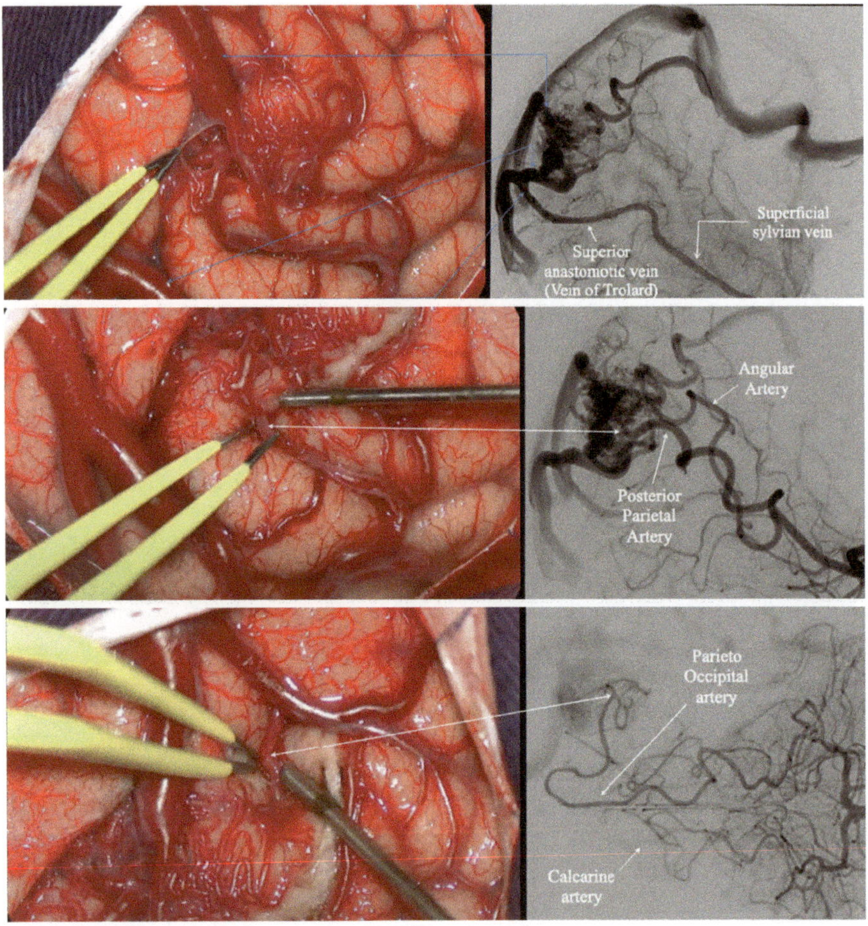

Fig. 3.6 Angiographic correlation of the venous and arterial vascularization of the AVM with intraoperative images

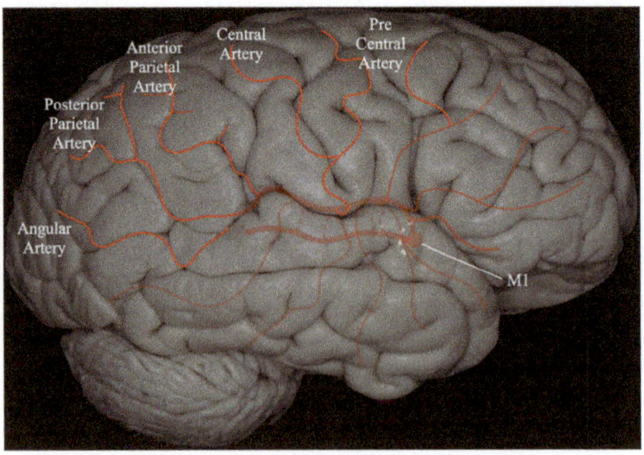

Fig. 3.7 Illustration of the middle cerebral artery cortical branches (M4 segment). The most common pattern involves the superior and inferior trunks having similar diameter (M2 segment), with the former giving rise to ascending groups and the latter to the descending branches. However, superior or inferior trunk domination may occur

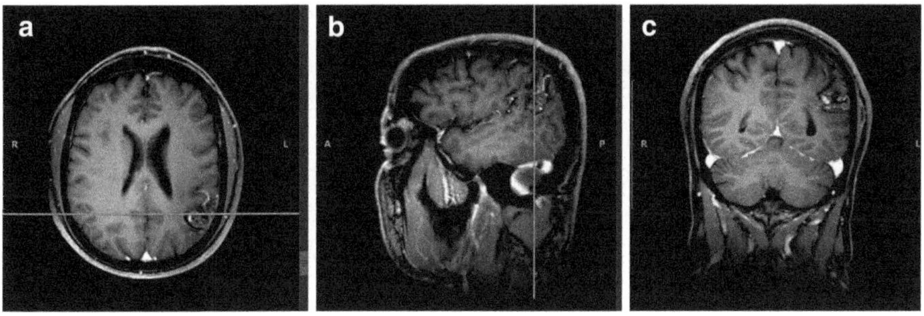

Fig. 3.8 T1 weighted brain angio MRI, obtained from the (**a**) axial, (**b**) sagittal, and (**c**) coronal planes, showing a left parietal AVM to be described below

of visual perception deficits as well as moderate deficits in executive functions, memory, and calculation. Figure 3.8 reveals the brain MRI of the patient.

3.3.1 Based on the Figure, What Is the Location of the AVM?

The AVM boundaries are superiorly the intraparietal sulcus, anteriorly the supramarginal gyrus, and posteroinferiorly the occipital lobe. Therefore, the AVM is located in the angular gyrus of the inferior parietal lobe.

On the lateral surface, the inferior parietal lobule is located below the intraparietal sulcus, which separates the superior and inferior lobules. The intraparietal sulcus is directed anteroposteriorly and has a low continuity rate. The intraparietal point is directed toward the roof of the atrium and occipital horn.

Two gyri form the inferior parietal lobule, i.e., the supramarginal gyrus anteriorly and the angular gyrus posteriorly, can be separated by a highly variable sulcus called the primary intermediate sulcus of Jensen, which represents the inferior vertical branch of the intraparietal sulcus.

The supramarginal gyrus forms a horseshoe shape, embracing the end of the lateral fissure. The inferior convolution of the supramarginal gyri merges with the superior temporal gyrus, forming the temporoparietal junction.

The angular gyrus lies posteriorly adjacent to the supramarginal gyrus. The most inferior portion is usually continuous with the middle temporal gyrus. The internal configuration of the angular gyrus is determined by the pattern of branching of the superior temporal sulcus. Usually, there are three terminal branches, a vertical branch that can reach the sulcus of Jensen, a horizontal middle one called the angular sulcus, and an inferior, less constant one. Therefore, a tip to identify the angular gyrus is to follow the posterior projection of the superior temporal sulcus. The angular gyrus ends at the occipital lobe.

3.3.2 Are There Any Craniometric Points Relevant to the Inferior Parietal Lobule?

The craniometric points of importance for the inferior parietal lobule are (1) the intraparietal point, which corresponds to the point of intersection between the intraparietal sulcus and the postcentral sulcus. It is located 6 cm anterior to the lambdoid suture and 5 cm lateral to the sagittal suture and halfway to the asterion. (2) Eurion: The eurion is located at the central point of the parietal eminence, anatomically related to the supramarginal gyri. The impression of the supramarginal gyrus protrusion on the parietal bone characterizes the parietal tuberosity or bossa, corresponding to the eurion.

3.3.3 Which Are the Deep Structures Related to the Inferior Parietal Lobule?

The posterior parietal cortex is formed by the superior and inferior lobules and the cortex within the region of the intraparietal sulcus. The inferior parietal lobule covers the ventral aspect of the posterior parietal cortex and is involved in several neurocognitive domains such as language processing, visuospatial orientation, memory, auditory function, arithmetic ability, praxis, and executive functions.

The inferior parietal lobule is anatomically connected mainly via association fibers to other cortical and subcortical areas. Two major fiber bundles dominate the architecture of the white matter underlying the inferior parietal lobule: the arcuate fasciculi, which provide connections to Broca's area and are therefore related to language, and the superior longitudinal fasciculus, which provides connections between the inferior parietal lobule and the frontal cortex.

The posterior portion of the lateral ventricle's body and atrium is located deeply at the level of the inferior portion of the postcentral gyrus and inferior parietal lobule, respectively.

3.3.4 Finally, Still Regarding Case 2, What Is the Vascularization of This AVM?

This AVM is irrigated by the angular and temporo-occipital arteries. The angular artery merges from the posterior end of the Sylvian fissure, coursing over the superior temporal gyrus until the occipital lobe. The temporo-occipital artery supplies the posterior half of the superior temporal gyrus and can share a common trunk with the angular artery (Fig. 3.9).

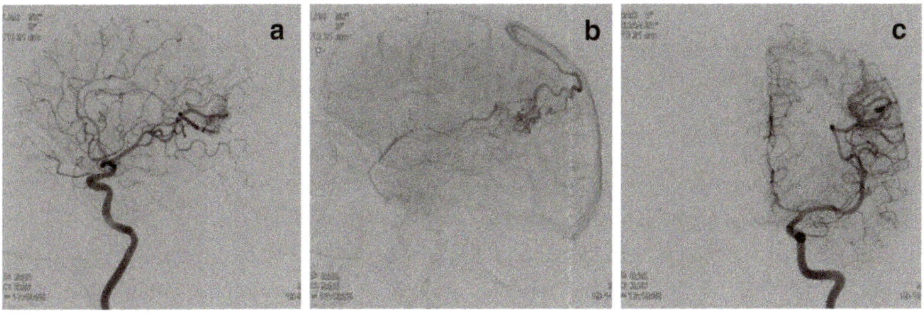

Fig. 3.9 Angiographic aspect of the parietal AVM. (**a**) shows the internal carotid artery at lateral view; (**b**) shows the early venous phase; (**c**) postero-anterior view of the left carotid artery injection, arterial phase

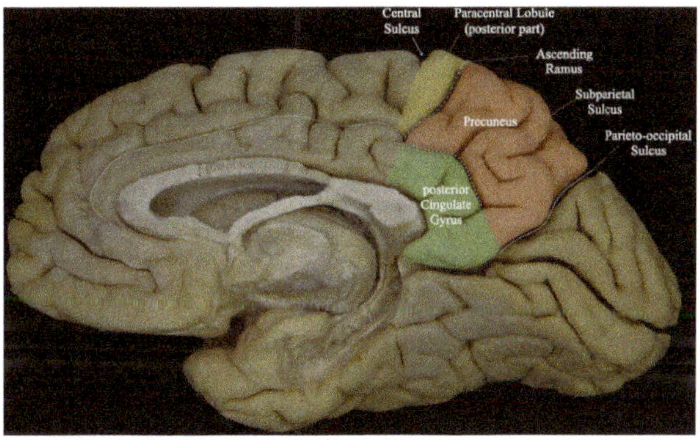

Fig. 3.10 Medial surface of the parietal lobe. It is composed by the precuneus (red), the posterior part of the cingulate gyrus (green), and the posterior paracentral lobule (yellow)

Drainage is accomplished through a parietal ascending vein to the superficial system through the superior sagittal sinus; there is also a small reflux to the superficial Sylvian vein (*another case example is also demonstrated in the* Video 3.1).

3.4 Medial Surface

On the medial aspect, the parietal lobe is delimited anteriorly by the central sulcus and posteriorly by the parieto-occipital sulcus. It includes the precuneus, posterior portion of the cingulate gyrus, and posterior paracentral lobule (Fig. 3.10).

The precuneus is a quadrilateral area that represents the medial extension of the superior parietal lobule, anteriorly delineated by the marginal branch of the cingulate sulcus, inferiorly by the subparietal sulcus, and posteriorly by the parieto-occipital sulcus. The posterior paracentral lobule represents the medial extension of the postcentral gyrus. The cingulate gyrus can be divided into three parts: an anterior part inferior to the superior frontal gyrus, a midcingulate cortex inferior to the paracentral lobule, and a posterior portion inferior to the precuneus. The last two can be considered in the study of the medial aspect of the parietal lobe (Fig. 3.10).

The sulci of importance on the medial surface are the ascending ramus of the cingulate sulcus, comprising the anterior boundary between the paracentral lobule and the precuneus; the cingulate sulcus, which separates the cingulate gyri from the paracentral lobule; and the subparietal sulcus, which separates the precuneus from the cingulate gyrus (Fig. 3.10).

Case 3. A 6-year-old male patient presented with a 6-month, left-side, pulsatile headache. The neurological examination showed no abnormalities (Figs. 3.11 and 3.12).

3.4.1 Where Is the AVM Located?

This AVM is located in the medial aspect of the parietal lobe, at the posterior part of the precuneus. Its nidus is limited posteriorly by the parieto-occipital sulcus and inferiorly by the subparietal sulcus, coinciding with the precuneus limits.

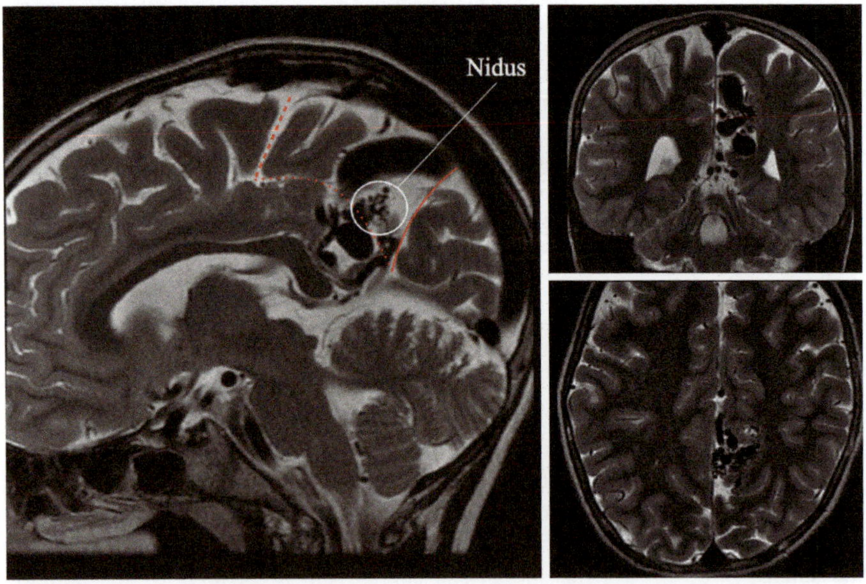

Fig. 3.11 Magnetic resonance imaging T2 acquisitions. Sagittal sequence (left) showing the AVM in the medial surface of the parietal lobe. The precuneus is delimited by the ascending ramus of the cingulate sulcus (red dashed line), subparietal sulcus (red dotted line), and parieto-occipital sulcus (red line). In the right, coronal (upper) and axial (lower) sequences

3.4.2 Which Arteries Supply This AVM?

The medial aspect of the parietal lobe is supplied: (1) anteriorly by the anterior cerebral artery distal branches (A5) and (2) posteriorly by the posterior cerebral artery branches. The distal branches of the middle, posterior, and anterior cerebral arteries may exhibit extensive anastomosis (Fig. 3.13).

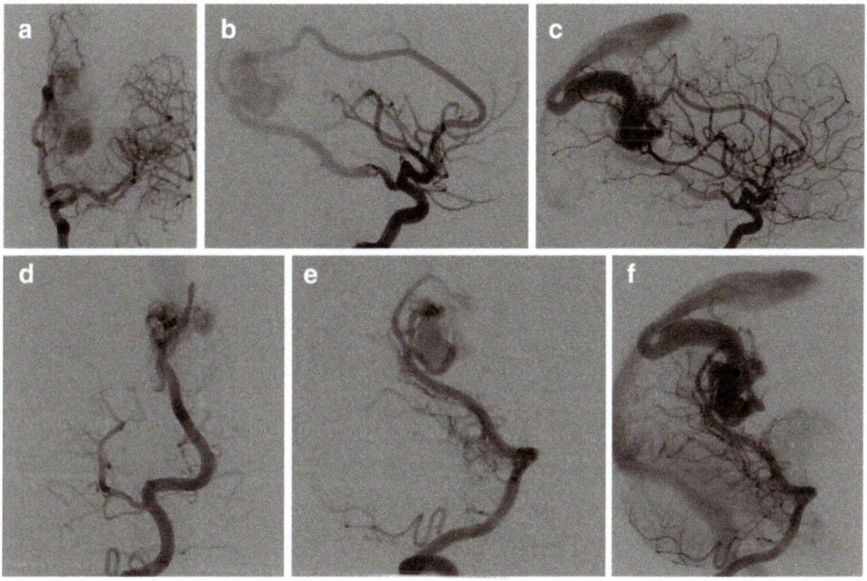

Figs. 3.12 Digital arteriography. (**a**) Postero-anterior view of left carotid artery injection; (**b**, **c**) postero-anterior and lateral view of the early anterior phase; (**d**) vertebral injection; (**e**) postero-anterior view of the vertebral injection in the early arterial phase; (**f**) postero-anterior view of the vertebral injection in the venous phase

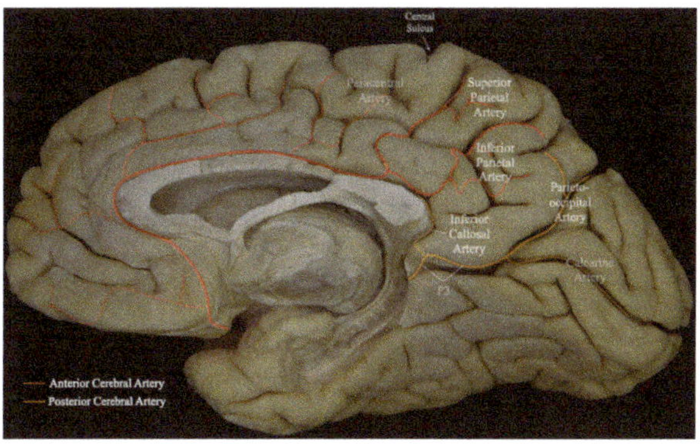

Fig. 3.13 Illustration of the arterial vascularization of the medial surface of the parietal lobe

The anterior cerebral artery branches that supply the medial parietal lobe are the paracentral, superior parietal, and inferior parietal arteries. The posterior cerebral artery branch that supplies the medial aspect of the parietal lobe is the parieto-occipital artery, which travels within the parieto-occipital fissure and supplies the precuneus and posterior cingulate gyrus (Fig. 3.13).

The AVM presents superficial venous drainage through an ectasic cortical vein connected to the superior sagittal sinus.

3.5 Sylvian Surface

The parietal lobe has a short Sylvian surface between the central sulcus and supramarginal gyrus, which faces the Sylvian surface of the temporal lobe and the insula, composing the parietal operculum. It is composed of the inferior portion of the postcentral and the supramarginal gyri, which form the posterior roof of the Sylvian fissure.

On such a surface, the supramarginal gyrus faces the gyri forming the posterior part of the planum temporale, and the postcentral gyrus faces Heschl's gyrus. Heschl's gyrus points to the atrium. The parietal operculum contains the secondary somatosensory cortex.

3.6 Occipital Lobe

The occipital lobe has no evident separation from the temporal and parietal lobe; thus, it is important to emphasize that the boundaries of the lobe are arbitrary. The anterior limit at the lateral surface is an imaginary line between the pretemporo-occipital notch and the impression of the parieto-occipital sulcus. The superomedial limit is the interhemispheric fissure, and inferior-lateral limit is the border that is parallel to the transverse sinus. It is important to note that the impression of the parieto-occipital sulcus in the lateral surface is marked in the cranial surface by the union of the lambdoid suture to the sagittal suture; this junction occurs 6–7 cm ahead of the inion, the external occipital protuberance. The intrinsic anatomy exhibits different patterns compared to the previously described lobes. To date, there are at least seven different classifications. Therefore, we will consider the two most prevalent forms, one that considers three gyri and two sulci—the superior, middle, and inferior occipital gyri separated by the interoccipital and lateral occipital sulci—and the other that considers two gyri separated by one sulcus, superior and inferior occipital gyri, and lateral occipital sulcus. However, in this case, the superior occipital gyrus can harbor the interoccipital sulci and transverse occipital gyri (Fig. 3.14).

The occipital lobe has its best definition in the medial surface (Figs. 3.15 and 3.16) because it is separated from the quadrangular lobule by the parieto-occipital sulcus or fissure that is very deep with 100% constant sulcus. The occipital lobe in the medial surface is composed of two structures separated by the calcarine sulcus, cuneiform, and lingual gyrus; the cuneiform is superior and triangular. The elongated lingual gyrus continues anteriorly as parahippocampal gyrus in the temporal lobe.

There is functional importance in recognizing the calcarine sulcus because the neighboring gyri are the primary visual area. However, the calcarine sulcus gives another essential information when performing the occipital transtentorial approach to the pineal region; the best way of finding the cistern of the quadrigeminal lamina is to follow the sulcus anteriorly. It also contains the posterior cerebral artery. The posterior projection of the parahippocampal gyrus is called the lingual gyrus, which makes up the occipital lobe and is an eloquent area.

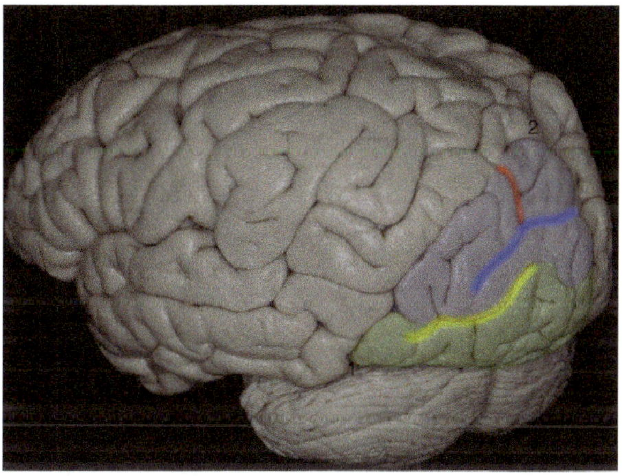

Fig. 3.14 This figure illustrates the posterior view of the lateral surface. The occipital lobe has many possible variations. Here, we show one type of presentation where the lateral occipital sulcus (yellow) separates the occipital lobe in two gyri, the superior occipital gyrus (purple) and inferior occipital gyrus (green). Inside the superior occipital gyrus, we can see the transverse occipital sulcus (blue) establishing a connection with the intra-occipital sulcus (red) that is the direct continuation of the intra-parietal sulcus. Number 1 represents the temporo-occipital notch that is the impression of the vein of Labbè at the point where the vein drains to the transverse-sigmoid sinus junction. Number 2 represents the impression of the parieto-occipital sulcus in the lateral surface. By connecting these two numbers as an imaginary line, we delineate the anterior limit of the occipital lobe once there are no clear separations between the occipital lobe and both the temporal and parietal lobe

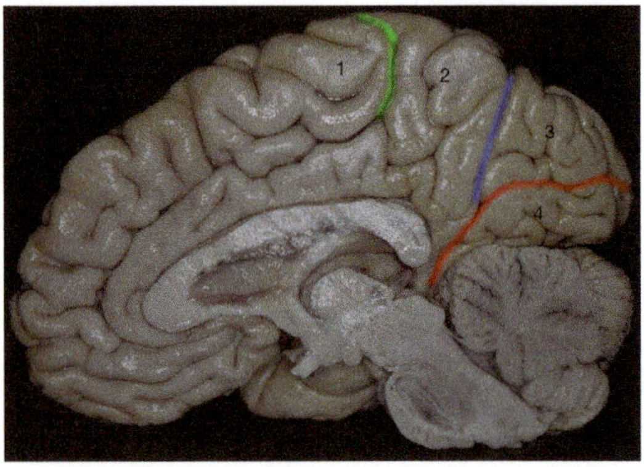

Fig. 3.15 The medial surface of the brain. The green sulcus is the marginal branch of the cingulate sulcus, which causes an impression of the postcentral gyrus named "pars marginalis" and separates the paracentral lobule from the quadrangular lobule (2) that is the representation of the parietal lobe in the medial surface; parieto-occipital sulcus (blue) is one of the major landmarks of the medial surface; it is 100% constant sulcus. Particularly, a deep sulcus is sometimes called as parieto-occipital fissure; its superior end causes an impression in the medial surface, which helps to delineate the parietal lobe from the occipital lobe. The cuneiform lobe (3) represents the occipital lobe together with the lingual gyrus (4); the cuneiform lobe and lingual gyrus as the margins of the calcarine sulcus (red) are also called the calcarine fissure (red), which also has a deep impression in the medial surface, and it is beyond being the representation of the visual primary area. This fissure has an important role in the occipital transtentorial approaches, as demonstrated in the figure if you follow anteriorly this sulcus; its anterior end reaches the pineal region or the quadrigeminal cistern

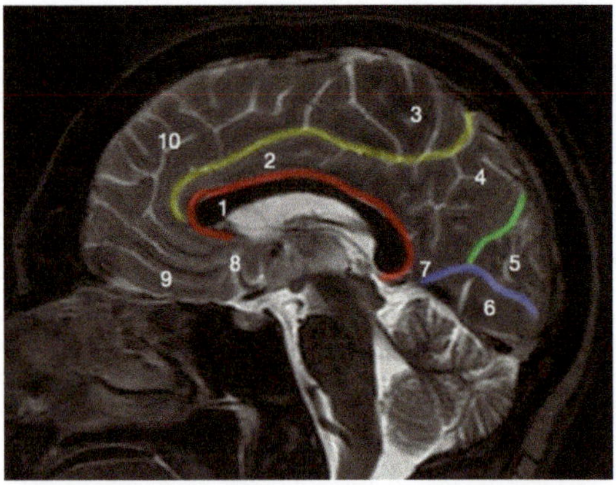

Fig. 3.16 This is a sagittal view of a Cube-FLAIR sequence of the magnetic resonance imaging of the brain that reviews all the major anatomic strictures of the medial surface. The corpus callosum (1) is encircled by the sulcus of the corpus callosum (red). The cingulate gyrus (2) is just above the sulcus of the corpus callosum (red) and, below its own sulcus, the cingulate gyrus sulcus (yellow). This sulcus has a marginal branch that delineates the paracentral lobule (3) from the quadrangular lobe (4). The last is separated by the parieto-occipital sulcus (green) from the cuneiform gyrus (5) that together with the lingual gyrus (6) represents the occipital lobe in the medial surface. They are separated by the calcarine sulcus (blue), which is deep and had an anterior pathway until the quadrigeminal cistern. The other numbered structures are isthmus of the cingulate gyrus (7), subcallosal area (8), rectus gyrus (9), and frontal medial gyrus (10)

Suggested Bibliography

Binkofski FC, Klann J, Caspers S. The neuroanatomy and functional role of the inferior parietal lobule and intraparietal sulcus. Neurobiol Lang. 2016. https://doi.org/10.1016/B978-0-12-407794-2.00004-3

Burks JD, Boettcher LB, Conner AK, Glenn CA, Bonney PA, Baker CM, et al. White matter connections of the inferior parietal lobule: A study of surgical anatomy. Brain Behav. 2017;7:e00640.

Caspers S, Amunts K, Zilles K. Posterior parietal cortex: multimodal association cortex. The human nervous system. . 3rd ed. 2012. https://doi.org/10.1016/B978-0-12-3742360.10028-8

Frigeri T, Paglioli E, de Oliveira E, Rhoton AL Jr. Microsurgical anatomy of the central lobe. J Neurosurg. 2015;122:483–98.

Koutsarnakis C, Liakos F, Kalyvas AV, Liouta E, Emelifeonwu J, Kalamatianos T, et al. Approaching the atrium through the intraparietal sulcus: mapping the sulcal morphology and correlating the surgical corridor to underlying fiber tracts. Oper Neurosurg (Hagerstown). 2017;13:503–16.

Martino J, De Lucas M. Subcortical anatomy of the lateral association fascicles of the brain: a review. Clin Anat. 2014;27:563–9.

Monroy-Sosa A, Jennings J, Chakravarthi S, Fukui MB, Celix J, Kojis N, et al. Microsurgical anatomy of the vertical rami of the superior longitudinal fasciculus: an intraparietal sulcus dissection study. Oper Neurosurg (Hagerstown). 2019;16:226–38.

Ono M, Kubik S, Abernathey CD. Atlas of the cerebral sulci. Stuttgart: Georg Thieme Verlag; 1990.

Ribas GC, Yasuda A, Ribas EC, Nishikuni K, Rodrigues AJ Jr. Surgical anatomy of microneurosurgical sulcal key points. Neurosurgery. 2006;59:ONS177–210, discussion ONS210–1.

Serrato-Avila JL, da Costa MDS, Stávale JN, Lima JVF, Carrasco-Hernandez JP, Alejandro SA, Chaddad-Neto F. Microsurgical Resection of a Left Supramarginal Gyrus AVM Causing Radionecrosis. World Neurosurg. 2020;138:317.

Vallar G. and Coslett HB eds. The parietal lobe. Handbook of clinical neurology, Vol. 151 (3rd series). https://doi.org/10.1016/B978-0-444-63622-5.00003-6.

Wild HM, Heckemann RA, Studholme C, Hammers A. Gyri of the human parietal lobe: volumes, spatial extents, automatic labelling, and probabilistic atlases. PLoS One. 2017;12:e0180866.

Chapter 4
Surgical Anatomy of the Temporal Lobe

4.1 Introduction

The anatomy of the temporal lobe can be divided into four surfaces: lateral, inferior (or basal), superior (or Sylvian), and medial (Fig. 4.1). The temporal lobe is located beneath the Sylvian fissure, behind the greater wing of the sphenoid bone, and rests on the lateral middle fossa of the skull base and the anterior part of the tentorium (the posterior part of the basal surface). Medially, the temporal lobe encircles the cerebral peduncles and their respective cisterns and cisternal structures. Posteriorly, the temporal lobe is limited by an imaginary vertical line that runs from the preoccipital notch, in the inferolateral surface, to the superior point of the parieto-occipital sulcus (the sulcus of the medial surface). Behind the posterior limit of the Sylvian fissure, the temporal lobe is separated from the parietal lobe by an imaginary horizontal linc that runs from that point to the vertical imaginary line.

External features of the cranium and craniometric points allow delimitation of the temporal lobe. The squamous suture (the suture between the temporal and parietal bones), from its most anterior point (the anterior squamous point, AntSqP) until its most superior point (the superior squamous point, SupSqP), runs at the same level as the Sylvian fissure. The SupSqP is located at the same plane of an imaginary vertical line passing through the anterior ear depression. These two craniometric points are, respectively, located at the same level as the anterior Sylvian point (AntSyP), the point at which the apex of the pars triangularis meets the Sylvian fissure, and the inferior Rolandic point (IRP), the midpoint of the subcentral gyrus facing the Sylvian fissure. The ASyP represents the position of the limen of the insula on the lateral surface of the brain, that is, the point of transition of the

Supplementary Information The online version of this chapter (https://doi.org/10.1007/978-3-030-82747-2_4) contains supplementary material, which is available to authorized users.

F. Chaddad-Neto, M. D. Silva da Costa, *Microneuroanatomy and Surgery*,
https://doi.org/10.1007/978-3-030-82747-2_4

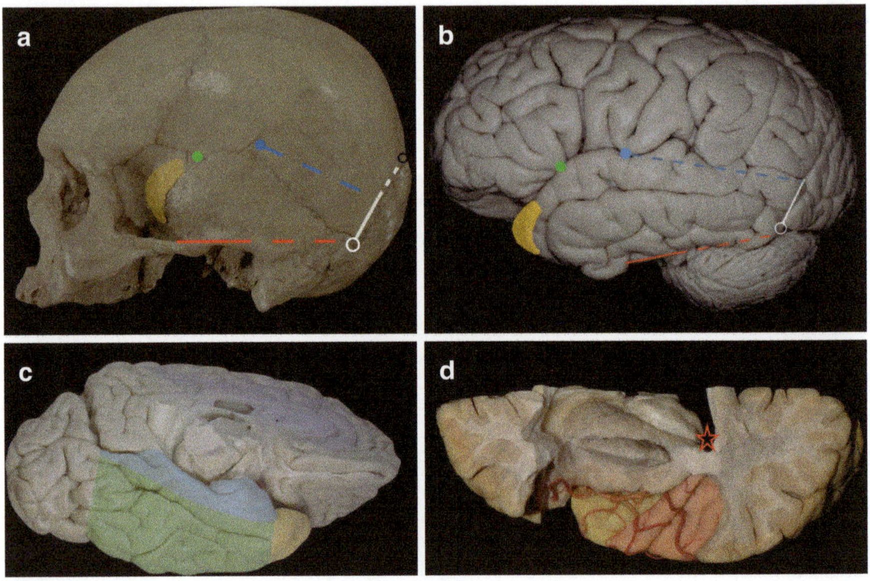

Fig. 4.1 Temporal lobe surfaces and cranial correlations. The delimitation of the lateral surface of the temporal lobe in the skull is illustrated in **a** and **b**. The yellow shading is located below the great wing of the sphenoid bone, which covers the temporal pole. The green dot corresponds to the AntSqP/AntSyP and the blue dot to the SupSqP/IRP. The blue dashed line is the superior limit of the posterior part of the temporal lobe. The red dashed line indicates the superior border of the zygomatic arch, which correlates to the inferior border of the inferior temporal gyrus. The red dashed line meets the asterion/preoccipital notch (white circle), the posterior limit of the inferior of the lateral surface of the temporal lobe. In A, the black circle indicates the angle between the sagittal and lambdoid suture, which coincides to the superomedial limit of the parietal lobe (not seen in **b**). The white line indicates the limit between the temporal and occipital lateral surfaces. (**c**) Basal surface highlighted in green, the medial surface in blue and the temporal pole in yellow. (**d**) Transversal cut of the left hemisphere at the level of the atrium (red star). The superior surface of the temporal lobe is divided into the *planum polaris* (yellow shading) and the *planum temporalis* (red shading). (Note that the anterior long transverse gyrus points to the atrium. *AntSqP* anterior squamous point, *AntSyP* anterior Sylvian point, *SupSqP* superior squamous point, *IRP* inferior Rolandic point)

M1 (medial to lateral trajectory) to M2 (anterior to posterior trajectory) segment of the middle cerebral artery (MCA). The greater wing of the sphenoid bone encompasses the temporal lobe pole. The superior border of the zygomatic arch is in the same line as the inferolateral border of the inferior temporal gyrus. The asterion (the meeting of the parietomastoid, occipitomastoid, and lambdoid sutures) corresponds to the preoccipital notch, and the angle point at the meeting of the lambdoid and sagittal sutures corresponds to the superior limit of the parieto-occipital sulcus. Thus, a line along the lambdoid suture permits the estimation of the anterior limit of the occipital lobe. In the same way, a horizontal line from the SupSqP to the lambdoid line estimates the superior limit of the posterior part of the temporal lobe.

4.2 Inferior Horn of the Lateral Ventricle and Boundaries

The temporal or inferior horn of the lateral ventricle is located at the level of the middle temporal gyrus, approximately 3 cm deep to its cortical surface and 3 cm from the anterior limit of the temporal pole. This cavity contains limbic structures, the head and body of the hippocampus and the amygdala, and is surrounded by white matter bundles of projection fibers (optic radiation), commissural fibers (corpus callosum – tapetum and anterior commissure), and association fibers (superior longitudinal fascicle, SLF; inferior longitudinal fascicle, ILF; and the inferior fronto-orbital fascicle, IFOF).

The anterior wall of the inferior horn is molded by the impression of the amygdaloid nucleus. The tapetum, the corpus callosum fibers extending to the temporal lobe, constitutes the lateral wall and the lateral part of the roof. The tail of the caudate nucleus complements the medial part of the roof and the superior part of the medial wall. The floor is composed of the collateral eminence, which is the bottom of the collateral sulcus on the basal surface of the temporal lobe. The head and body of the hippocampus, the fimbria, and the choroid plexus delimit the inferior part of the medial wall. These structures comprise the ventricular part of the medial temporal lobe.

Important white matter bundles enter and exit the temporal lobes, some of which surround the temporal horn of the lateral ventricle. Outside the tapetum, the optic radiation bundles run along the lateral wall and roof of the temporal horn. Outside the optic fibers, IFOF runs along the middle and posterior parts of the superolateral aspect of the temporal horn. The ILF fibers course below the temporal part of the IFOF, connecting the anterior temporal lobe to the occipital lobe. In the superior and middle temporal regions, outside the IFOF and ILF fibers, the SLF runs, which can be divided into three bundles: the temporoparietal vertical-oriented bundle, the fronto-temporal arch-like bundle (called the arcuate fasciculus), and the frontoparietal horizontal-oriented bundle, which is not related to the temporal lobe (Fig. 4.2).

4.3 Temporal Stem

Projection, association, and commissural fiber bundles pass over the temporal horn of the lateral ventricle and cross the inferior limiting sulcus of the insula (ILS), from the level of the amygdala anteriorly to the lateral geniculate body (Fig. 4.3). These white matter bundles attach the temporal lobe to the insula and are clinically relevant in relation to the spread of tumors and epileptic seizures. Therefore, they are important as a surgical route to the inferior horn and mesial structures. These fibers can be described according to an anterior to posterior, lateral to medial, and cranial to caudal arrangement, as follows:

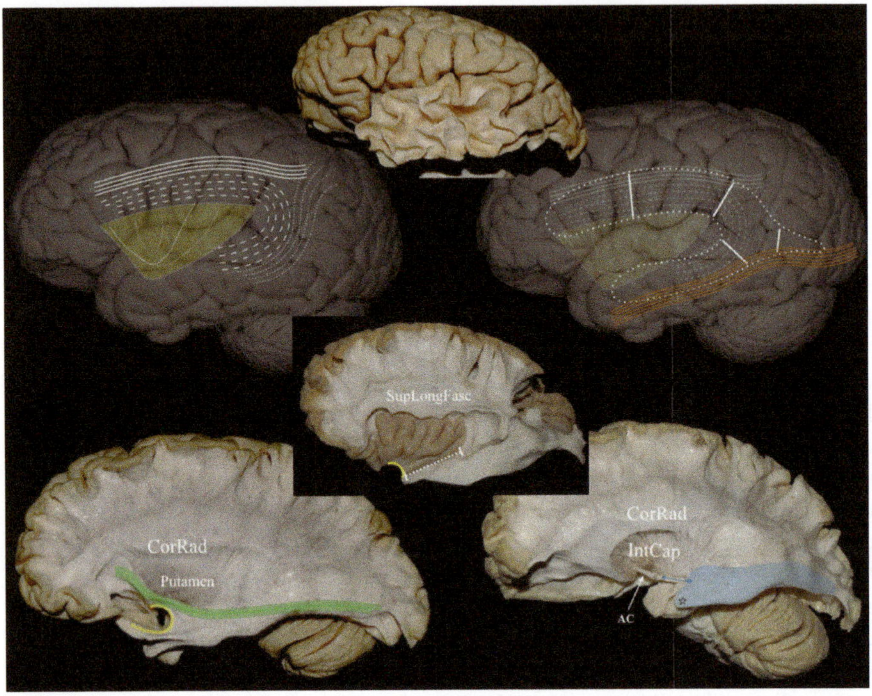

Fig. 4.2 Temporal lobe from outside to inside. Middle and superior image – the superior, middle, and inferior temporal gyrus cortex was resected through Klingler's technique, exposing the subcortical short U fibers. Left and superior image – projection of the most superficial white matter bundles of the temporal lobe: the temporal part of the arcuate fasciculus (dashed lines) and temporoparietal component (dotted lines) of the SLF. The arcuate fasciculus surrounds the posterior insula (illustrated in yellow). Right and superior image – the ventricular cavity (light dotted line) is projected in the lateral surface of the left hemisphere. The ILF (red lines) runs along the inferior part of the middle temporal gyrus and the temporal horn and is deep and inferior to the SLF bundles. Middle and inferior image – the inferior limiting sulcus of the insula (white-dotted line) is the path of the temporal stem. Left and inferior image – the uncinate fascicle (yellow) and IFOF (green) topography is projected. They comprise the anterior and lateral layers of the temporal stem. Right and inferior image – the anterior commissure fibers (white arrow) cross the inferior limiting sulcus just medial and behind the uncinate fascicles and IFOF. The OR (highlighted in blue) is the most deep and posterior fibers of the temporal stem. The Meyer's loop (star) comprises the anterior bundle of the OR, connecting to the LGB (projected as the blue circle), which connects to the optic tract (blue arrow). The OR passes under the posterior part of the lentiform nucleus, contributing to the sublenticular part of the internal capsule. SLF superior longitudinal fascicle, ILF inferior longitudinal fascicle, IFOF inferior fronto-occipital fascicle, OR optic radiation, LGB lateral geniculate body, AC anterior-commissure

- The uncinate fasciculus (UF) is the most anterior fibers from the limen of the insula, coming from the temporal pole, passing through the extreme/external capsule, ventrally to the claustrum, at the anteroinferior part of the insula, reaching the orbitofrontal region.

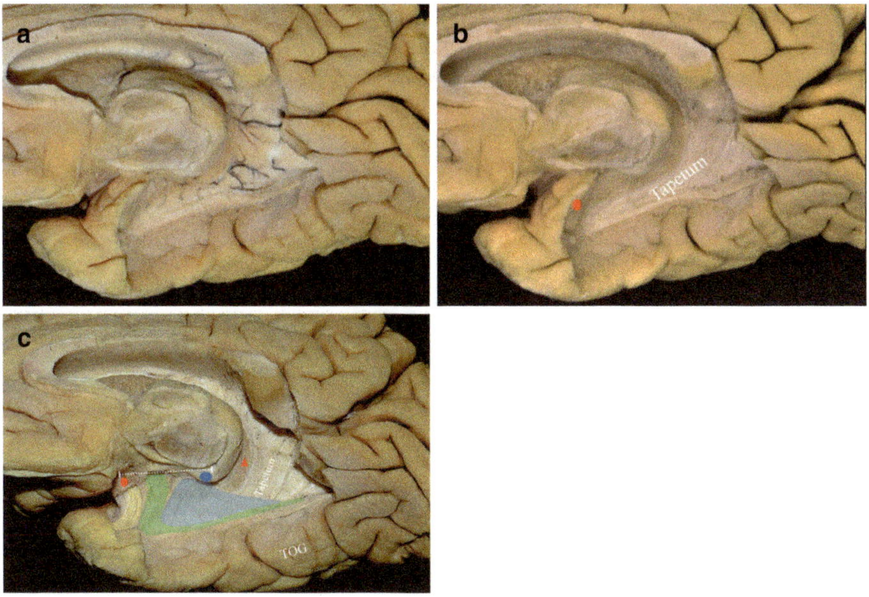

Fig. 4.3 Temporal lobe from inside to outside. The uncus, PHG, part of the amygdala, the hippocampus, fornix, and choroid plexus were removed. (**a**) The ependymal layer and ependymal veins are visualized at the roof, floor, and lateral wall of the temporal horn and atrium. (**b**) The ependyma was removed. The vertical-oriented fibers of the tapetum are identified. (**c**) The tapetum fibers at the temporal horn and part of the tail of the caudate nucleus (red triangle) were resected. The ILS of the insula is projected (dotted line) from the level of the amygdala (red circle) to the LGB (blue circle). Part of the OR (blue) can be identified just lateral to the tapetum. Observe the different orientations of the fibers. The IFOF is highlighted in green and the UF in yellow. (*TOG* temporo-occipital gyrus, *PHG* parahippocampal gyrus, *ILS* inferior limiting sulcus, *OR* optic radiation, *IFOF* inferior fronto-occipital fascicle)

- The inferior fronto-occipital fascicle (IFOF) runs superiorly and posteriorly. These association fibers come from the inferior frontal lobe, cross the ILS in the extreme/external capsule, and run ventrally to the claustrum, going posteriorly to the posterior temporal, parietal, and occipital lobes.

- The anterior commissure (AC) crosses the ILS behind and deep to the UF and IFOF. The AC fibers cross the midline just anteriorly to the anterior fornix column, and then some of them turn anteriorly to connect to the olfactory system, and the others go straight, pass below the globus pallidus, and cross the ILS to reach and connect the temporal lobes, ending in the amygdaloid nucleus.

- The optic radiation (OR) contributes to the temporal stem in more posterior and deeper trajected fibers. It originates in the lateral geniculate body (LGB) and projects to the calcarine cortex. From the metathalamus, the OR passes above the stria medullaris and the tail of the caudate nucleus, contributing to the sublenticular part of the internal capsule. Along its path, the corpus callosal fibers at the roof and lateral wall of the inferior horn and atrium, and the lateral wall and part of the roof of the occipital horn of the lateral ventricle. The OR is divided into three bundles. The anterior bundle, or Meyer's loop, assumes an anterior direction from its origin, along the roof of the inferior horn, going approximately 5 mm beyond the anterior wall and distancing approximately 30 mm from the temporal tip. At this point, it turns sharply and runs posteriorly with the central and posterior bundle as part of the sagittal stratum (Fig. 4.4). The anterior bundle fibers end at the lingual gyrus along the inferior lip of the posterior part of the calcarine fissure. The central bundle fibers end at the lateral part of the temporal pole. The posterior bundle fibers end along the superior lip of the posterior part of the calcarine fissure at the cuneus gyrus.

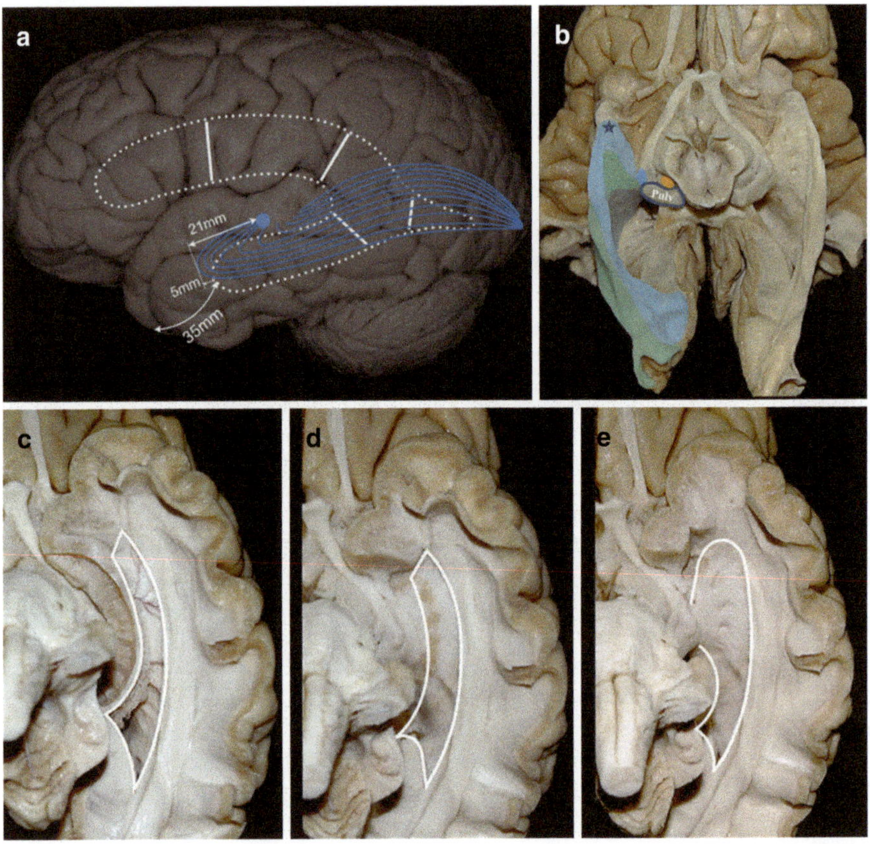

Fig. 4.4 Optic radiation (OR) and correlation to the temporal horn (TH). (**a**) The OR course along the lateral wall of the occipital horn and roof and lateral wall of the atrium and TH. The anterior limit of Meyer's loop (anterior bundle) goes approximately 5 mm beyond the anterior limit of the TH and about 21 mm from the geniculate body (blue circle). The anterior limit of the TH is about 35 mm from the temporal tip. (**b**) The anterior (light blue), central (green), and posterior (dark blue) bundles of the OR running from the LGB (blue circle) and pulvinar of the thalamus (dark blue line) toward the occipital lobe. The MGB is highlighted in orange. The blue star indicates Meyer's loop. (**c**, **d**, and **e**) The ependymal layer, tapetum, and OR, respectively, from inside to outside covering the lateral wall and roof of the TH. (*TH* temporal horn, *OR* optic radiation, *LGB* lateral geniculate body, *MGB* medial geniculate body)

4.4 Sagittal Stratum

The sagittal stratum (SS) is the set of sagittal white mater bundles running laterally to the atrium. These fibers are contiguous with the fibers of the temporal stem, from the posterior end of the ILS of the insula. The ILF is the most superficial layer, and the IFOF is the intermediate component. The OR is the deepest layer and is the main component of the SS.

4.5 Functions of the Temporal Lobe

The superior temporal gyrus is related to the function of semantic language, auditory comprehension, and naming living objects. The middle temporal gyrus is related to phonological function. Finally, the inferior temporal gyrus relates the phonological interface to semantics, naming non-living objects (colors, places). Some studies suggest that an area may also be related to numerical recognition and face perception (Fig. 4.5).

The IFOF is related to semantic association, reading, and judgment. The ILF participates in reading phonology, semantics, and naming objects. The arcuate fasciculus works on articulatory processing and phonology.

4.5.1 Lateral Surface of the Temporal Lobe

Approaches to the lateral surface of the temporal lobe pathologies must be planned in an attempt to give the surgeon a perpendicular and direct view of the lesion. For this purpose, head position, skin incision line, and extension of craniotomy depend on the topography of the lesion, adjacent cortex, and vasculature.

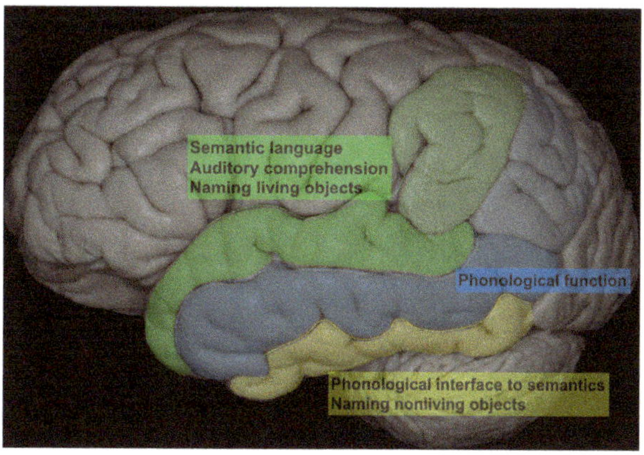

Fig. 4.5 Simplified illustration of the cortical functions of the temporal lobe. Speech functions correspond to the dominant side

4.5.1.1 Surface Anatomy and Delimitation

The lateral surface of the temporal lobe is formed by three horizontal gyri, separated by two sulci. The superior temporal sulcus (STS) is frequently continuous, but the inferior temporal sulcus (ITS) is always discontinuous.

The superior temporal gyrus (STG) is located between the Sylvian fissure above and the STS below. Posteriorly, the STG runs superiorly behind the posterior end of the Sylvian fissure and continues as the supramarginal gyrus (SMG) in the parietal lobe.

The STS runs between the STG and middle temporal gyrus (MTG). It is parallel to the Sylvian fissure, and, in the posterior end, it bifurcates into an ascending branch, posteriorly to the SMG, and a horizontal branch that ends in the angular gyrus (AG).

The MTG runs between the STS and the ITS. The inferior temporal gyrus (ITG) is located below the ITS and also presents an inferior surface, which is the most lateral gyrus of the basal surface of the temporal lobe.

The lateral temporal surface can be correlated to the lateral surface of the skull and the respective craniometric points (Fig. 4.6):

- The inferolateral border of the inferior temporal gyrus is in the same plane as the superior aspect of the zygomatic arc.

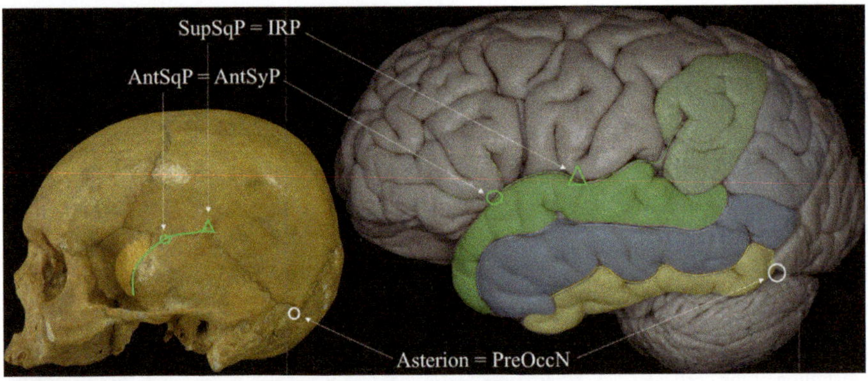

Fig. 4.6 Lateral surface of the temporal lobe and skull point correlations. The anterior ascending part of the squamous suture (green line) coincides with the Sylvian fissure. The greater wing of the sphenoid bone (yellow shading) covers the temporal tip. The asterion (the meeting point of the squamous, parietomastoid, and occipitomastoid sutures) is identified with a white circle. At this level is the PreOccN, the posterior limit of the temporal lobe. The superior, middle, and inferior temporal gyri are highlighted in green, blue, and yellow shading, respectively. The first is contiguous to the SMG (light green shading) and the second to the AG (light blue) at the parietal lobe. The last is contiguous to the lateral surface of the occipital lobe. The AntSqP is the meeting point of the squamous, sphenoparietal, and sphenotemporal sutures and corresponds to the AntSyP (green circle). The most superior point of the squamous suture is the SupSqP, corresponding to the IRP (green triangle), the Sylvian fissure point at the level of the central sulcus. (*PreOccN* preoccipital notch, *SMG* supramarginal gyrus, *AG* angular gyrus, *AntSqP* anterior squamous point, *AntSyP* anterior Sylvian point, *SupSqP* superior squamous point, *IRP* inferior Rolandic point)

- The anterior part has an oblique direction, where the temporal pole projects below the lesser sphenoid wing. This change in the direction of the anterior lateral surface of the temporal lobe starts at the anterior Sylvian point (AntSyP) at the Sylvian fissure, located below the tip of the pars triangularis of the frontal operculum.
- AntSyP correlates to the anterior squamous point (AntSqP) in the skull and coincides with the deep part of the Sylvian cistern to the limen of the insula. The brainstem is localized posteriorly to this point in the coronal plane and is covered by the midpart of the medial temporal lobe.
- Anteriorly to the AntSyP/AntSqP, thick external structures interpose as surgical obstacles: the sphenoid ridge, temporal muscle, and zygomatic arch. Additionally, this region usually contains rich venous drainage to the sphenoparietal sinus, beneath and parallel to the sphenoid ridge.

4.5.2 Vascularization of the Lateral Temporal Surface

4.5.2.1 Arterial Supply

The MCA is responsible for the blood supply to the lateral surface of the temporal lobe through branches of the M4 segment. The extension of the contribution of these arteries may range from the middle temporal gyrus to the fusiform gyrus at the basal surface. Therefore, the inferior temporal arteries from the PCA may contribute to the arterial supply to the inferior temporal gyrus (Fig. 4.7).

The MCA presents four segments: each one is defined according to its respective direction. The M1 segment (sphenoidal segment) extends from medial to lateral, parallel to the lesser sphenoid. It begins at the ICA bifurcation, courses in the carotid cistern (from the lateral edge of the optic nerve until the lateral limit of the uncus), passes over the anterior segment of the uncus, and then reaches the basal part of the Sylvian cistern. At the level of the limen of the insula, the MCA turns 90° posteriorly, forming the genu and becoming the M2 segment (insular segment). The MCA usually bifurcates at the M1–M2 transition point into the superior or frontal trunk and inferior or temporal trunk, but it may occur early at the M1 segment. In this case, M1 is divided into pre- and post-bifurcation. Rarely, the MCA bifurcates at the M2 segment. Trifurcation may also occur very uncommonly. The M2 segment runs from anterior to posterior, within the hemispheric part of the Sylvian cistern, along the insular cortex, which is supplied by small branches of M2. The frontal and temporal trunks branch into arteries that reach the superior and inferior limiting sulci of the insula, which turns in a medial to lateral direction to run under the frontoparietal and over the temporal operculum, respectively. These branches comprise the M3 segment (opercular segment). When they reach the opercular border, the frontal branches turn upward and the temporal branches turn downward, where the M4 segment (cortical) begins, that is, hemispheric branches coursing from inferior to

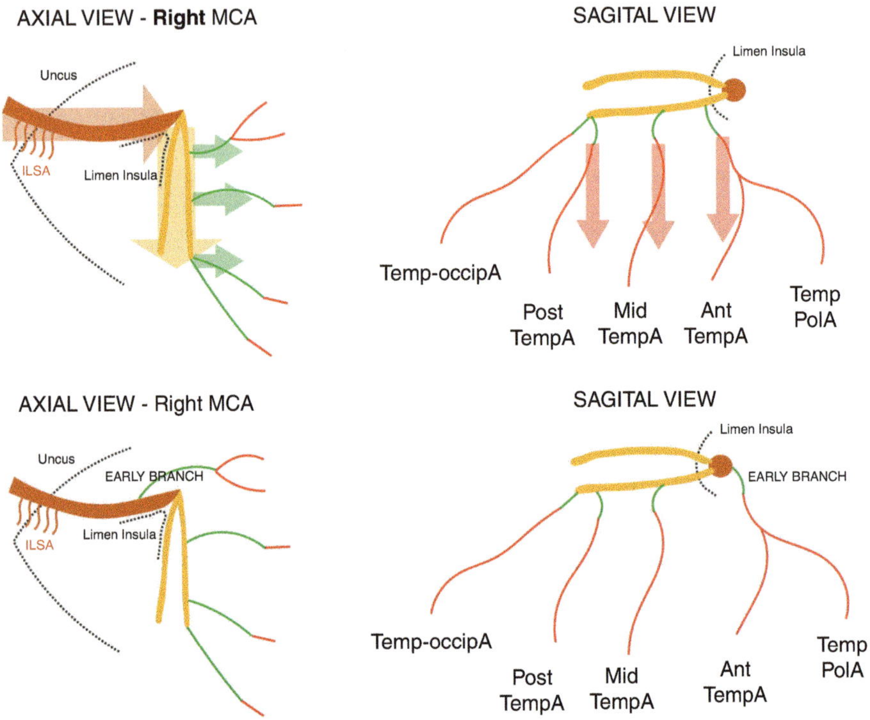

Fig. 4.7 Illustration of the MCA segments. The bifurcation of the main trunk is drawn at the M1–M2 transition at the level of the limen of the insula. However, it occurs very commonly at the M1 segment or, uncommonly, at the M2 segment. In A, the arrows indicate the blood flow direction, respectively, to each segment. In B, the illustration of an early branch toward the anterior temporal lobe, temporal pole, and/or the uncus. (*lLSA* lateral lenticulostriate arteries)

superior (frontoparietal arteries) and from superior to inferior (temporal arteries) (Fig. 4.7).

The temporal arteries from the MCA can be classified from anterior to posterior as the uncal, temporopolar, anterior, middle, posterior, and temporo-occipital arteries. They usually arise from the inferior trunk of M2. However, there is a variable superior or inferior trunk dominance range, which determines the contribution to the temporoparietal region arterial supply. In the same way, a less developed inferior or temporal trunk limits its contribution to the inferior temporal lobe and reflects the extension of the PCA supply to the temporal lateral surface by inferior temporal arteries. Additionally, the uncal, temporopolar, or anterior temporal artery may originate from M1 as an early branch. The same may occur in the orbitofrontal artery, which usually arises from the M2 superior trunk (Fig. 4.7).

4.5.2.2 Venous Drainage

The venous network of the lateral surface of the temporal lobe is composed of ascending and descending veins. The former are composed of temporosylvian veins, responsible for drainage of the superior temporal gyrus and the basal aspect of the temporal pole toward the superficial Sylvian vein (SSV). The last group drains the middle and inferior temporal gyri and eventually the angular gyrus through the anterior, middle, and posterior temporal veins. The descending veins converge toward the preoccipital notch to communicate with the lateral tentorial sinus, together with the temporobasal veins, or to the transverse-sigmoid sinus (TSS) (Fig. 4.8).

The vein of Labbé (VL) or inferior anastomotic vein usually starts at the midportion of the Sylvian fissure and always drains to the TSS. However, its inferior trajectory may occur given the location of the anterior temporal until the posterior temporal veins. The VL connects to the posterior end of the SSV and the superior anastomotic vein, the vein of Trolard. These three major veins form an anastomotic communication between the superior sagittal sinus (SSS), the transverse-sigmoid sinus, and the sphenoparietal (SPS) or cavernous sinus (CS), depending on the outflow direction of the SSV. However, there may be a dominance of one or another (Fig. 4.8).

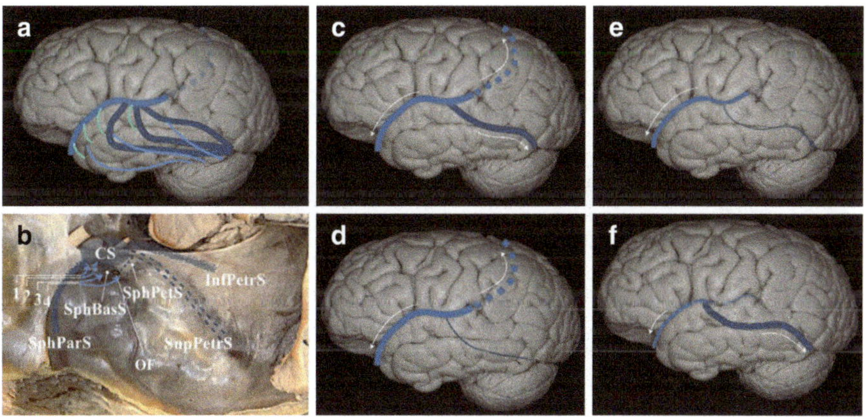

Fig. 4.8 (**a**) Different Labbé positions (dark blue). Ascending temporal veins illustrated in light green and descending anterior, middle, and posterior temporal veins in light blue. (**b**) Possible SSV outflow: (1) toward SphParS; (2) toward CS; (3) toward SphBasS; (4) toward SphPetS. (**c–f**) Scheme of variation of the lateral surface superficial anastomotic venous network. This is a very simple illustration. The most important message is that there is a wide variety of combinations, since absence of anastomosis until a rich anastomotic patency permitting occlusion of one of the large veins with no damage. (**c**) SupSyV, Trolard, and Labbé are similar. (**d**) SupSyV and Trolard dominance. (**e**) SupSyV is dominant. (**f**) SupSyV and Labbé are dominant. (*CS* cavernous sinus, *InfPetrS* inferior petrous sinus, *OF* oval foramen; *SphBasS* sphenobasal sinus, *SphParS* sphenoparietal sinus, *SSV* superficial Sylvian vein, *SupPetrS* superior petrous sinus, *SupSyV* superficial Sylvian vein)

The SSV receives tributaries from the superior temporal and inferior frontal and parietal regions and usually drains anteriorly toward the SPS, along the sphenoid ridge, or turns inferiorly toward the CS. Less commonly, it may surround the anterior surface of the temporal lobe to communicate with the pterygoid sinus through the sphenobasal sinus or to the superior petrosal sinus (SPS) through the sphenopetrosal sinus (Fig. 4.8).

4.5.3 Approaches to the Lateral Surface of the Temporal Lobe

4.5.3.1 Anterior Part of Lateral Surface

This patient presented with a grade III Spetzler-Martin AVM at the temporal pole reaching the lateral surface of the anterior temporal lobe. The main arterial supply was from the inferior trunk of the M1 segment (early MCA bifurcation) and multiple venous outflows: superficial drainage was through the frontal and anterior temporal veins, and deep drainage was through the BVR (Fig. 4.9). She underwent microsurgical resection after two sessions of embolization.

4.6 Planning the Surgery

Surgery must be planned according to the topography of the lesion, adjacent cortex and subcortical functional areas, and relevant vascularization. Thereafter, craniotomy is performed to expose the whole region of interest. For this, anatomic craniometric correlation can provide a more accurate approach.

Fig. 4.9 (**a** and **b**) Pre-embolization magnetic resonance imaging scan. The T2 axial image (**a**) demonstrates the nidus (white arrow) at the temporal pole, anteriorly to the rhinal incisure, illustrated in orange. The T2 sagittal image (**b**) shows the correlation of the temporal pole/AVM (white arrow) to the sphenoid ridge (in red) and the superior (blue) and inferior (yellow) temporal gyri. Observe that the superficial Sylvian fissure (green) localizes at the level of the sphenoid ridge. (**c–f**) Digital arteriography. Right internal carotid artery injection. (**c** and **d**) Arterial phase. (**e** and **f**) Venous phase. (**c**) Pre-embolization. AVM nidus in red dashed circle receiving feeders from the enlarged temporal trunk (MCA bifurcation at M1 segment). (**d**) Post-embolization. Note the reduction of the temporal trunk due to reduced AVM flow. Early MCA bifurcation is evident, at the M1 segment, as identified. (**e**) Pre-embolization, perfil acquisition. Superficial venous drainage through the frontal and anterior temporal vein and deep venous drainage toward the basal vein of Rosenthal. (**f**) Post-embolization, anteroposterior acquisition. (AVM nidus draining to the anterior temporal vein was excluded. *Ant temp vein* anterior temporal vein, *BVR* basal vein of Rosenthal, *SSS* superior sagittal sinus., *SupTempG* superior temporal gyrus, *MidTempG* middle temporal gyrus, *InfTempG* inferior temporal gyrus)

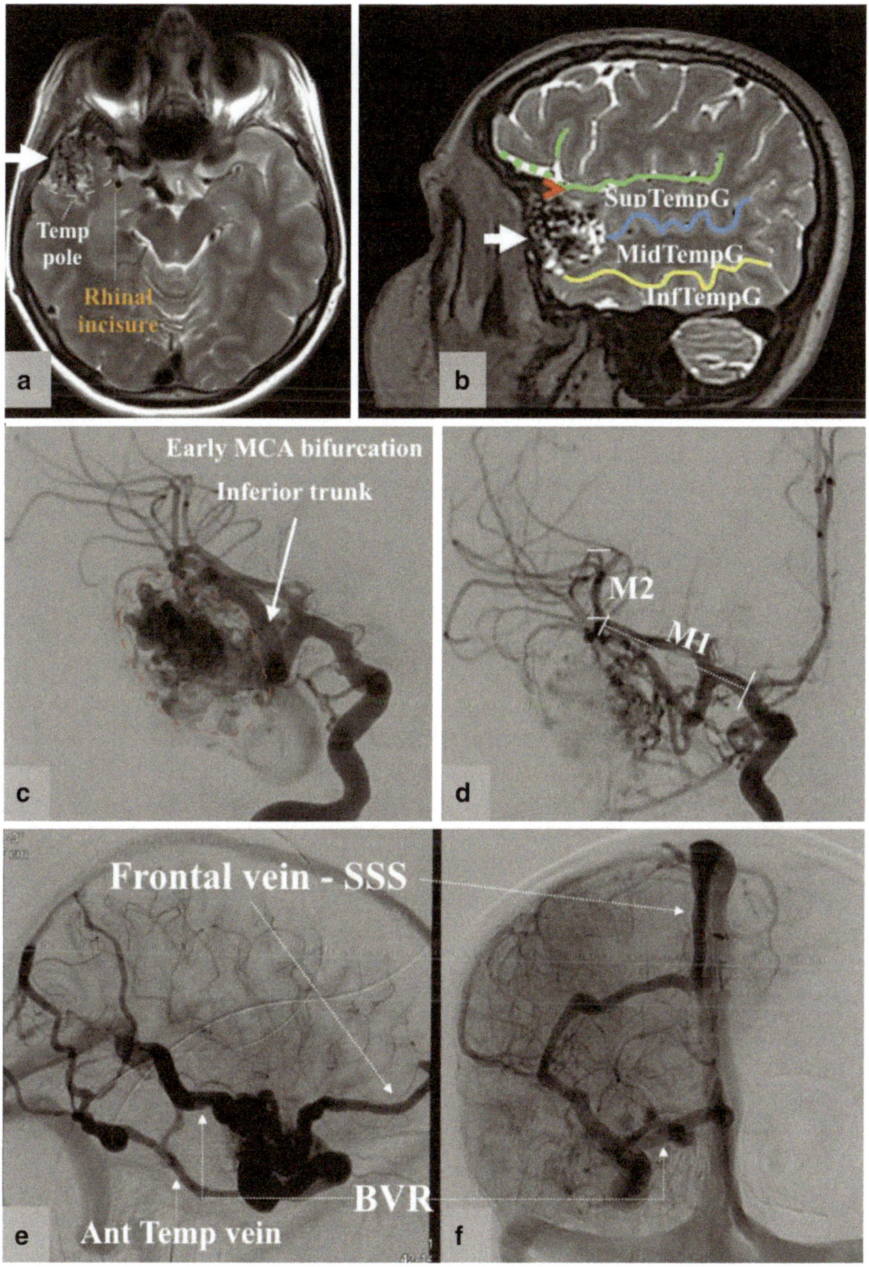
Temp pole
Rhinal incisure
a
SupTempG
MidTempG
InfTempG
b
Early MCA bifurcation
Inferior trunk
c
M2
M1
d
Frontal vein - SSS
BVR
Ant Temp vein
e
f

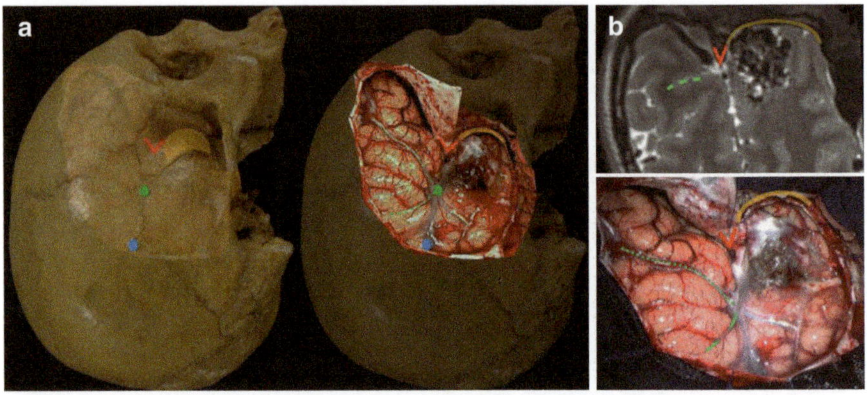

Fig. 4.10 Correlation of the cranium and craniometric points to the AVM, temporal lobe, and Sylvian fissure. (**a**) Lateral view of the skull and the surgical field translucent overlaid. The green and blue spots correspond to the anterior and superior squamous points, respectively. The red V shape demarcates the sphenoid ridge. The greater sphenoid wing covering the temporal pole is highlighted in yellow. (**b**) The surgical field superimposed on the translucent skull. The green and blue spots correspond now to the anterior Sylvian point and inferior Rolandic point, respectively. Of note, the frontal vein draining the AVM courses along the anterior ramus of the Sylvian fissure (green dotted line), which delineates with the ascending ramus (green dashed line) the *pars triangularis*. The greater sphenoid wing was drilled to expose the temporal pole

Craniometric points can delineate the temporal lobe externally. In this case, the MRI shows that the nidus was anterior to the AntSyP (the point where the apex of the pars triangularis meets the Sylvian fissure), which can be estimated by the AntSqP. The anterior temporal lobe is beneath the lesser sphenoid wing, and the greater sphenoid wing covers the temporal pole (Fig. 4.10).

The temporal pole is covered by the greater sphenoid wing. For good AVM exposure, a pretemporal craniotomy was performed. The pretemporal craniotomy was centered on the temporal fossa and gave transsylvian, temporopolar, and subtemporal access. In this case, this approach allowed the exposure of the whole lateral surface of the AVM. The superficial venous drainage could be identified initially and deep venous drainage after opening the Sylvian fissure (Fig. 4.11). At the Sylvian cistern, the feeding arteries from the inferior trunk at the M1 segment were identified, coagulated, and cut.

4.6.1 Posterior Part of Lateral Surface

This male patient presented with a small posteroinferior AVM. The lesion was located at the posteroinferior limit of the lateral surface of the temporal lobe, at the level of the preoccipital notch, that is, at the confluence point of the temporal veins (Fig. 4.12).

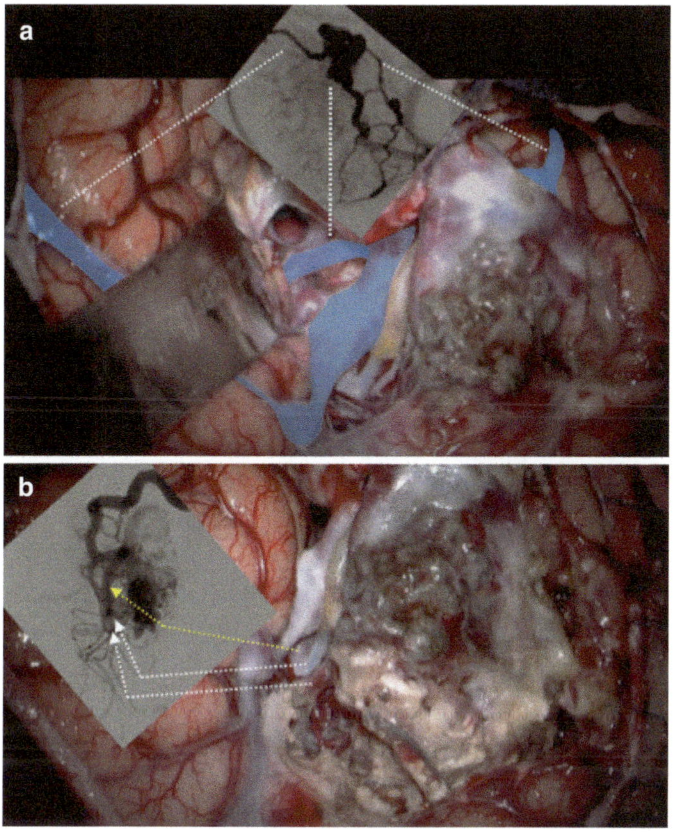

Fig. 4.11 Surgical exposition and identification of the AVM angioarchitecture. (**a**) Sylvian cistern opened. Drainage veins are highlighted in blue and correlated to the angiographic venous phase. (**b**) AVM partially dissected. The arterial feedings are identified. The enlarged MCA inferior trunk (yellow dotted arrow) giving of a previously embolized (superior white dotted arrow) and still patent (inferior white dotted arrow) feeders

4.7 Planning the Surgery

The preoccipital notch is located at the level of the asterion, the meeting point of the parietomastoid, occipitomastoid, and parieto-occipital (lambdoid) sutures. The arterial supply to the lateral surface of the posteroinferior temporal gyrus is made by the temporo-occipital artery (MCA branch) but may also receive PCA supply through the interior temporal branch. In this case, the AVM feeders arise from both territory and venous drainage by the posterior temporal vein (Fig. 4.12).

The surgical approach was then centered on the asterion, with the patient in the park-bench position. The feeder from the MCA artery was found and coagulated at the lateral surface. A transsulcal (inferior temporal sulcus) and transgyrus approach led to the discovery of the basal feeder from the PCA. The superficial drainage vein was finally coagulated (Fig. 4.13) (*case example is also demonstrated in the* video 4.1).

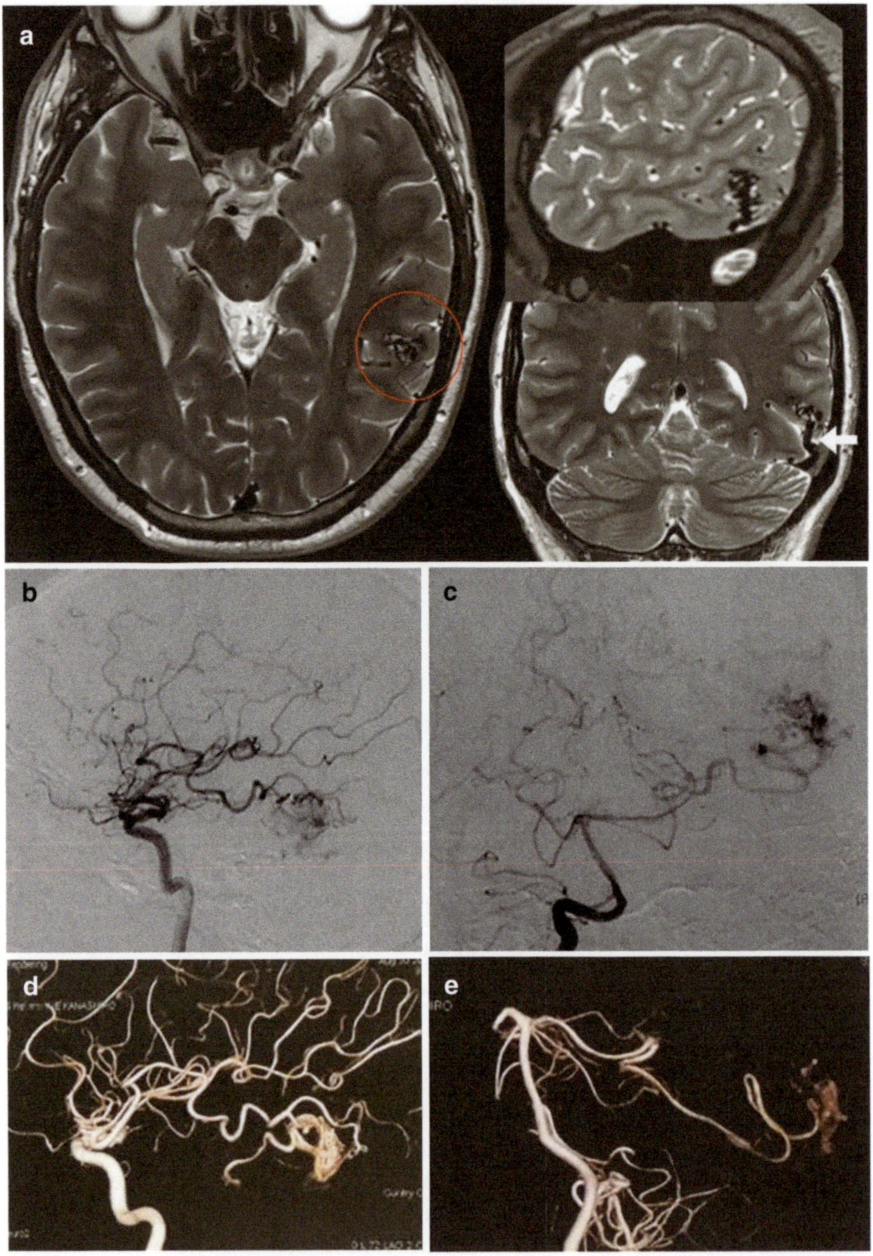

Fig. 4.12 MRI T2 sequences (**a**), In the left, axial image showing the small nidus at the postero-inferior lateral surface of the temporal lobe (red circle). In the right, sagittal (above) and coronal (below) images showing the nidus at the preoccipital notch. The white arrow points to the vein of drainage toward the transverse-sigmoid sinus. (**b**) Perfil view of left carotid artery injection. The temporo-occipital branch of the MCA gave rise to a feeder to the AVM. (**c**) Anteroposterior view of right vertebral injection. The posterior inferior artery from the PCA contributed to the AVM's blood supply. (**d**) 3D reconstruction of the digital angiography of the internal carotid injection and (**e**) 3D reconstruction of the vertebrobasilar injection

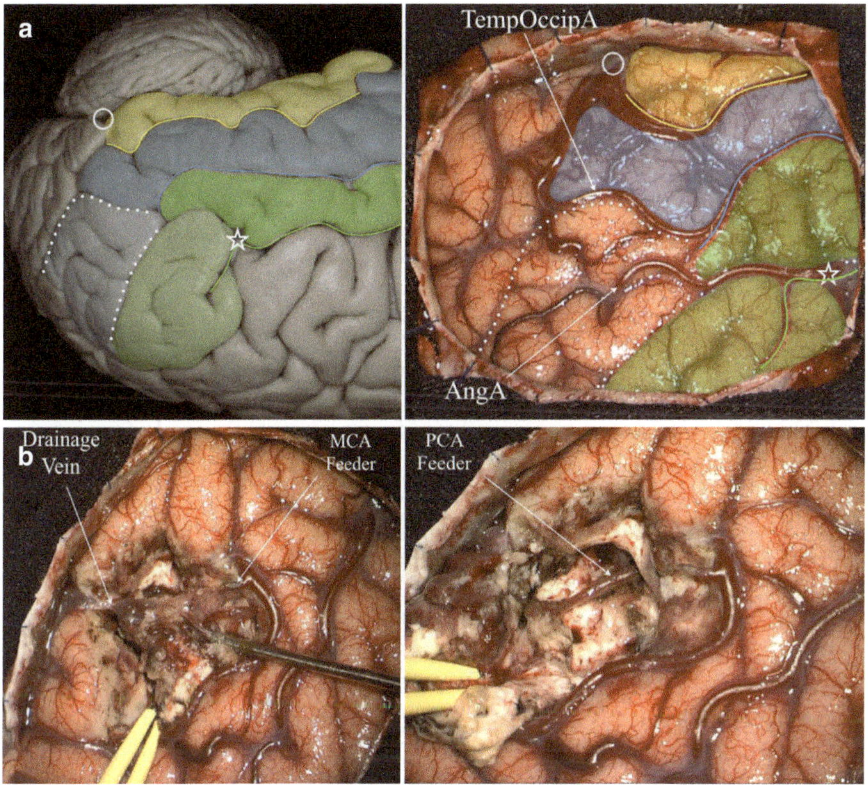

Fig. 4.13 (**a**) Surgical exposure (right) and anatomical correlation (left). The white circle is located at the preoccipital notch, the posterior limit of the inferior temporal gyrus (yellow shading). Note the venous confluence at this point (right image). The white star indicates the posterior Sylvian point, the posterior limit of the Sylvian fissure (green line). The AngA courses over the supramarginal gyrus (light green shading) toward the angular gyrus, delimitated by the white dotted lines. These gyri are the parietal continuation of the superior temporal (dark green shading) and the middle temporal gyri (blue shading), respectively. The TempOccipA passes at the posterior limit of the middle temporal gyrus and gives rise to an AVM feeder. (**b**) Transsulcal and transgyrus (posterior end of the middle temporal gyrus) resection of the AVM. Left: The temporo-occipital (MCA) feeder was coagulated and cut. The single vein of drainage is also identified. Right: AVM retracted is possible to identify the inferior temporal (PCA) feeder, which was coagulated and cut before the vein. (*AngA* angular artery, *TempOccipA* temporo-occipital artery)

4.8 The Medial Surface of the Temporal Lobe

The medial surface of the temporal lobe is composed of limbic structures and phylogenetic older neuronal formation. The human cerebral cortex cytoarchitecture is divided into neocortex (six cell layers), which represents 90%, and allocortex, which is divided into paleocortex (three cell layers) and archicortex (four cell layers). The archicortex includes the hippocampus, the indusium griseum, and the paraterminal gyrus. The paleocortex comprises structures of the olfactory circuit, such as the olfactory bulb and piriform cortex.

The medial temporal lobe encircles the lateral aspect of the mesencephalon and thalamus and maintains a straight correlation to cisternal structures, representing important landmarks for microsurgical approaches.

4.8.1 Surface Anatomy

The uncus is the most anterior structure. It is laterally limited by the rhinal sulcus, which separates it from the temporal pole. Sometimes the rhinal sulcus is continuous with the collateral sulcus in the basal surface, separating the entorhinal cortex of the uncus from the temporo-occipital (fusiform) gyrus. At the level of the inferior choroidal point, the entorhinal cortex of the uncus continues posteriorly as the parahippocampal gyrus (PHG).

The uncus can be divided into an anterior segment, which contains the amygdala and posterior segment related to the head of the hippocampus. These structures are located in the superior part of the uncus and form the anterior wall and anterior part of the medial wall of the inferior horn of the lateral ventricle, respectively. Between them in the ventricular space is located the uncal recess, which projects medially forming the apex of the uncus. The posterior part of the uncus is divided in a longitudinal direction by the uncal notch.

The surface of the uncus features intimately related intrinsic and surrounding structures. At the anterior segment, the semilunar gyrus represents the cortical amygdaloid nuclei. Laterally, it connects to the lateral olfactory stria and lateral olfactory cortex. Posterolaterally, it is separated from the anterior perforated substance and optic tract by the entorhinal sulcus. Anteromedially, the semiannular sulcus separates it from the ambient gyrus, constituted by the entorhinal cortex, and is directly related to the ambient cistern. The anterior choroidal artery passes over the superior surface of the uncus in its trajectory toward the inferior choroidal point. The ambient gyrus is contiguous to the entorhinal cortex of the inferior part of the anterior and posterior segment of the uncus, the last located below the uncal notch. The superior part of the posterior segment of the uncus is occupied by the head of the hippocampus. Its surface presents, from anterior to posterior, the uncinate gyrus, continuous to the PHG, the band of Giacomini, the anteromedial aspect of the dentate gyrus, and the intralimbic gyrus, which is continuous with the fimbria. The posterior limit of the uncus is the inferior choroidal point.

The inferior choroidal point (ICP) represents the inferolateral end of the choroidal fissure, and the foramen of Monro corresponds to the superior choroidal point. The ICP represents the posterior limit of the uncus and the head of the hippocampus. Therefore, it is the anterior limit of the body of the hippocampus. This is the entry point of the anterior choroidal artery, which supplies the choroid plexus of the inferior horn and the exit point of the inferior ventricular vein. The fimbria fornix and the choroid plexus arise from here, which surround the thalamus all along the choroidal fissure. Of note, the choroid plexus rests on the head of the hippocampus but does not adhere to it. At the temporal horn, the choroid plexus is attached laterally to the fimbria fornix by the tenia fornix and medially to the thalamus by the tenia thalami (or tenia choroidea). The topographic correlation of the fornix and thalamus changes along the choroidal fissure, as the fornix runs medially along the atrium and then superiorly along the body of the lateral ventricle. The choroidal fissure surgical route must be opened through the tenia fornix, once the tenia choroidea contains thalamic vessels.

The ICP is an important surgical landmark. It is located at the same level as the lateral border of the cerebral peduncle. Anteriorly is the head of the hippocampus and uncus. The anterior limit of the PHG is located posteriorly, and from here the posterior cerebral artery runs into the hippocampal sulcus. The ICP also demarcates the limit between the crural, anterior, and ambient cistern posteriorly. At the cisternal roof just posterior to the level of the ICP, one finds the lateral geniculate body (LGB) and the basal vein of Rosenthal (BVR). The LGB is located at the level of the lateral border of the cerebral peduncle in the coronal plane. Another important structure below the ICP is the fourth cranial nerve entry point into the free edge of the tentorium. Therefore, it is possible to cut the free edge of the tentorial posteriorly to the ICP safely with preservation of the trochlear nerve. A schematic illustration of the ICP as a landmark is shown in Fig. 4.14.

The fornix is an arch-like structure that integrates the limbic system, running from the hippocampus to the mammillary bodies. It is formed by the fimbria, crus, body, and column. The fimbria is formed at the level of the ICP from the collected fibers of the alveus, a thin white matter layer of myelinated afferent and efferent neurons at the superior surface of the ventricular aspect of the head of the hippocampus. It runs backward as the most cranial structure along the medial wall of the

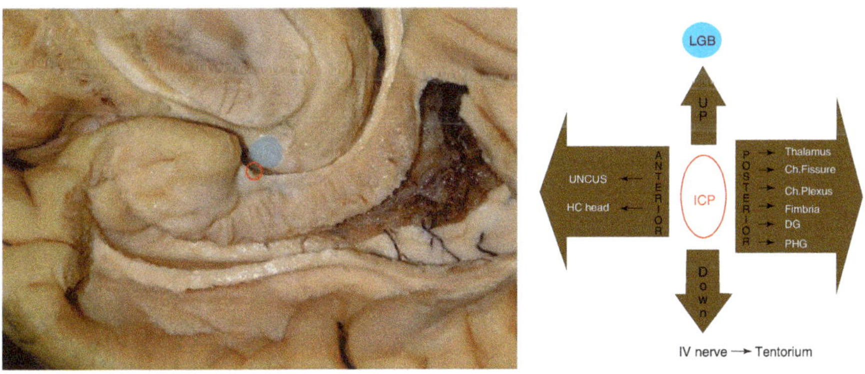

Fig. 4.14 Inferior choroid point (ICP) landmark. At the top, the specimen of the medial temporal lobe, PHG removed, showing the ICP (red circle) and surrounding structures. Above: lateral geniculate body (LGB), highlighted in blue. Anteriorly: uncus and hippocampus (HC) head. Posteriorly, from top to bottom: thalamus, choroidal fissure, choroidal plexus, fimbria, dentate gyrus (DG), and parahippocampal gyrus (PHG). Below: fourth cranial nerve entry point into the free edge of tentorium (not shown)

temporal horn, just above the dentate gyrus and below the lateral geniculate body. The tenia fornix attaches the fimbria to the choroid plexus superomedially at the choroidal fissure. Opening the tenia fornix at the temporal horn gives access to the crural cistern (Fig. 4.15).

Posteriorly, the fimbria fornix is posterolateral to the pulvinar of the thalamus, where it becomes the crus fornix, a diencephalic commissure also referred to as the hippocampal commissure. At this point, these fibers are attached to the posterior part of the inferior aspect of the corpus callosum, the great telencephalic commissure. From here, the crus becomes the body of the fornix, which course a lateral to medial and posterior to anterior direction to reach the superomedial aspect of the thalamus and meet the inferior border of the septum pellucidum. The body of the fornix forms the superior layer of the roof of the third ventricle, passes above the foramen of Monro, and turns sharply downward to become the column of the fornix, which course just posterior to the anterior commissure, reaching the mammillary bodies (Fig. 4.15).

At the medial surface of the temporal lobe, behind the uncus, the fimbria is the most superior structure. The DG, which continues posteriorly and superiorly as indusium griseum, surrounds the corpus callosum. There is a sulcus between them, the fimbrodentate sulcus. Beneath the DG, the hippocampal sulcus separates it from the subiculum, the medial aspect of the PHG, which constitutes the most inferior structure of the medial temporal lobe. The PHG comprises the medial and basal surfaces of the temporal lobe. Posteriorly, it continues to the mediobasal occipital lobe, where it becomes the lingual gyrus beneath the calcarine fissure (Fig. 4.15).

The long anteroposterior axis of the medial temporal lobe requires different surgical approaches according to different regions. Anatomic correlation understanding is required for good surgical planning.

Fig. 4.15 Medial to lateral view of the medial temporal lobe structures. (**a**) Left: Medial temporal lobe. The red circle indicates the ICP. Vertical white dashed line at the level of the apex of the uncus, between the anterior and posterior segments. Right: Uncus surface. Green shading – entorhinal cortex; dark blue shading – semilunar gyrus; red line – semiannular sulcus; light blue shading – ambient gyrus.; in pink, orange, and light gray, the uncinate gyrus, band of Giacomini, and intralimbic gyrus, respectively. (**b**) PHG removed, exposing the temporal horn and trigone. Left: Blue shading demonstrates the LGB, at the cisternal roof, above the ICP (red circle) and lateral to the lateral border of the cerebral peduncle. White star indicates the choroid plexus of the atrium. Right: In red shading, the amygdala nucleus; in blue shading, dentate gyrus of hippocampus; in dark gray, yellow, and green shading, the fimbria, crus, and body of fornix, respectively. (**c**) Hippocampus, fornix, and choroid plexus removed, exposing the ependyma of the roof and lateral wall of the temporal horn. Left: blue shading = LGB. Right: Red shading at the amygdala nucleus; purple shading illustrates the caudate nucleus tail; intense yellow represents the collateral eminence, the projection of the collateral sulcus along the floor of the temporal horn; and the faint yellow the collateral trigone, at the floor of the atrium. (ICP inferior choroidal point, LGB lateral geniculate body)

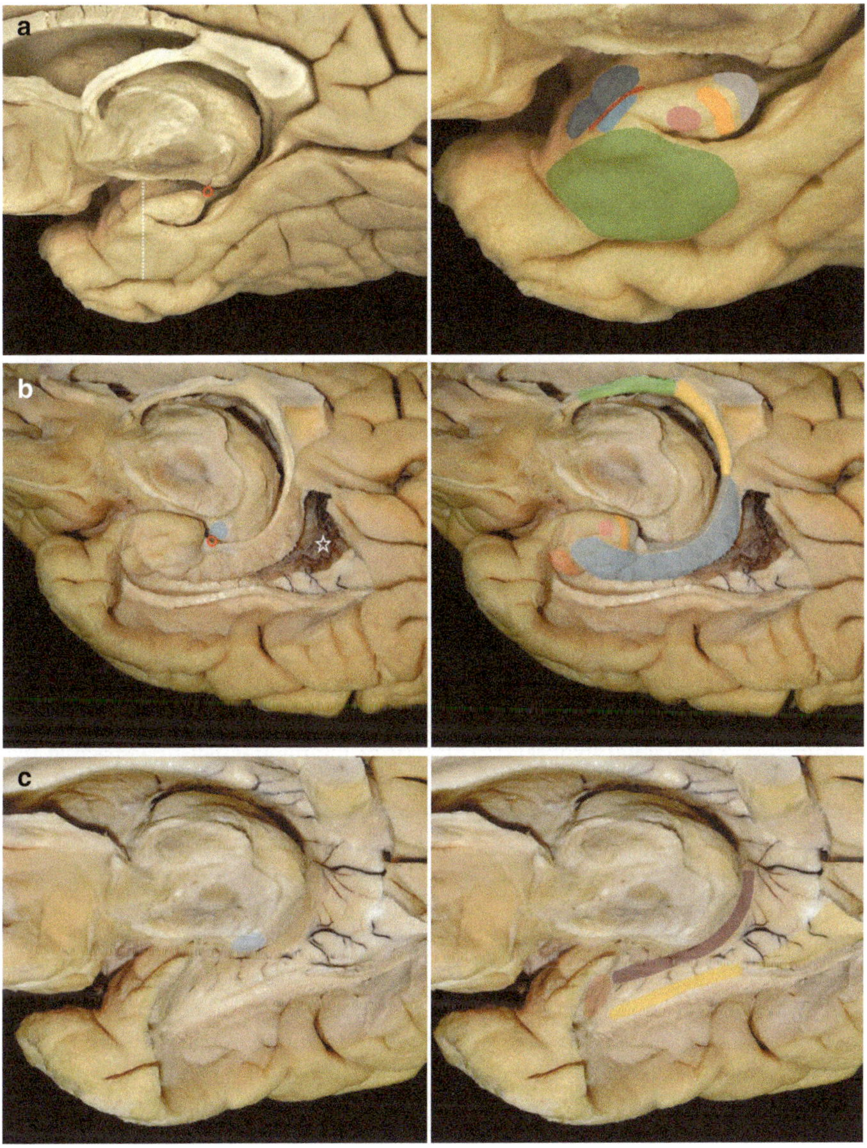
a
b
c

The proximal Sylvian fissure runs behind the lesser sphenoid wing. Therefore, the lesser sphenoid wing is a landmark below which is the temporal pole. The medial projection of the lesser wing ends in the anterior clinoid process, which constitutes the anterior part of the roof of the cavernous sinus. The **medial aspect of the temporal pole**, anterior to the rhinal incisura, covers the lateral wall of the cavernous sinus. Therefore, the lateral wall of the cavernous sinus can be accessed by the temporopolar approach through the pretemporal craniotomy, which includes resection of the greater sphenoid wing.

The **anterior segment of the uncus** occupies the lateral extension of the carotid cistern, limited laterally by the rhinal incisura. This segment faces the internal carotid and anterior incisural spaces. Thus, it is located anteriorly to the cerebral peduncles and pons in the coronal plane. The medial border stays above the oculomotor triangle, and its anterior part is parallel to the optic chiasm/optic nerve. The proximal M1 courses above the anterior segment and reaches the basal portion of the Sylvian cistern as soon as it crosses the rhinal incisure. Hence, pterional craniotomy for the transsylvian approach allows exposure of the proximal Sylvian fissure and basal Sylvian cistern, and consequently the proximal M1 artery and the anterior part of the uncus, that is, the amygdala.

The **posterior segment of the uncus** faces the anterior part of the P2 segment (P2A) of the posterior cerebral artery (PCA). The anterior choroidal artery runs along the superior aspect of the posterior segment before reaching the inferior choroidal point. The posterior segment is lateral to the cerebral peduncle and forms the anterior part of the middle incisural space (the space surrounded by the tentorial incisure along the lateral midbrain). The roof of the middle incisural space at the level of the posterior segment of the uncus is composed of the optic tract.

The medial surface of the temporal lobe behind the uncus (three limbic layers formed by the subiculum, DG, and fimbria) delimitates the lateral wall of the posterior part of the middle incisural space. The roof is composed of the LGB anteriorly, the medial geniculate body, posterior to the superior limit of the lateral mesencephalic sulcus, and the inferior surface of the pulvinar of the thalamus. The posterior part of the P2 segment (P2P) begins at the point of change of direction of the PCA, at the most lateral aspect of the cerebral peduncle. This point coincides with the posterior end of the uncus. From here, the P2P turn slightly cranially to run along the hippocampal sulcus above the subiculum.

4.8.2 Vascularization of the Medial Temporal Surface

4.8.2.1 Arterial Supply

The arterial supply of the medial temporal lobe originates from branches of the anterior choroidal artery (AChA), ICA, MCA, and PCA. There is a wide anastomosis network of small cortical branches from those arteries, which can be easily understood by segment division based on the name of the respective structure. Each of these arteries contributes in a variable manner as they pass around the medial lobe regions.

The **anterior uncal arteries** from the AChA, MCA, and/or ICA supply the anterior segment of the uncus. The posterior segment of the uncus is mainly supplied by AChA branches, the **posterior uncal artery(ies)**, which can reach the uncal sulcus to supply the cisternal surface of the hippocampal head. The entorhinal area, at the inferior surface of the uncus, can be supplied by the **uncoparahippocampal artery** (AChA, PCA, PComA), **hippocampal-uncoparahippocampal artery** (AChaA, PCA, PComA), **anterior parahippocampal artery** (MCA, PCA, ICA), or **anterior hippocampal-parahippocampal artery** (PCA) (Fig. 4.15). All these arteries, as the respective names propose, are responsible for a variable percentage of the arterial supply of each structure. The posterior part of the medial temporal lobe receives arterial supply from the PCA branches (Fig. 4.16).

The PCA arteries originate from the basilar bifurcation, run from medial to lateral crossing over the third nerves, and then turn backward to surround the brainstem heading for the occipital lobe. The basilar bifurcation usually lies in the interpeduncular cistern at the level of the sulcus pontomesencephalic but may range from below the mammillary bodies at the third ventricle floor upward to the midprepontine space downward. The first segment (P1) always runs above and perpendicular to the third nerve. Thus, in the low-lying basilar bifurcation, the P1 courses in an upward oblique direction, and a high basilar bifurcation causes P1s to run downward.

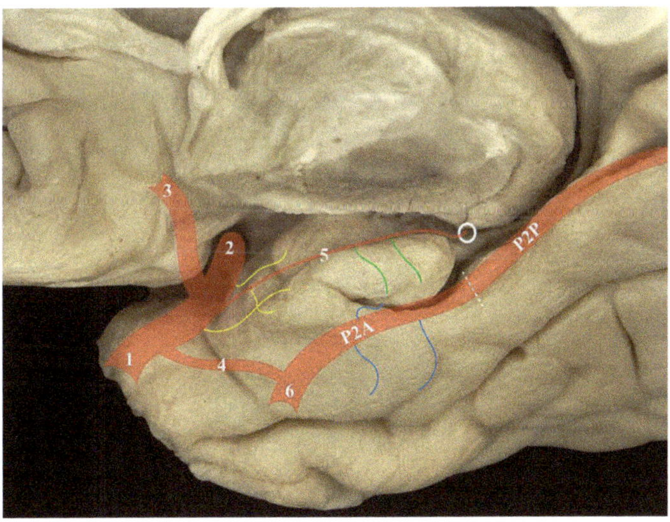

Fig. 4.16 Simplified illustration of the arterial supply to the uncus. (1) right ICA; (2) MCA; (3) ACA; (4) PCoA; (5) AChA; (6) PCA. In yellow, the anterior uncal arteries, which may arise from the ICA, AChA, and/or MCA. In green, the posterior uncal arteries from the AChA, which enters in the inferior choroidal point (white circle) to supply the choroid plexus. In blue, the uncoparahippocampal arteries from the P2A, which are separated from the P2P segment in the level of posterior border of the cerebral peduncle (projection of the white dashed line). However, the posterior uncus and the cisternal surface of the hippocampal head may also be supplied by the AChA and PCoA, and the anterior parahippocampal gyrus/entorhinal area may be supplied also by branches from the MCA and ICA

The main trunk of the PCA is divided into four segments and gives rise to perforating, choroidal, and cortical branches along its trajectory. P1 ends at the posterior communicating artery (PComA). The second segment of PCA (P2) is divided into two parts according to the change in the direction of the trajectory. The anterior part of P2 (P2A) is located in the crural cistern and is lateral to the cerebral peduncle surface and medial to the uncal sulcus, in a medial to lateral posterior oblique direction. The P2A ends at the lateral border of the cerebral peduncle. At this point, the posterior part of P2 (P2P) begins, where its oblique backward course changes to lateral to medial. This segment is located at the ambient cistern and courses along the hippocampal sulcus. Thus, it maintains a straight relation to the pulvinar of the thalamus, choroidal fissure, fimbria, and dentate gyrus. The posterior part of P2 (P2P) ends at the posterior edge of the quadrigeminal plate, where the third segment (P3) arises at the quadrigeminal cistern. Both P3 converge medially, and the most medial point, which faces one to another, is called the angiographic collicular point. At this point, the P3 gives off the posterior pericallosal or splenial artery, which surrounds the splenium to meet the anterior pericallosal artery. Then, the P3 segment ran posteriorly in the medial surface of the occipital lobe inside the anterior part of the calcarine sulcus toward the meeting point of the parieto-occipital to the calcarine sulci. Here, the fourth segment (P4), where the PCA bifurcation usually occurs in the calcarine and parieto-occipital arteries. The first course along the distal part of the calcarine sulcus and the last course inside the parieto-occipital sulcus.

4.9 PCA Branches to the Medial Temporal Lobe

Along its course, the PCA segments give rise to perforating, choroidal, and cortical branches.

Perforating branches:

P1

- Superior surface: Posterior thalamoperforating arteries (anterior thalamoperforating arteries arise from the posterior communicating artery) (Fig. 4.17).
- Inferior surface: Long circumflex arteries, predominantly from the inferior surface of P1, surround the brainstem medially between the PCA and superior cerebellar artery (SCA), originating from perforating branches to the cerebral peduncles, pulvinar, and MGB, and reach the quadrigeminal cistern, which supplies the quadrigeminal plate through a network anastomosis with branches of the SCA (Fig. 4.17).

P1/P2A

- Short circumflex arteries to medial geniculate bodies.
- Perforating arteries: to the posterior part of the cerebral peduncles, optic tract, and midbrain tegmentum.

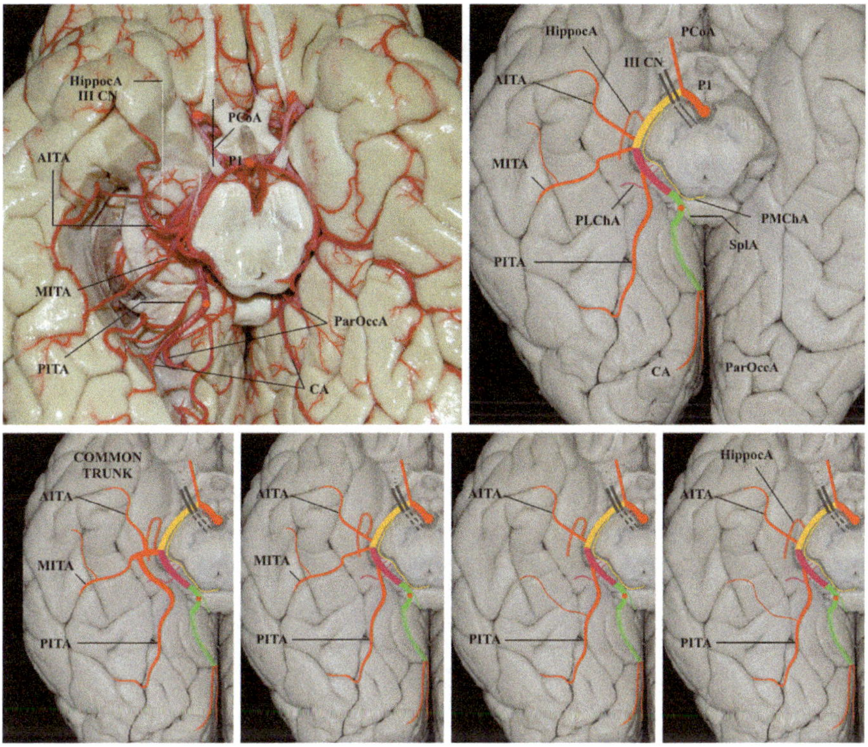

Fig. 4.17 PCA segments and inferior temporal arteries schematic illustration. (*AITA* anterior inferior temporal artery, *MITA* middle inferior temporal artery, *PITA* posterior inferior temporal artery, *CA* calcarine artery, *HippocA* hippocampal artery, *ParOccA* parieto-occipital artery, *PCoA* posterior communicating artery, *PLChA* posterolateral choroidal artery, *PMChA* posteromedial choroidal artery, *SplA* splenial artery)

P2P

- Thalamogeniculate arteries (initial part of P2P) to the LGB and pulvinar, which arise at the superior or medial surface of the P2P.
  Choroidal branches:

P2A

- Posteromedial choroidal artery: It surrounds the mesencephalon medially to the P2A and P2P to reach the choroidal fissure posteromedial to the pulvinar and then the velum interpositum from its posterior aspect (Fig. 4.17).

P2P

- Posterolateral choroidal artery: runs a straight lateral course of a path to reach the choroidal plexus of the atrium (Fig. 4.17).
  Cortical branches:

P2A

- Hippocampal artery: This is the first cortical branch arising from the PCA. In some cases, it may originate from the inferior common trunk or anterior inferior temporal artery. The hippocampal artery courses through the fimbriodentate sulcus to supply the anterior PHG, uncus, dentate gyrus, and fimbria.

P2A/P2P

- Inferior temporal arteries (anterior, middle, and posterior) arise from the P2A/P2P junction or from the distal P2A or proximal P2P. Sometimes, there is a single inferior trunk giving off those branches, and less usually one of them may be absent. They supply the basal surface of the temporal lobe and contribute to the medial temporal lobe supply. A schematic illustration of the variabilities of inferior temporal arteries in the five groups proposed by Rhoton is shown in Fig. 4.17.

P3

- Splenial or posterior pericallosal arteries. These run in the inferior and posterior surfaces of the splenium and anastomose to the anterior pericallosal artery from the anterior cerebral artery.

P4

- Parieto-occipital and calcarine arteries. Usually, the PCA bifurcation at the occipital lobe occurs at the junction of the anterior and posterior segments of the calcarine fissure, that is, the meeting point of the parieto-occipital sulcus and calcarine fissure. However, the bifurcation may be more proximal along the P3 segment (Fig. 4.17).

4.9.1 Venous Drainage

The BVR is the main path of the tributary veins from the medial temporal lobe. The BVR is divided into segments according to the brain structures. It originates from the junction of the anterior cerebral vein and deep middle cerebral vein, at the lateral margin of the midpart of the optic tract (OT), and then crosses below the midpart of the OT to course medially to it underneath the anterior perforating substance – *striate segment*. From here, the BVR course is backward along the roof of the crural cistern and parallel to the posterior segment of the uncus – *anterior peduncular segment*. At the lateral border of the cerebral peduncle, its backward trajectory changes from medial to lateral to lateral to medial direction, coursing along the roof of the ambient cistern – *posterior peduncular segment* (posterior surface of the cerebral peduncle) and *proximal mesencephalic segment* (lateral surface of the mesencephalic tegmentum) (Fig. 4.18). Then, the BVR crosses the lateral part of the quadrigeminal cistern (*distal mesencephalic segment*) to reach the great vein of Galen. Along its course, the BVR receives tributaries from the medial temporal lobe structures and temporal horn of the lateral ventricle, thalamus, atrium, and occipital

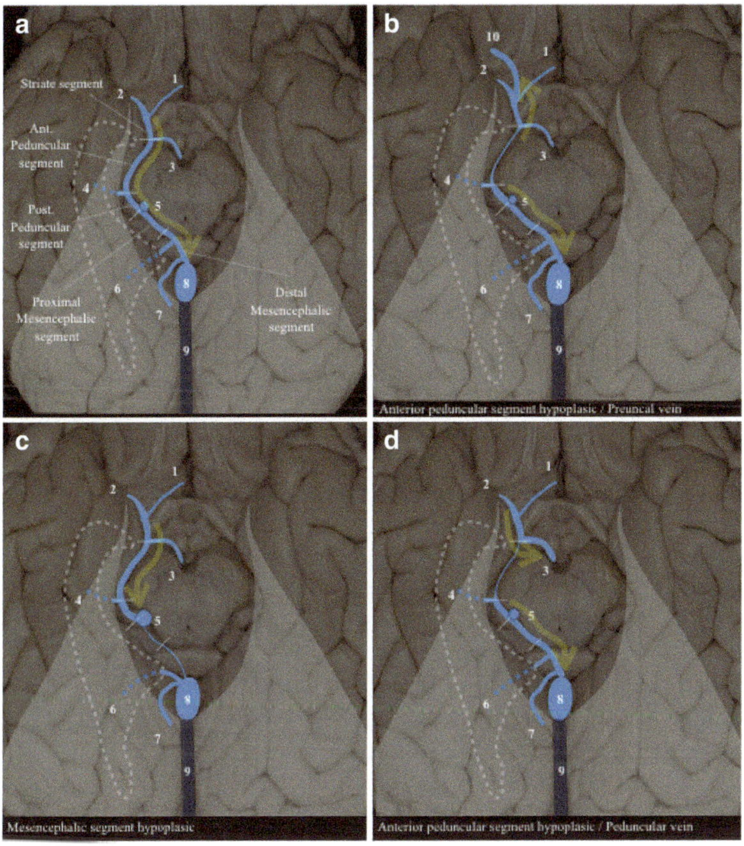

Fig. 4.18 Schematic illustration of patterns of drainage of the BVR segments. (1) Anterior cerebral vein; (2) deep middle cerebral vein; (3) peduncular vein; (4) inferior ventricular vein; (5) lateral mesencephalic vein; (6) lateral atrial vein; (7) internal occipital vein; (8) vein of Galen; (9) straight sinus; (10) (**b**) = preuncal vein. The yellow arrows indicate the outflow drainage direction. (**a**) All segments drain toward the vein of Galen. (**b**) Anterior outflow. The striate segment drains forward through the preuncal vein. (**c**) Inferior outflow. The striate and peduncular segments drain through the lateral mesencephalic vein. (**d**) Medial outflow. The striate segment drains into the peduncular vein

lobe: *anterior uncal vein* (cortical surface of the anterior uncus); *inferior ventricular vein* (anterior hippocampal vein from the hippocampal head amygdala vein, ependymal and choroid plexus veins of the temporal horn, and anterior and posterior longitudinal hippocampal vein from the hippocampal sulcus), coming out through the ICP; *pulvinar veins*, coming from above into the ambient cistern; *lateral atrial vein*; *medial temporal veins*; and the *internal occipital vein* (or *calcarine vein*), from the anterior part of the calcarine sulcus (Fig. 4.18). These last two veins may drain to the internal cerebral vein directly into the vein of Galen.

The BVR drainage system, as in other cerebral venous systems, is involved in an anastomotic venous network. Thus, the segments of BVR may present different patterns other than the most common posterior drainage toward the vein of Galen. There may be an anterior, medial, or inferior venous outflow predominance of the anterior and midpart of the BVR. In such cases, one or more segments of the BVR may be hypoplastic or absent.

The anterior part of the BVR may drain anteriorly through the *preuncal vein*, which arises from the anterior junction point of the anterior and deep middle cerebral veins, toward the sphenoparietal or directly toward the cavernous sinus. In these cases, the anterior peduncular segment is hypoplasic.

The striate and anterior peduncular segments may drain inferiorly through the *lateral mesencephalic vein*, which runs along the lateral mesencephalic sulcus and connects the BVR to the superior petrosal sinus or to a lateral tentorium sinus. In these cases, the mesencephalic segment is hypoplasic.

In a few cases, the striate segment drains medially into the peduncular vein when the anterior peduncular segment is hypoplasic. The peduncular vein runs medially from the medial surface of the BVR along the anterior border of the cerebral peduncle toward the anterior pontomesencephalic vein (which drains to the basilar venous plexus).

In rare cases, all segments may be independent: the striate segment draining through the uncal vein; the peduncular segment outflowing through the lateral mesencephalic vein: and the mesencephalic segment toward the vein of Galen. Alternatively, all segments of the BVR may present an anterior drainage path through a prominent preuncal vein.

4.10 Basal Surface of the Temporal Lobe

Most of the basal surface of the temporal lobe is located on the skull base middle fossa. The posterior part, which continues with the basal surface of the occipital lobe, lies on the supratentorial surface of the tentorium. As the lateral surface, the inferior surface presents three longitudinal gyri separated by two sulci, although there are fewer constants.

4.10.1 Surface Anatomy

The inferior surface of the uncus, entorhinal cortex, and parahippocampal gyrus (PHG) comprises the most medial gyrus. Posteriorly, the PHG continues as the lingual gyrus of the occipital lobe and posterosuperiorly meets the isthmus of the cingulate gyrus behind the splenium of the corpus callosum. The lingual gyrus is separated from the cuneus (occipital lobe) by the posterior part of the calcarine fissure. The parahippocampus is separated from the isthmus of the cingulum by the anterior part of the calcarine fissure (Fig. 4.19).

The collateral sulcus (ColS) is lateral to the uncus and PHG and medial to the temporo-occipital or fusiform gyrus. It is constant, but not always a continuous sulcus, and may be continuous or not to the rhinal sulcus. The ColS runs along the axis of the temporal horn and atrium of the lateral ventricles. Because of this, its fundus bulges in the floor of the ventricles: the collateral eminence at the temporal horn and the collateral trigone at the atrium (Fig. 4.19).

The temporo-occipital or fusiform gyrus is between the ColS and the temporo-occipital sulcus at the basal surface of the temporal and occipital lobes. It is not always a well-defined gyrus because of the inconstancy of the basal intermediate sulci. Sometimes the collateral and temporo-occipital sulcus meets each other at the anterior part. In these cases, the fusiform gyrus is more evident and easier to identify (Fig. 4.19).

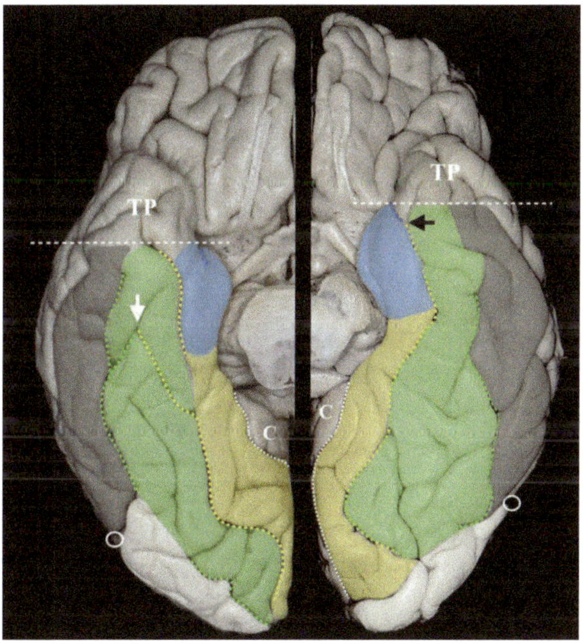

Fig. 4.19 Temporo-occipital basal surface. The white circle corresponds to the preoccipital notch. In gray, the inferior temporal gyrus. In green, the temporo-occipital gyrus. In yellow, the parahippocampal gyrus continuing as lingual gyrus (occipital lobe). In blue, the uncus and entorhinal cortex. The rhinal incisure (blue dotted line) separates the uncus from the TP. In the right basal surface (left in the figure), the collateral sulcus (yellow dotted line) meets (white arrow) the temporo-occipital sulcus (green dotted line). In this case, the temporo-occipital gyrus presents a fusiform aspect. In the left basal surface (right in the figure), the collateral sulcus does not meet the temporo-occipital sulcus but the rhinal incisure (black arrow). (*C* Cingulum (istmo), *TP* temporal pole)

The most lateral gyrus of the basal surface of the temporal lobe is the inferior aspect of the inferior temporal gyrus, located lateral to the temporo-occipital sulcus and anterior to the preoccipital notch. Posteriorly, the inferior aspect of the inferior occipital gyrus is the basal surface of the occipital lobe.

4.10.2 *Vascularization of Basal Temporal Surface*

4.10.2.1 Arterial Supply

The inferior surface of the temporal and most of the inferior surface of the occipital lobes are supplied by the inferior temporal arteries and cortical branches from the P2 segment of the PCA (Fig. 4.17). They are divided into anterior, middle, and posterior inferior temporal arteries. However, they are not constant, and there may be a common trunk giving rise to the inferior arteries. According to the dominance pattern, the PCA arteries may contribute to the inferior part of the lateral surface of the temporal lobe. Alternatively, the lateral part of the basal surface may receive arterial supply from the MCA temporal branches.

The anterior inferior temporal arises from the distal part of the P2A segment to supply the anteroinferior surface of the temporal lobe. Sometimes, it may reach the temporal pole, contributing to its blood supply.

The middle inferior temporal artery is the less frequent branch. It originates at the P2A–P2P transition to supply the midpart of the basal surface.

The posterior inferior temporal artery is the most frequent branch and usually the larger one, which arises from the P2P segment and courses posterior and lateral to supply the basal surface of the temporal and occipital lobes.

The common temporal trunk, when present, originates from the vicinity of the P2A–P2P segment transition. It arises from branches to the inferior cortex of the temporal and most of the inferior occipital lobe.

4.10.2.2 Venous Drainage

The inferior surface of the temporal lobe presents superficial and deep venous drainage. The temporo-occipital and inferior surface of the inferior temporal gyrus are drained by temporobasal veins toward the lateral tentorial sinuses, at the level of the preoccipital notch. The anterior, middle, and posterior temporobasal veins converge to connect to the sinuses at the lateral aspect of the tentorium, together with the occipitobasal veins. The inferior surface of the uncus and PHG is drained by the BVR through the medial temporal veins, anterior hippocampal vein, and anterior uncal vein (Fig. 4.20).

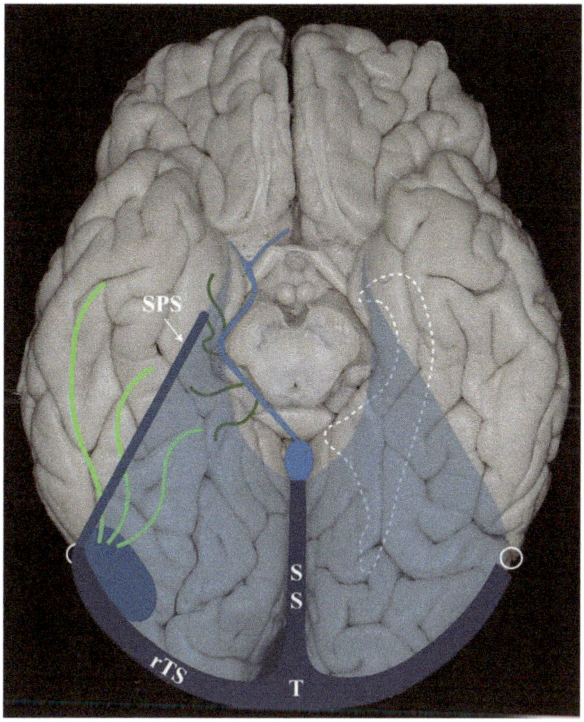

Fig. 4.20 Illustration of the veins of the basal temporal surface. The light green lines represent the venous drainage of the temporo-occipital and inferior temporal gyri toward a tentorial venous lake or the transverse sinus. The parahippocampal gyrus veins (dark green lines) drain into the basal vein of Rosenthal (light blue). The blue shade corresponds to the tentorial area. The white dashed line represents the projection of the ventricular cavity (temporal and occipital horn and atrium). (*rTS* right transverse sinus, *SPS* superior petrous sinus, *SS* straight sinus, *T* torcula)

4.10.3 Approaches to the Medial and Basal Temporal Lobe

The medial temporal lobe can be accessed through the transsylvian, temporopolar, transventricular, subtemporal, and supracerebellar transtentorial routes. The basal temporal lobe can be accessed through the subtemporal and supracerebellar transtentorial routes.

The pterional craniotomy gives access to the transsylvian and transsylvian transventricular routes. The pretemporal craniotomy provides access to the transsylvian, temporopolar, transcortical, transsulcus, transventricular, and subtemporal routes. A more restricted temporal craniotomy may be used for transcortical and transsulcus transventricular and subtemporal approaches. The median or paramedian suboccipital craniectomies provide access to the supracerebellar transtentorial route.

The transsylvian approach permits the surgeon to reach the anterior segment and apex of the uncus, while the temporopolar permits the surgeon to reach the posterior segment. The medial temporal lobe posterior to the uncus may be accessed through the subtemporal or supracerebellar transtentorial approach (Fig. 4.21).

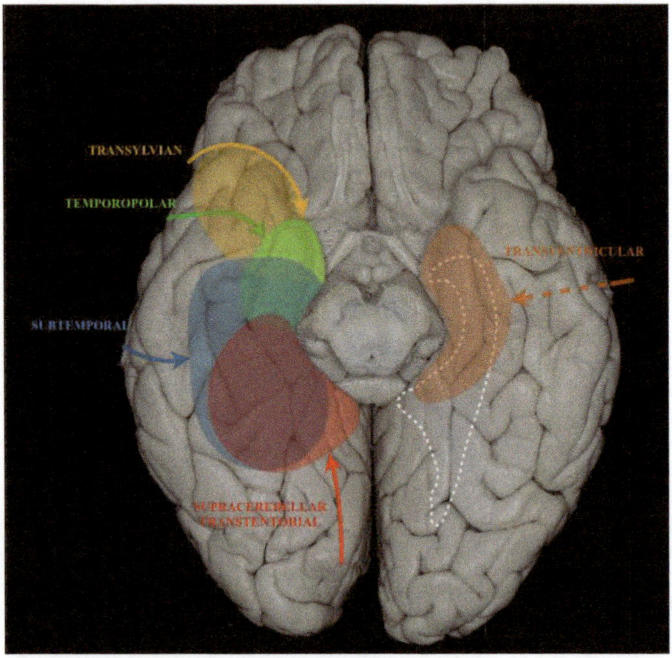

Fig. 4.21 Illustration of the routes to access the basal surface. The transsylvian approach (yellow) allows access until the anterior part of the entorhinal cortex and the temporopolar (green) reaches the posterior part of the uncus. The subtemporal approach (blue) may allow to reach from the posterior part of the uncus up to the parahippocampus gyrus at the level of the quadrigeminal cistern. The vein of Labbé may limit the temporal retraction. The supracerebellar transtentorial (red) (median or paramedian suboccipital craniotomy) leads to access the parahippocampal gyrus up to the crural cistern. The entorhinal and posterior parahippocampal gyrus may be approached also by the transventricular route (light orange)

4.10.3.1 Transcortical Transventricular Approach to the Temporal Horn

A 13-year-old male patient presented with spontaneous subarachnoid hemorrhage and subsequent hydrocephalus. Further investigation demonstrated an AVM in the medial temporal lobe. The MRI (Fig. 4.22) showed that the involved structure was the anterior part of the hippocampus. This was concluded by the localization of the ICP, which marks the transition of the head and body of the hippocampus. The

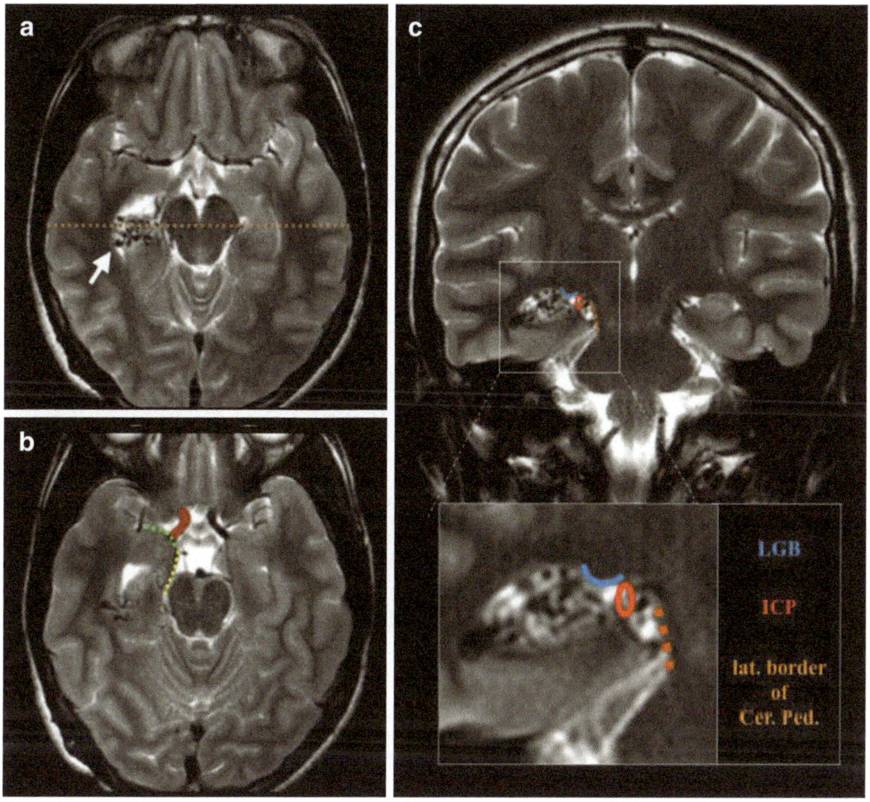

Fig. 4.22 T2-weighted magnetic resonance imaging sequences. (**a**) Axial image demonstrates the AVM (white arrow) just behind the imaginary line (orange dot line) crossing the lateral border of cerebral peduncles, which coincides with the ICP. (**b**) A 5 mm lower (than in **a**) axial plane illustrating the uncus – anterior segment border with a green dotted line and posterior segment border with a yellow dotted line. The posterior segment contains the hippocampus head. An important microsurgical anatomy aspect is the correlation of the internal carotid artery (highlighted in red shading on the right) to the border of the anterior segment of the uncus. The middle cerebral artery crosses over it (not shown). The posterior segment of the uncus is correlated to the posterior cerebral artery (P2A segment). (**c**) Coronal plane at the level of the lateral border of the cerebral peduncle (orange dotted line). At the same plane is located the ICP (red circle) and the lateral geniculate body at the roof of the cistern (blue circular line). The anterior limit of the AVM nidus is identified. (*ICP* Inferior choroidal point)

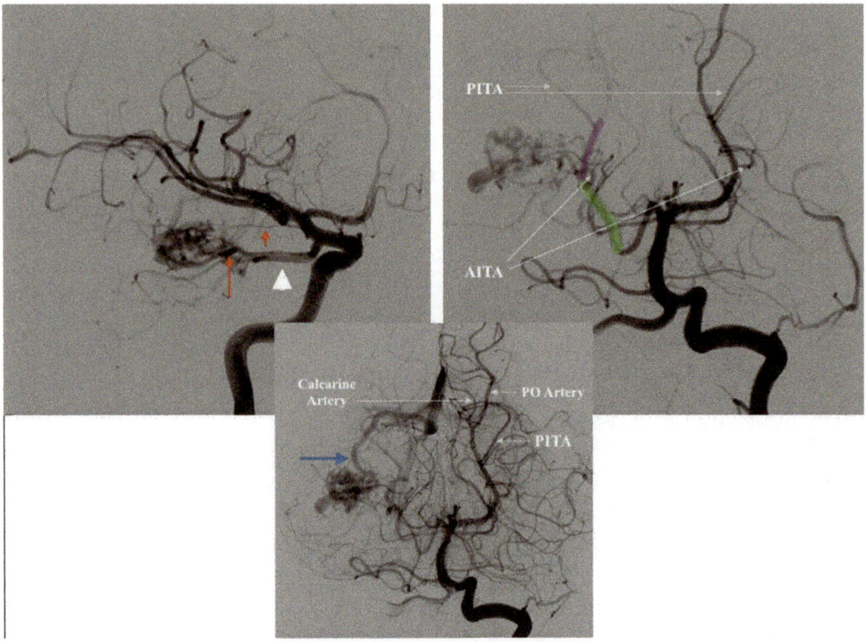

Fig. 4.23 The oblique right ICA injection (upper left image) shows the arterial feeders from the hippocampal artery, which arises from the AITA (long red arrow), and from the anterior choroidal artery (small red arrow). This patient showed a fetal posterior communicating artery pattern (white arrowhead). The upper right image (left vertebral injection, early arterial phase) illustrates the P2A in green shading and the P2P in purple shading. The AITA arises from P2A/P2P and the PITA from P2P, as indicated by the arrows. The inferior central image (left vertebral injection, late arterial phase) demonstrates the venous drain through the basal vein of Rosenthal (blue arrow). The contralateral PITA and P4 segment are identified by the white arrows. (*AITA* Anterior inferior temporal artery, *PITA* posterior inferior temporal artery)

digital subtraction angiography (Fig. 4.23) demonstrated that the AVM feeders came from the PCA and ICA. The patient initially underwent insertion of a ventriculoperitoneal shunt followed by microsurgical resection of the AVM 1 year later.

Surgery was performed through a question mark incision, with the head turned to the contralateral side and extended, which allowed a better angle of visualization to the medial structures. A temporal craniotomy exposed the middle temporal gyrus for transcortical access to the temporal horn of the lateral ventricle. Once the lesion was on the nondominant side, approximately 4–5 cm from the temporal tip is the limit from the approach (Fig. 4.24).

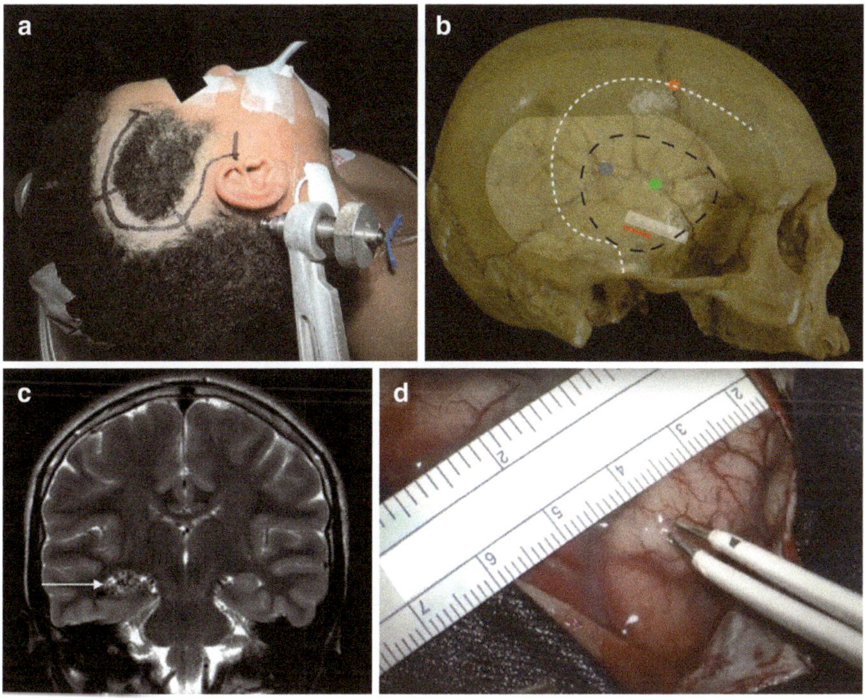

Fig. 4.24 (**a**) Patient's head position and skin incision mark. (**b**) Skull projection of the skin incision (white dashed line), craniotomy area (black dashed line), and middle temporal gyrus corticectomy (red line). Craniometric points and a translucent lateral surface of the right brain hemisphere were projected for anatomic correlation: the green circle indicates the anterior squamous point/anterior Sylvian point and the blue circle corresponds to the superior squamous point/inferior Rolandic point. The orange circle identifies the stephanion (meeting point of the superior temporal line and the coronal suture), which correspond to the meeting point of the inferior frontal sulcus and the precentral sulcus. (**c**) Coronal T2-weighted magnetic resonance imaging indicating the transcortical route (white arrow) toward the AVM at the right temporal horn of the lateral ventricle. (**d**) Intraoperative delimitation of the posterior limit of the corticectomy

Aspiration of the T2 gyrus leads to the temporal horn, where it is possible to identify the medial structures: hippocampus head, ventricular surface of the amygdala, choroidal plexus, and the choroidal point. The AVM is identified just posteriorly to the head of the hippocampus. After dissection of the nidus, the main feeder from the hippocampal artery (which course into the fimbrodentate sulcus) and then the choroidal feeder was dissected, coagulated, and cut. The drainage vein was identified at the posterolateral aspect of the nidus, as previously identified in the arteriography, and coagulated for resection of the AVM (Fig. 4.25).

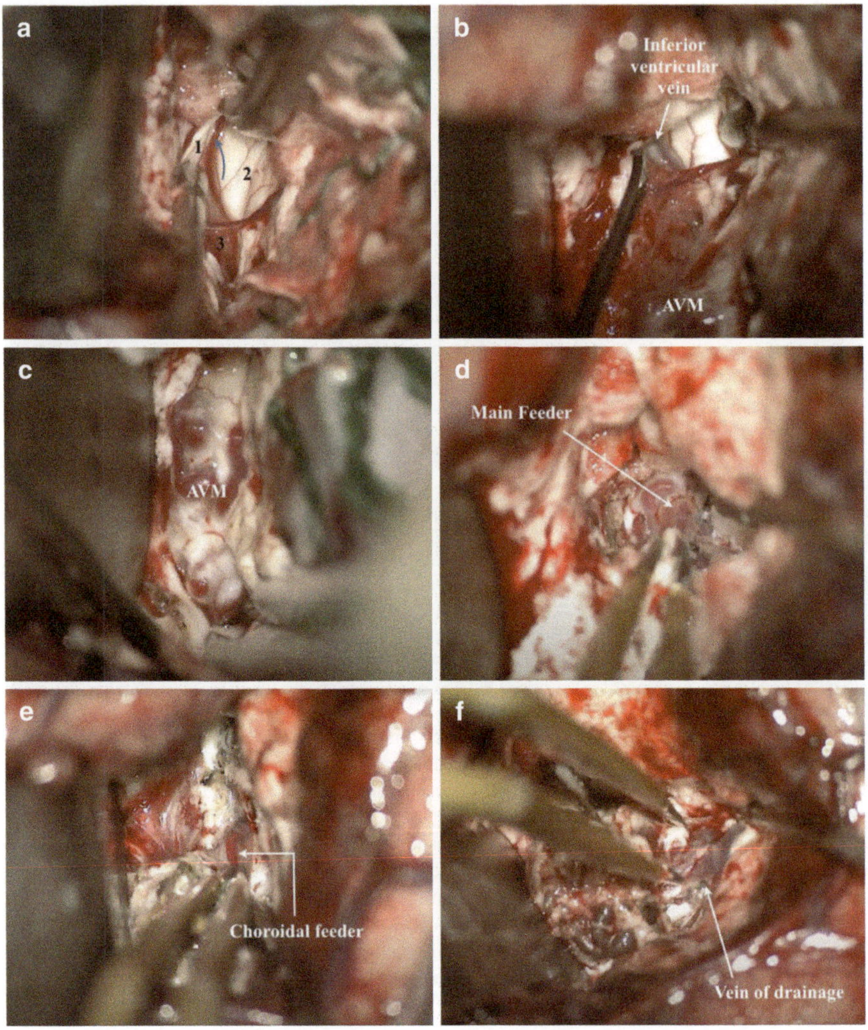

Fig. 4.25 (**a**) Temporal horn: (1) ventricular surface of the amygdala; (2) hippocampus head; (3) choroidal plexus. (**b**) The AVM is identified posteriorly to the ICP, where comes out the inferior ventricular vein. (**c**) The AVM nidus was dissected. (**d**) The main feeder (hippocampal artery) from the anterior inferior temporal artery coming from the fimbrodentate sulcus. (**e**) The choroidal feeders are coagulated. (**f**) Vein of drainage. (*ICP* Inferior choroidal point)

4.10.3.2 Supracerebellar Infratentorial Approach to a Posterior Parahippocampus AVM

A 24-year-old female patient developed a spontaneous hemorrhage in the left medial temporal region and intraventricular hemorrhage 3 years ago. Further MRI investigation demonstrated an AVM nidus (Fig. 4.26), and arteriography showed arterial feeders from a branch of the inferior temporal artery arising from the P2P segment and venous drainage through the BVR (Fig. 4.27).

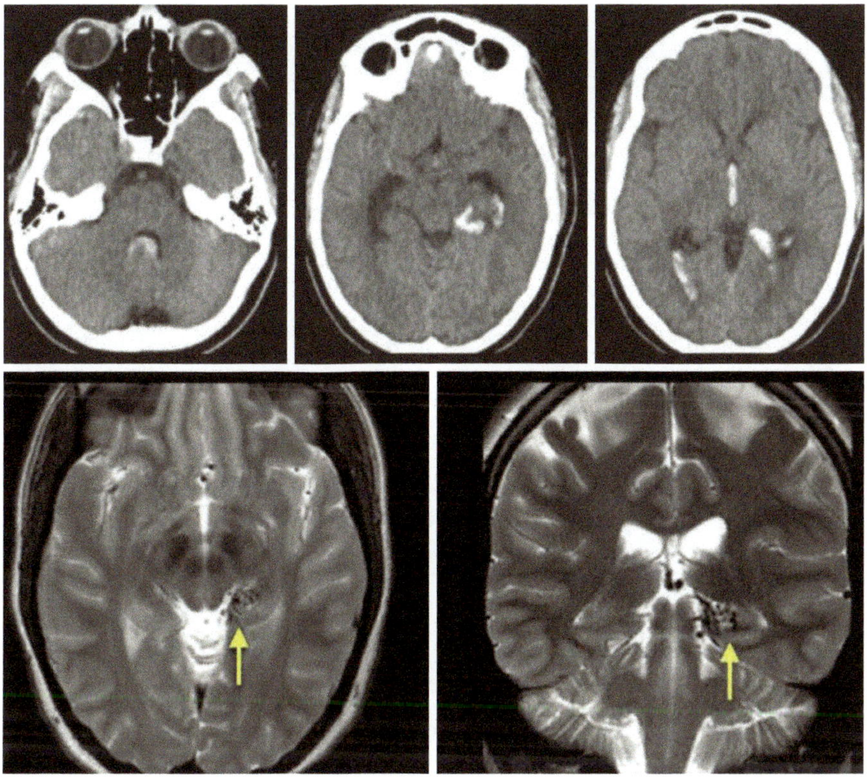

Fig. 4.26 Upper: computed tomography scan showing intraventricular hemorrhage and hyperdensity in the medial temporal lobe posterior to the uncus (central picture). Lower: T2-weighted magnetic resonance imaging axial (left) and coronal (right) slices demonstrate the small nidus of the AVM (yellow arrows) at the posterior parahippocampus gyrus. (*AVM* Arteriovenous malformation)

This lesion was located along the hippocampal sulcus at the left hippocampus body/posterior parahippocampal gyrus, lateral to the ambient cistern and tegmentum of the midbrain and beneath the pulvinar of the thalamus. The posterior segment of the posterior cerebral artery (P2P) coursed along the medial border of the AVM, giving off its feeder, a branch from the posterior inferior temporal artery.

Surgical approaches to this segment of the medial temporal region may be performed through (1) a transventricular transcortical or transsulcal approach; (2) a subtemporal approach; or (3) a supracerebellar transtentorial approach. The transventricular approach gives a straight view to the surgeon but would implicate damage to the optic radiation and, in cases of dominant side, to important speech cortical area and fascicles. The subtemporal approach allows for reaching the medial temporal lobe, but there are two important limitations: (1) the caliber and trajectory of the vein of Labbé and (2) the superior oblique direction of the basal surface of the posterior half of the temporal lobe along the tentorial plane. Because of the superior oblique direction from lateral to medial, the temporal lobe must be retracted for subtemporal approaches. A prominent vein of Labbé, as in the case of this patient

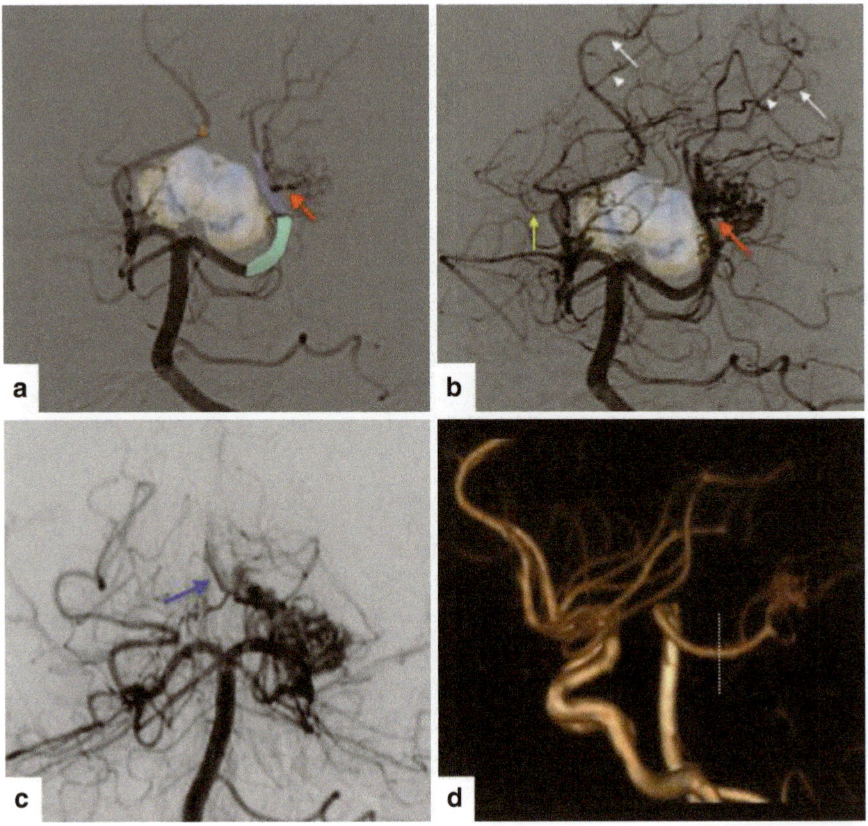

Fig. 4.27 (**a**, **b**, and **c**) Arteriography – left vertebral injection. (**a**) Green and purple shading illustrates the P2A (lateral and parallel to cerebral peduncle) and P2P (lateral and parallel to the tegmentum) segments of PCA, respectively. Red arrow points to the posterior inferior temporal feeder to the AVM. The collicular point at P3 segment is indicated in orange shading. (**b**) Yellow arrow = posterior inferior temporal artery. White arrowheads = parieto-occipital arteries. White arrows = calcarine arteries. Red arrow = posterior inferior temporal feeder to the AVM. (**c**) Blue arrow = venous drainage through basal vein. (**d**) 3D reconstruction. The white dashed line indicates the transition of P2A to P2P, at the lateral margin of the cerebral peduncle. *PCA* – posterior cerebral artery. P2A – anterior part of P2 segment of posterior cerebral artery. P2P – posterior part of P2 segment of posterior cerebral artery.

(see Fig. 4.28), would make this a risky surgical approach. Therefore, it was opted for an infratentorial supracerebellar approach. Although there is a possibility of large tentorial venous lakes, the drainage of this AVM was through the BVR, as shown in the arteriography illustrated in Fig. 4.27.

The approach was performed through a left paramedian suboccipital craniotomy, which included the superior nuchal line to expose the transverse sinus. This allows maximum exposure of the superior surface of the cerebellum for the supracerebellar route (Figs. 4.29 and 4.30) shows the initial view of the paramedian supracerebellar

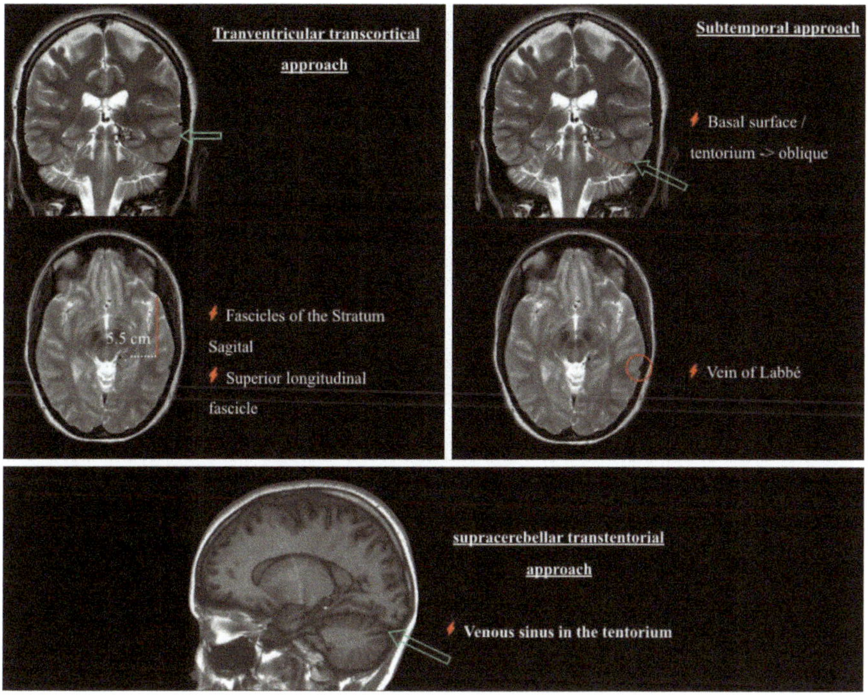

Fig. 4.28 Routes to access the parahippocampus along the region of the ambient cistern. The green arrows indicate the approaches. The describes the main limitations of the respective route

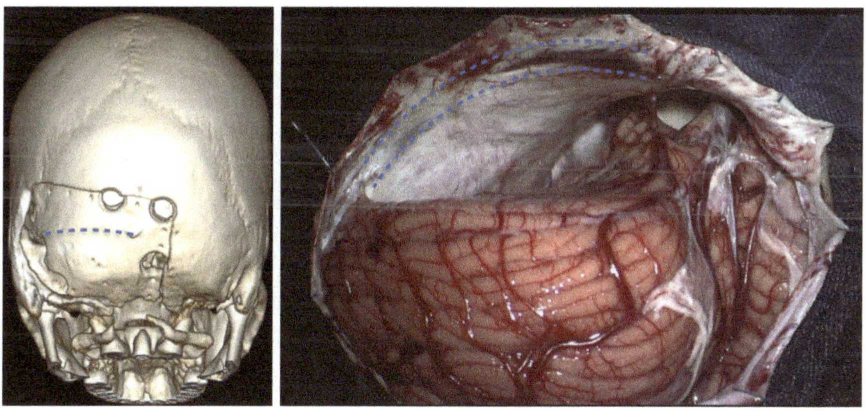

Fig. 4.29 Postop 3D computed tomography scan (left) demonstrating the craniotomy area exposing the transverse sinus. The blue dashed line indicates the superior occipital line, which corresponds to the transverse sinus topography. Surgical field (right) photo showing the superior surface of the cerebellum and the tentorium. The blue dashed lines delimitate the transverse sinus

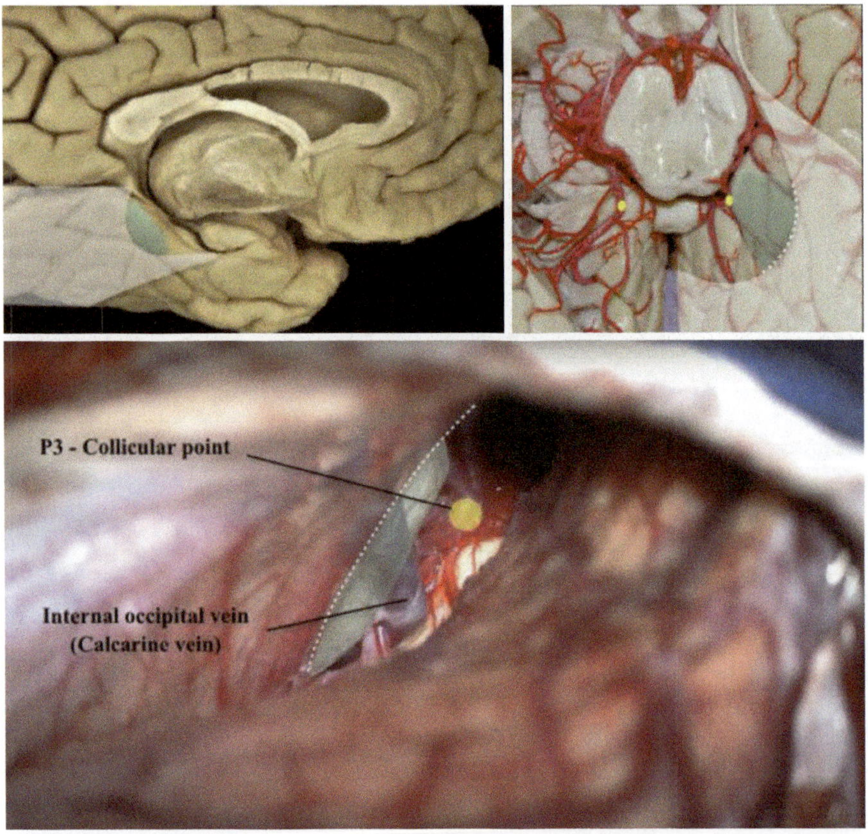

Fig. 4.30 The green shading represents the temporal lobe visualization. The white shading localizes the tentorial area of the basal temporal lobe surface. It is possible to see the medial surface of the parahippocampal gyrus (green shading) and the free edge of the tentorium (white dashed line) at the posterior incisural space. The structures will be more or less evident according to the extent of the posterior incisural space. The collicular point of the P3 segment (yellow shading) is identified, just above the inferior colliculi of the quadrigeminal plate. In this case, it is possible to see the calcarine vein running medial and inferiorly to the basal vein of Rosenthal

approach. The resection of the tentorium widens the visualization of the basal surface of the temporal lobe (Fig. 4.31). The AVM was accessed through a corticectomy of the posterior parahippocampal gyrus (Fig. 4.32).

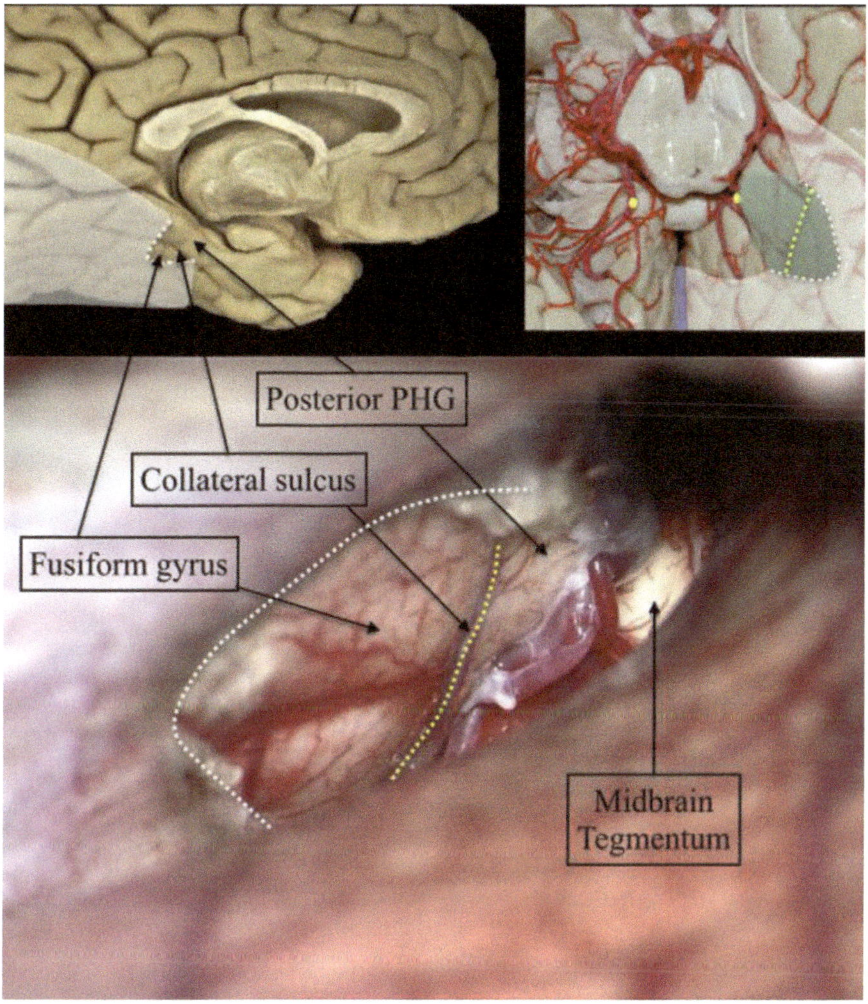

Fig. 4.31 The tentorium resection widens the basal surface view, exposing the fusiform gyrus

4.11 Sylvian Surface of the Temporal Lobe

The Sylvian or superior surface of the temporal lobe presents an anterior concave part, called *planum polare* (PP), and a posterior part, which has three temporal transverse gyri, called *planum temporale* (PT), which is located in the primary auditory area.

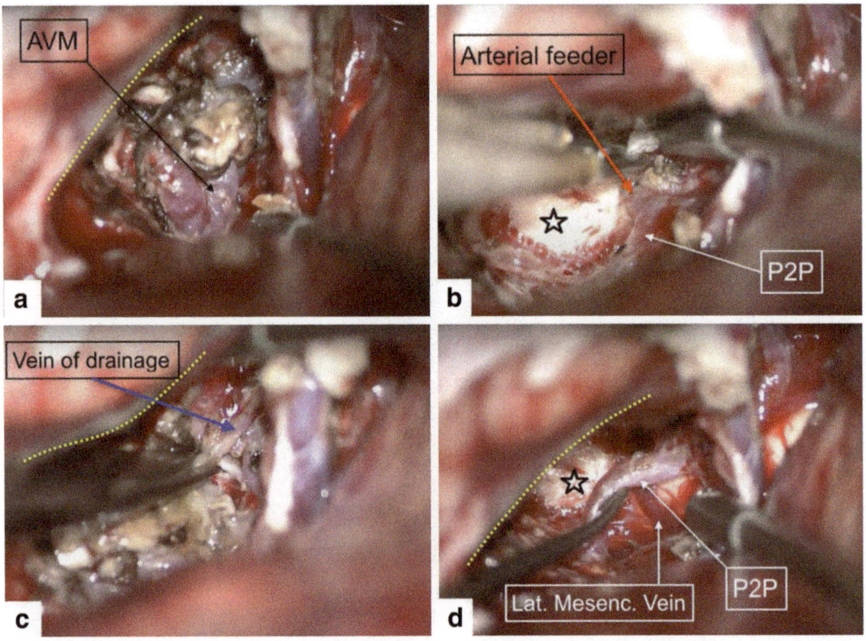

Fig. 4.32 (**a**) After corticectomy of the posterior parahippocampal gyrus, the arteriovenous malformation (AVM) was identified. The yellow dotted line indicates the collateral sulcus. (**b**) The nidus was retracted by the aspirator, and the P2P feeder was identified and cut after coagulation. The star indicates the pulvinar. (**c**) The nidus was retracted laterally, and the vein of drainage was identified, coagulated, and cut. (**d**) The AVM was resected. The lateral mesencephalic vein running along the lateral mesencephalic sulcus was identified at the ambient cistern, medial to P2P segment

4.11.1 Surface Anatomy

The PP faces the insular cortex surface, whose long and short gyri impression over the Sylvian temporal surface leads to a concave shape. From the AntSyP forward, the PP makes an oblique lateral to medial trajectory underneath the sphenoid ridge along the sphenoidal part of the Sylvian fissure. Consequently, in the coronal plane, the PP appears oblique, while PT is horizontal (Fig. 4.33).

PT is the superior surface of the temporal lobe posterior to the insular cortex. Thus, it does not present a concave but a convolutional shape. There are usually three transverse gyri: anterior, middle, and posterior. The anterior temporal transverse gyrus is the most prominent and is also known as Heschl's gyrus (HG). Brodmann area 41 is part of the primary auditory cortex (Fig. 4.33).

The lateral surface aspect of the HG identifies the anterior limit of the PT. It is located at the same coronal plane of the postcentral gyrus. Additionally, the external acoustic meatus is also on the same coronal plane, serving as an external reference.

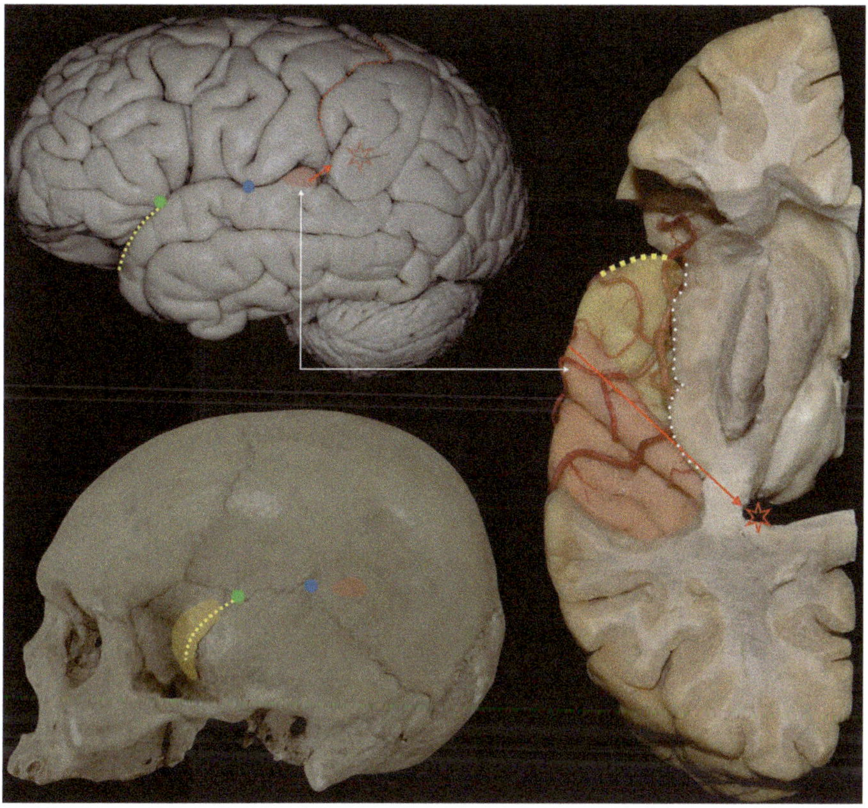

Fig. 4.33 Sylvian surface of the temporal lobe correlation to lateral surface of the brain and skull. In the right, superior view of a left hemisphere cut in an axial plane along the inferior frontal gyrus (at the level of the lateral ventricle). Left superior figure, lateral view of another cadaver brain. Left inferior figure, left skull view. The PP is highlighted in yellow and the PT in red. The PP is related to the insular cortex surface (white dotted line). The anterior part of the PP curves medially (yellow dotted line) from the AntSyP (green circle) forward. The PT is posterior to the inferior circular limiting sulcus of the insula. The HG points (red short arrow in the left superior figure and red long arrow in the right figure) to the atrium (red star). In the left superior figure, the HG lateral surface aspect is highlighted in red. It is located below the postcentral sulcus (red dotted line) and posterior to the IRP (blue circle). The atrium is projected in the lateral surface at the level of the supramarginal gyrus (smooth red star). In the left inferior figure, the HG is projected (red shading) posterior to the SupSqP (blue circle). (*AntSyP* Anterior Sylvian point, *SupSqP* superior squamous point, *IRP* inferior Rolandic point, *HG* Heschl's gyrus, *PP* planum polare, *PT* planum temporale)

The deep end of the gyrus is located just posterior to the meeting point of the superior and inferior limiting circular sulcus of the insula and points to the atrium of the lateral ventricle. However, what does this mean?

The insula is the superficial aspect of the central core, the deep region of the brain encompassing the basal ganglia, thalamus, and internal, external, and extreme capsules. Therefore, from HG back, there is no insula, no basal ganglia, and no thalamus. It can be a useful landmark for deep brain surgical planning.

4.11.2 Vascularization of the Sylvian Surface of the Temporal Lobe

4.11.2.1 Arterial Supply

Branches from the M3 segment of the temporal trunk of the MCA. These branches run over the frontoparietal and temporal operculum in a medial to lateral direction.

The most distal artery turns laterally at the top of the insula, usually the angular artery, which curves over the deep end of the HG. It is well identified on angiography as the most superior and medial M2–M3 artery in AP view and as the most superior and posterior M2–M3 artery in the perfil view. This is the angiographic Sylvian point (Fig. 4.34).

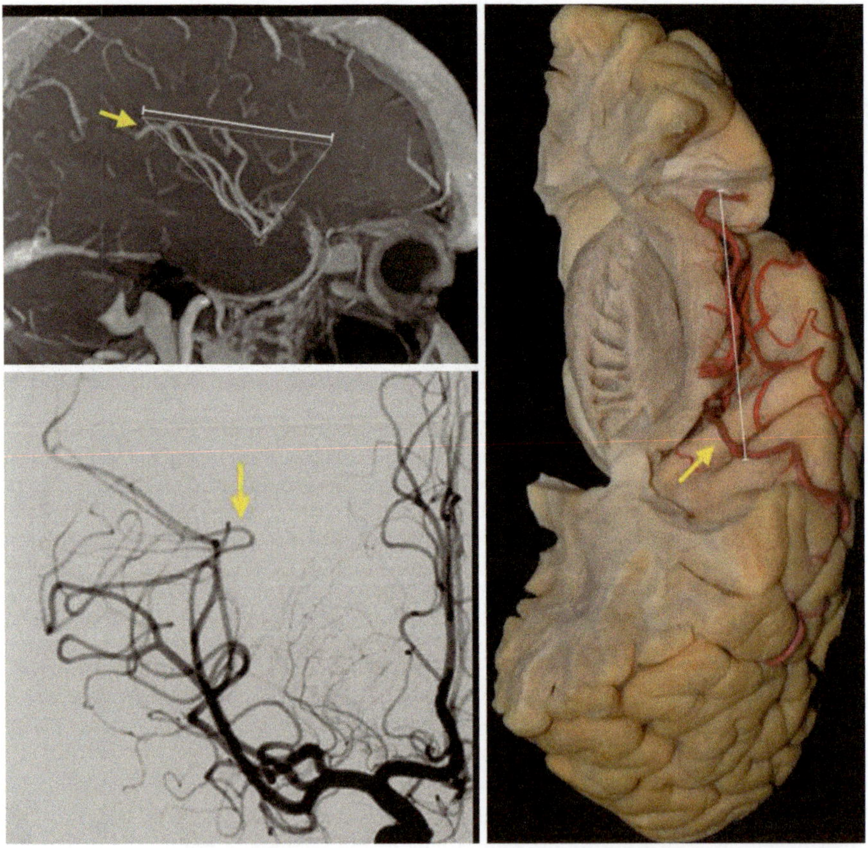

Fig. 4.34 The sagittal T1-weighted magnetic resonance imaging (MRI) demonstrates the arterial (M2 segment) delimitation of the insula. The white triangle lines correspond to the anterior, inferior, and superior limiting insular sulci. The right internal carotid injection angiography identifies the posterior insular point (yellow arrow), characterized by the most posterior M2 branch coursing over the deep part of the transverse temporal gyrus (Heschl's gyrus). This is also illustrated in the MRI figure and in the right hemisphere cut at the level of the superior limiting insular sulcus (white bar). Note that the deepest part of the Heschl's gyrus and the posterior Sylvian point are just posterior to the meeting point of the inferior and superior limiting sulcus of the insula

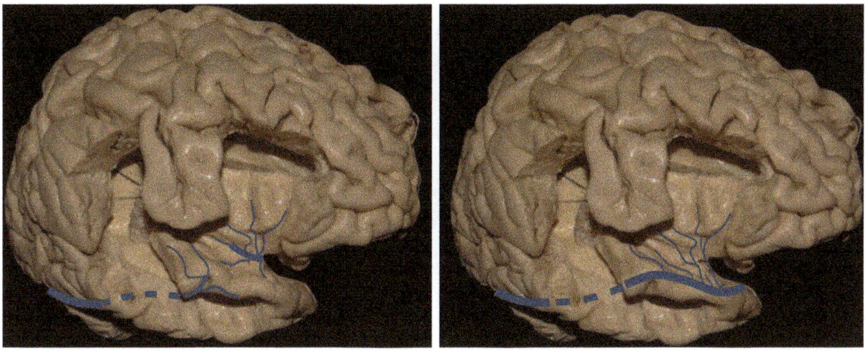

Fig. 4.35 Venous drainage of the superior surface of the temporal lobe. The pars triangularis, pars opercularis, inferior part of the SM gyrus, and middle temporal gyrus were resected to expose the anterior insular cortex, PP, PT, and lateral ventricle. Left: The deep Sylvian vein receives insular and PP tributaries. The PT veins drains toward the Labbé vein. Right: The PT, PP, and insular veins drain toward the superficial Sylvian vein. (*PP* Planum polaris, *PT* planum temporalis, *SM* supramarginal)

4.11.2.2 Venous Drainage

Venous drainage of the superior surface of the PT is usually through the superficial Sylvian/Labbé vein. The PP veins may contribute to the insular veins as tributaries to the deep Sylvian vein. In some cases, the insular veins drain toward the well-developed superficial Sylvian vein (Fig. 4.35).

4.11.3 Approaches to the Sylvian Surface of the Temporal Lobe

A brain AVM in the deep half of Heschl's gyrus on the right side was identified in a 25-year-old female patient with a recent history of headache (Fig. 4.36).

The arterial supply to the AVM was confirmed as feeders from the M2 artery at the angiographic Sylvian point. An enlarged deep Sylvian vein drained the AVM toward the cavernous sinus.

The patient underwent microsurgical resection through a wide posterior extended fronto-temporal craniotomy (Figs. 4.37 and 4.38).

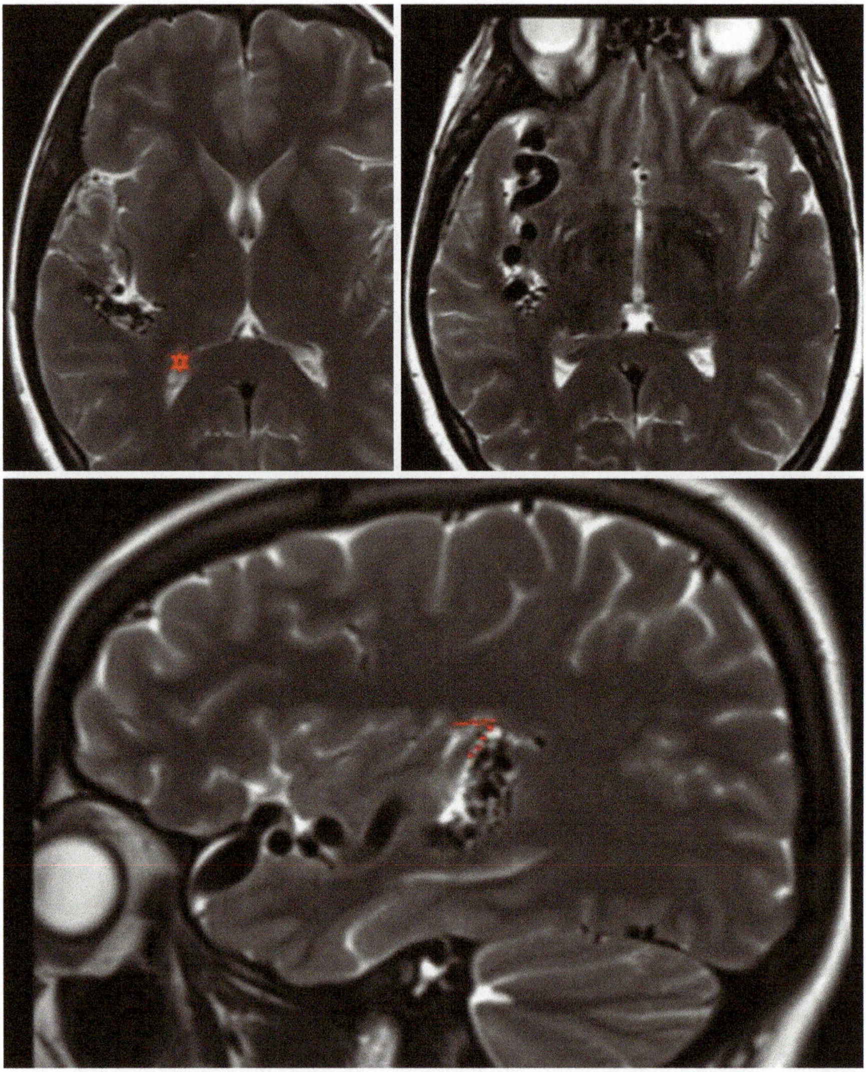

Fig. 4.36 T2-weighted magnetic resonance imaging sequences. The nidus at the Heschl's gyrus points to the atrium (red star). The deep middle cerebral vein is enlarged and tortuous. In the sagittal plane, it is possible to observe the location of the nidus (Heschl's gyrus) just behind the posterior insular point (the red v shape)

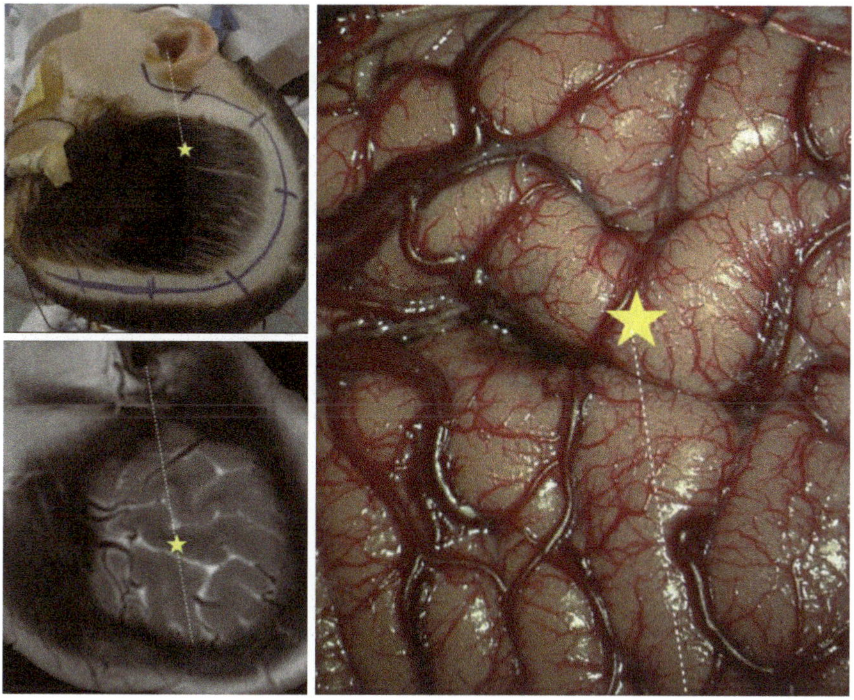

Fig. 4.37 The patient's head was turned to the left and parallel to the floor to keep HG (yellow star) at the highest point of the surgical field. HG is located above the external acoustic meatus and beneath the postcentral gyrus. (*HG* Heschl's gyrus)

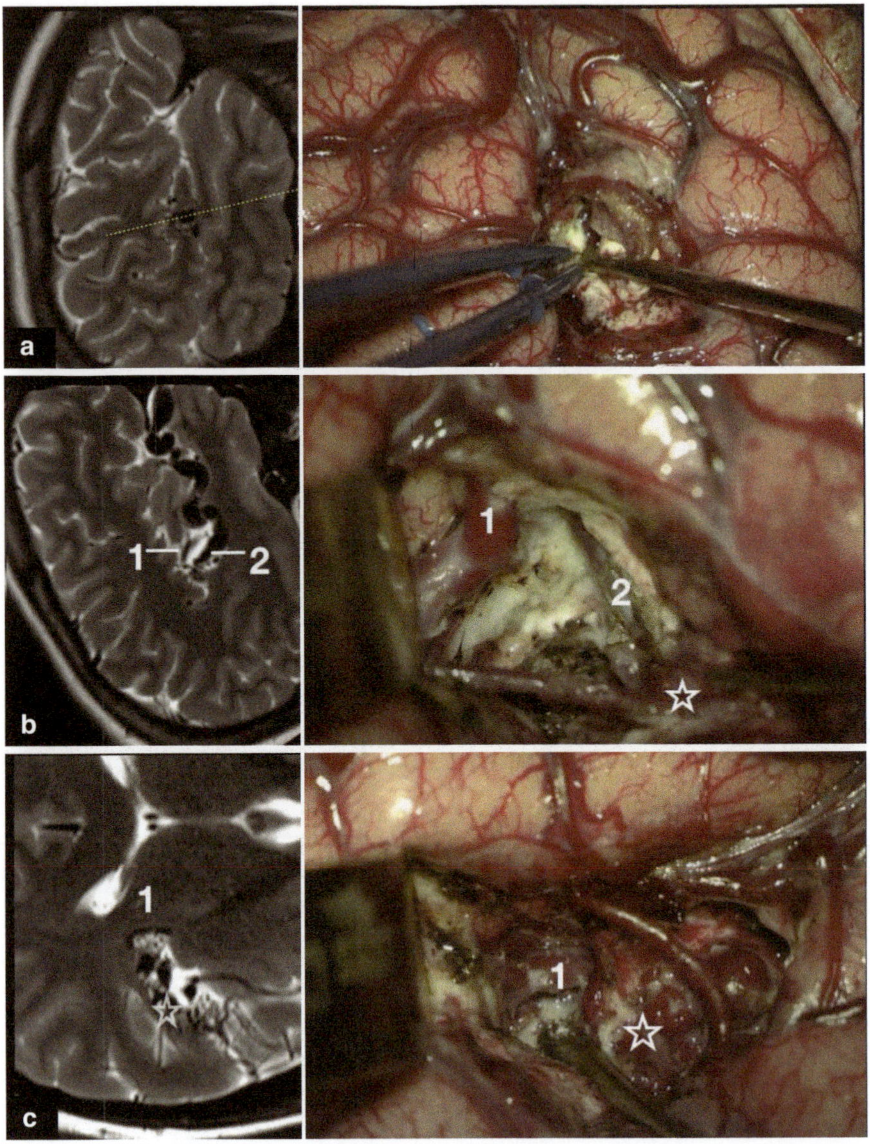

Fig. 4.38 (**a**) Corticectomy at HG. (**b**) The MCA superior trunk feeder (1) and the enlarged deep MCV (2) are identified anteriorly to the AVM (white star). (**c**) Axial T2-weighted magnetic resonance imaging showing the relation of the M2 branch running over the AVM at the deep part of the transverse temporal gyrus (the angiographic Sylvian point). The nidus (white star) was isolated from the artery (1) for further venous disconnection and en bloc resection. *AVM = Arteriovenous malformation; (HG* Heschl's gyrus, *MCA* middle cerebral artery, *MCV* middle cerebral vein)

Suggested Bibliography

Bozkurt B, da Silva CR, Chaddad-Neto F, da Costa MD, Goiri MA, Karadag A, Tugcu B, Ovalioglu TC, Tanriover N, Kaya S, Yagmurlu K, Grande A. Transcortical selective amygdalohippocampectomy technique through the middle temporal gyrus revisited: an anatomical study laboratory investigation. J Clin Neurosci. 2016;34:237–45.

Chaddad-Neto F, da Costa MDS, Caramanti RL, Costa BL, Silveira CF, Echeverria KG, Campos Filho JM, Centeno RS. Transtentorial approach for parahippocampal gyrus arteriovenous malformation resection: 3-dimensional operative video. Oper Neurosurg (Hagerstown). 2019;16(3):E83–4.

Chaddad-Neto F, Dória-Netto HL, Campos-Filho JM, Reghin-Neto M, Oliveira E. Pretemporal craniotomy. Arq Neuropsiquiatr. 2014;72(2):145–51.

De Witte E, Satoer D, Robert E, Colle H, Verheyen S, Visch-Brink E, et al. The Dutch Linguistic Intraoperative Protocol: a valid linguistic approach to awake brain surgery. Brain Lang. 2015;140:35–48.

de Oliveira JG, Párraga RG, Chaddad-Neto F, Ribas GC, de Oliveira EP. Supracerebellar transtentorial approach-resection of the tentorium instead of an opening-to provide broad exposure of the mediobasal temporal lobe: anatomical aspects and surgical applications: clinical article. J Neurosurg. 2012;116(4):764–72

Echeverría KG, da Costa MDS, Costa BL, Ortiz RZ, Paganelli SL, Granados OS, Doria-Netto HL, Centeno RS, Chaddad-Neto F. Microsurgical treatment of ruptured spetzler-martin grade 3 right hippocampal arteriovenous malformation: 3-dimensional operative video. Oper Neurosurg (Hagerstown). 2020;18(1):E17.

Herbet G, Duffau H. Revisiting the functional anatomy of the human brain: toward a meta-networking theory of cerebral functions. Physiol Rev. 2020;100:1181–228.

Ono M, Kubik S, Abcrbathcy C. Atlas of the cerebral sulci. Germany: Thieme; 1989.

Ribas GC. The cerebral sulci and gyri. Neurosurg Focus. 2010;28(2):E2.

Chapter 5
Surgical Anatomy of the Central Core of the Brain

5.1 Introduction

The central core is located within each cerebral hemisphere between the Sylvian fissure laterally and the third ventricle medially, constituting a single block composed externally by the insula and internally by the basal nuclei and thalamus. The term "central core" simplifies the topographic identification of several structures, including the thalamus, caudate nucleus, internal capsule, lenticular nucleus (putamen and globus pallidus), external capsule, claustrum, extreme capsule, and insular cortex, in addition to a group of short and long association, commissural, and projection fibers (Fig. 5.1).

For an easier comprehension, each component of the central core of the brain is studied separately; however, understanding all components as a whole, together with their anatomical correlations in 3D, remains challenging.

Supplementary Information The online version of this chapter (https://doi.org/10.1007/978-3-030-82747-2_5) contains supplementary material, which is available to authorized users.

F. Chaddad-Neto, M. D. Silva da Costa, *Microneuroanatomy and Surgery*,
https://doi.org/10.1007/978-3-030-82747-2_5

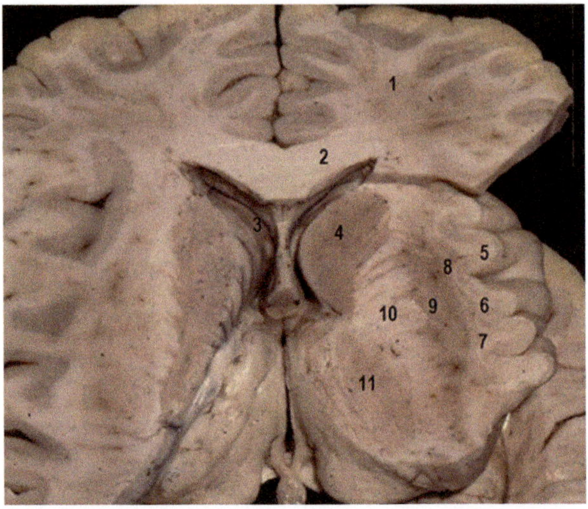

Fig. 5.1 Cross-sectional image illustrating the central core of the brain and its components. Left frontal, parietal, and temporal opercula were removed. The surface of the insula clearly represents the external shield of the anatomically well-defined cerebral core, which comprises gray and white matter in each hemisphere. The gray matter is composed of the insula itself, the claustrum, the basal nuclei (caudate nuclei, globus pallidus, and putamen), and the thalamus. The white matter is composed of the extreme, external, and internal capsule between the gray matter structures. (1) Frontal lobe, (2) corpus callosum, (3) anterior horn of lateral ventricle, (4) head of caudate nucleus, (5) insula, (6) extreme capsule, (7) claustrum, (8) external capsule, (9) lentiform nucleus, (10) internal capsule, (11) thalamus

5.2 Insular Cortex

The insula (or island of Reil) is a pyramidal shape region of the cerebral cortex located deep within the lateral sulcus. It is covered by the frontal, temporal, and parietal opercula (in Latin, curtains) and thus not visualized on the lateral surface of the brain. The insula forms the base of the Sylvian fissure and is only visible when the opercula are widely open. Removal of the opercula exposes the entire insula, which forms the external shield of the basal nuclei. This is important for surgical planning, since the basal nuclei are completely covered by the insula.

Understanding the anatomy of opercular and insular surfaces allows for planning of deeper surgical approaches. The anterior insular surface is covered by the fronto-orbital operculum, consisting of the posterior portion of the posterior orbital gyrus and the orbital part of the inferior frontal gyrus. Its lateral surface is covered by the frontoparietal operculum (triangular and opercular parts of the inferior frontal gyrus, subcentral gyrus, and superior part of the supramarginal gyrus). The temporal operculum is composed of the superior temporal gyrus, temporal pole, and supramarginal gyrus (Fig. 5.2).

The opercular surfaces delineate the superficial portion of the Sylvian fissure, especially its posterior branch. The anterior limit of the insula corresponds to a

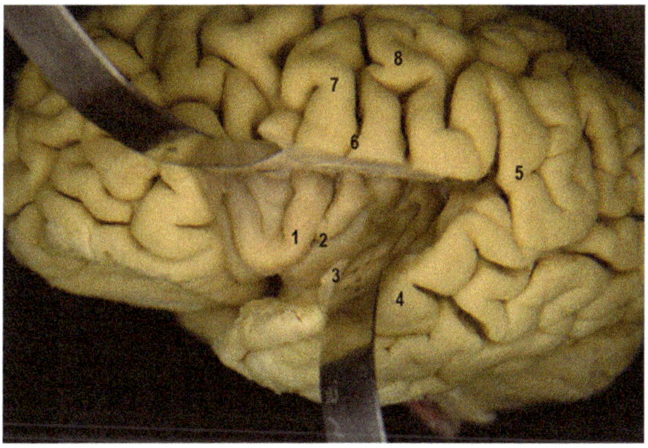

Fig. 5.2 Left lateral hemisphere after removal of the fronto-orbital, frontoparietal, and temporal opercula, visualization of the insula. Note correspondence between the central sulcus of Rolando and the central insular sulcus. The posterior limit of the insula is related to Heschl's gyrus. (1) Insula, (2) central insular sulcus, (3) Heschl's gyrus, (4) superior temporal gyrus, (5) supramarginal gyrus, (6) central sulcus of Rolando, (7) precentral gyrus, (8) postcentral gyrus

projection from the anterior Sylvian point, at the level of the triangular part of the inferior frontal gyrus. The posterior limit of the insula is delimited by the medial end of Heschl's gyrus, i.e., the transverse temporal gyrus, which is oblique and moves from lateral to medial in the superior temporal surface of the temporal lobe.

The anterior, superior, and inferior periinsular sulci (also known as limiting or circular sulci) separate the insular surface from the opercula. It is divided in two parts, the lateral and the anterobasal. The insular stem is located in the anterobasal portion, at the depth of the Sylvian fissure, and contains the limen insulae, a prominent bundle of fibers connecting the frontal and temporal lobes, which forms the lateral limit of the anterior perforated substance. It also represents the level at which the cerebral artery changes its direction axis (transition M1–M2). During surgery, it is necessary to locate this change in the direction of the middle cerebral artery, from medial to lateral and from anterior to posterior (M1 to M2), in order to ensure localization of the limen insula. At the most anteroinferior aspect of the insula lies the insular pole, which is formed by the transverse and accessory insular gyri.

The lateral surface of the insula is composed of short and long insular gyri. The central insular sulcus, the main and deepest sulcus of the insula, intersects it from the limen insulae to the superior periinsular sulcus, dividing it into anterior (larger) and posterior (smaller) portions. The anterior insula is composed of three short insular gyri, and their fusion forms the insular apex, which is the most prominent laterally projecting area and the tip of the pyramid shape. The posterior insula consists of the anterior and posterior long insular gyri. The anterior insula is connected to the frontal lobe, whereas the posterior insula connects to both the parietal and temporal lobes.

When viewing from the lateral surface of the cerebral hemisphere, the *pars opercularis* of the inferior frontal gyrus covers the superior portion of the anterior and middle

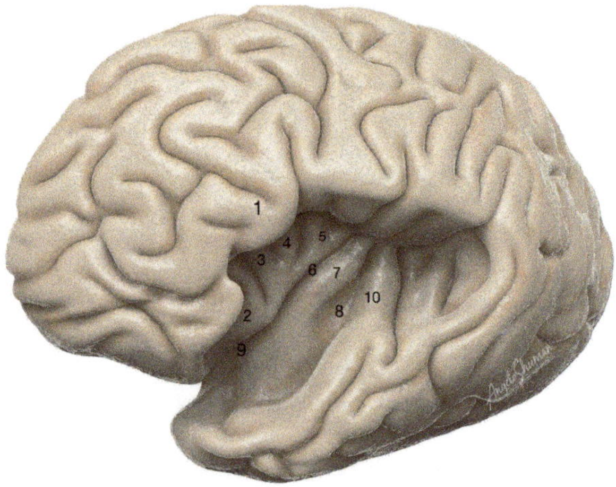

Fig. 5.3 Retraction of the opercula exposing the left insula. (1) Inferior frontal gyrus, (2) anterior periinsular sulcus, (3) anterior short gyrus, (4) middle short gyrus, (5) posterior short gyrus, (6) central insular sulcus, (7) long gyrus, (8) inferior limiting sulcus, (9) limen insulae, (10) Heschl's gyrus. (Illustration created by Angelo Shuman)

short insular gyri. Posteriorly, the supramarginal gyrus overlies the superior limiting sulcus and the superior portion of the posterior long gyri. The limen insula covers the uncinate fasciculus. The central insular sulcus courses obliquely following a similar direction and in the same projection of the central sulcus of Rolando (Fig. 5.3).

The insula is one of the paralimbic structures known as the mesocortex, and a variety of functions are attributed to it, although not yet fully understood. For example, memory, emotion, autonomic control, olfaction, and gustation are complex functions related to the insular cortex. Assessment by direct cortical stimulation links the insular cortex to autonomic regulation and visceral sensation, with descriptions of nausea, chewing, salivation, and changes in heart and respiratory rhythm. Functional neuroimaging investigations have revealed that the insula is activated in functions such as language, sensorimotor processing, taste and flavor perception, pain processing, and associated autonomic responses.

Anatomical and functional studies suggest a regional subspecialization of the insula. The anterior insula is related to attention processing and speech. It is also involved in the integration of autonomic-visceral information and emotional responses. The more anterior regions have connections to limbic structures, including the anterior cingulate cortex, prefrontal cortex, amygdala, and ventral striatum. In contrast, the posterior insula receives inputs from the sensory thalamus, and the association between the temporal, parietal, and occipital cortices plays a role in somatosensory, vestibular, and motor integration. An important distinction is the difference between right and left hemisphere insula functions. The right insula is probably more related to memory, emotional processing, and theory of mind, while the left is more related to speech.

Illustrative Case 1 Female, 29 years old, presenting seizures for half a year. Standard neurological examination did not reveal any deficits; meanwhile, neuropsychological assessment indicated a decrease in working memory, processing speed, and verbal fluency. MRI revealed a non-enhanced lesion suggestive of a left low-grade insular glioma (Fig. 5.4a, b, and c). Surgical resection was performed considering both oncological and functional indications. During surgery and while

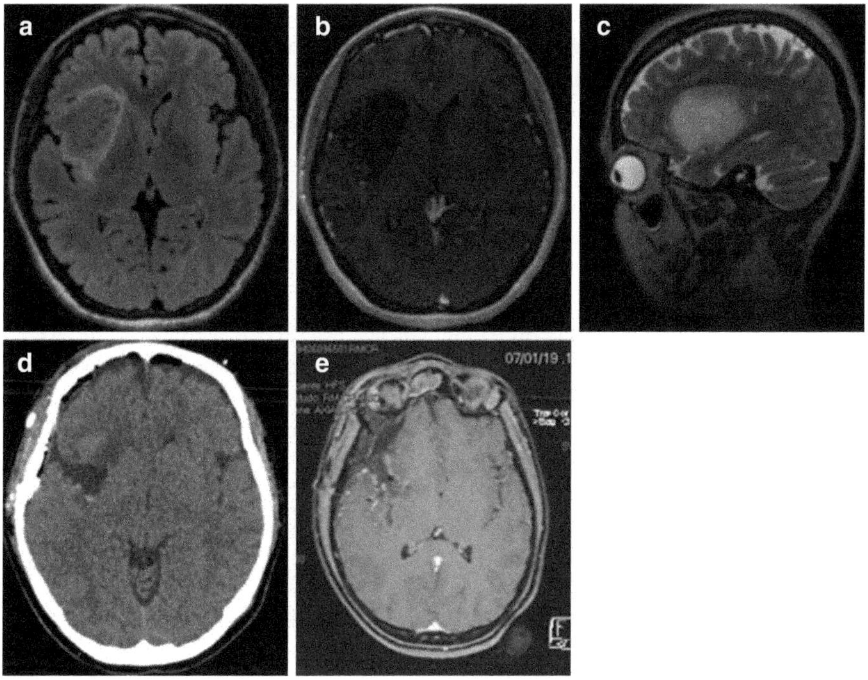

Fig. 5.4 (**a–c**) Preoperative brain MRI evidencing an insular glioma. (**d**) First day postoperative CT scan. (**e**) Late postoperative brain MRI showing the complete resection of the tumor

the insula was being manipulated, the patient started to present bradycardia. This fact limited the degree of lesion resection, requiring another step to completely remove the brain tumor (Fig. 5.4d and e).

Surgical Considerations The pterional approach offers a gateway to expose the Sylvian fissure and allows its microsurgical dissection (Fig. 5.4). The opercular lips that form the Sylvian fissure are slightly wider at the anterior Sylvian point, at the level of the pars triangularis. Therefore, starting microsurgical dissection at that point is favorable, thus exposing the insular apex, which is an important landmark for surgical orientation. Opening the Sylvian fissure is essential to access the circle of Willis and the anterior limits of the insula. The junction of the anterosuperior border of the anterior short insular gyrus with the periinsular sulci, namely, the anterosuperior insular point, is a landmark that indicates the anterior limit of the anterior limb of the internal capsule. Wide dissection of the posterior and more superficial branch of the Sylvian fissure allows the exposure of the polar plane of the superior surface of the temporal lobe until the beginning of the anterior transverse temporal gyrus (Heschl's gyrus). This gyrus runs posteriorly to the insula, with an oblique direction from anterior to posterior and from lateral to medial, and its medial end point toward the atrium of the lateral ventricle. The posterior insular point is the confluence of the superior and inferior periinsular sulci and indicates the direction to approach the posterior limb of the internal capsule, as well as the atrial portion of the lateral ventricle.

An important aspect of surgically approaching the insula is awareness of its arterial vascularization. Irrigation of the insula comes from the middle cerebral artery (MCA), classically divided into four segments. The first segment (M1-sphenoidal) moves laterally to the Sylvian vallecula and forms a genu at the level of the limen insulae, perpendicularly changing direction to another superior and posterior one. From the MCA, the genu forms the insular segment (M2). In approximately 15% of the cases, the MCA bifurcation is at the level of this genu, but it can also occur at an earlier or later level. Therefore, the MCA genu in the limen insulae and not necessarily this bifurcation is the transition point between M1 and M2.

The M1 segment is especially important for the irrigation of the basal nuclei and internal capsule through the lateral lenticulostriate arteries, as well as for the irrigation of the temporal pole. The M2 segment is usually formed by two trunks (superior and inferior) from the limen insulae, covering the insular surface along its route. Other bifurcations can also occur along its course, generating intermediate trunks, giving the impression of MCA trifurcations or quadrifurcations. The insula receives its blood supply predominantly from the M2 segment, which runs over the inferior limiting sulci and extends from the limen insulae to the periinsular sulci, in a region named posterior Sylvian point. Most of the insular arteries are short and supply the insular cortex and extreme capsule (Fig. 5.5).

The venous drainage of the insula is predominantly through the deep middle cerebral vein (DMCV) or deep Sylvian vein, a part of the deep venous system. Union of the insular veins forms the DMCV near the limen insulae, which run to the basal vein of Rosenthal. Some superficial insular veins, the sphenoidal group, send blood to the superficial middle cerebral vein (SMCV) of the superficial venous system.

Illustrative Case 2 A 22-year-old female presenting with intense sudden headache accompanied by weakness in the left limbs (grade III) and ipsilateral facial paresis. The patient was assessed in our service 1 month after the onset of clinical manifestations. Radiological investigation (Fig. 5.6) showed an arteriovenous malformation (AVM) in the insular territory. Initial non-contrast CT scan (Fig. 5.6a) showed a

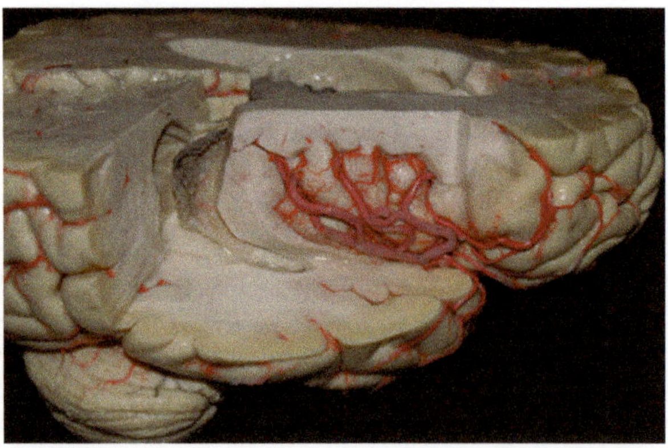

Fig. 5.5 Cross section of the right cerebral hemisphere at the level of the anterior horn, body, and atrium of the lateral ventricle. The temporal lobe is almost completely removed, and the right insula is exposed. The arteries of the insula originate from the M2 segment

hypodense area around the right basal nuclei, from the caudate head to near the ventricular atrium. MRI scan (Fig. 5.6b) indicated an enhancing lesion in the insular territory. Preoperative angiography showed an AVM irrigated by the M2 segment of the MCA (Fig. 5.6c) with superficial drainage (Fig. 5.6d). The surgical approach involved a pterional craniotomy. The patient was positioned with a slight contralateral rotation and deflection of the head (Fig. 5.7a). Neuronavigation was especially useful in defining the posterior limit of the lesion (Fig. 5.7b). Opening of the Sylvian

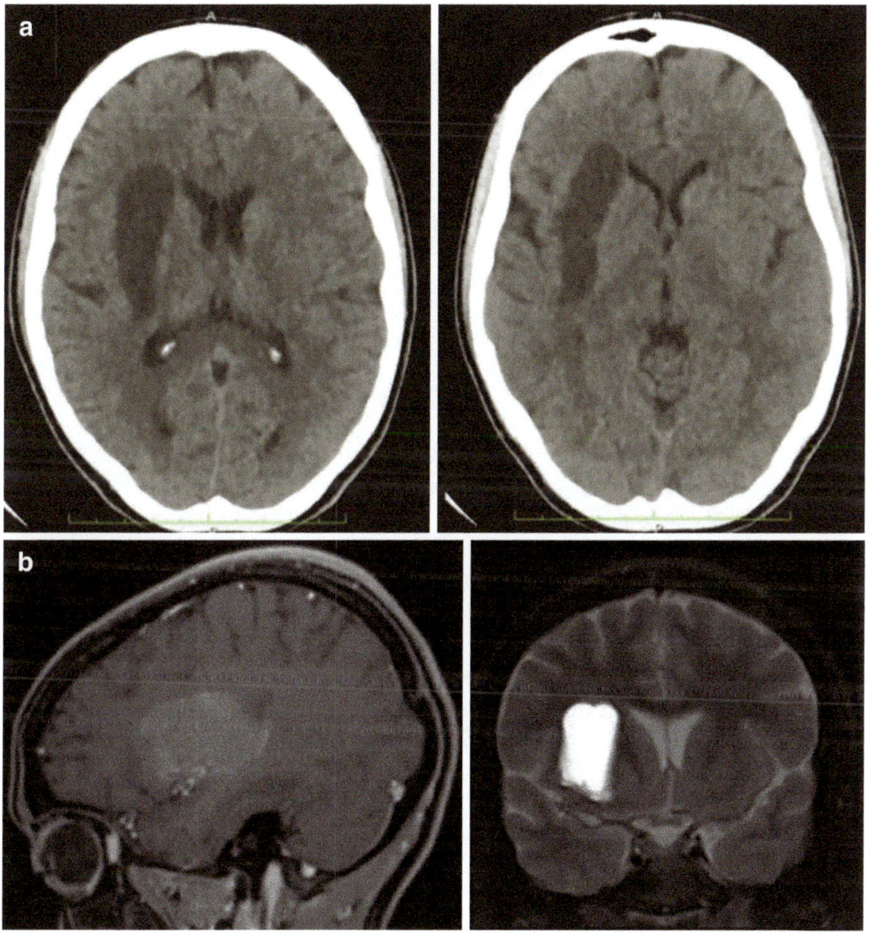

Fig. 5.6 (**a**) Initial non-contrast CT scan shows extensive hypodensity in the anteroposterior direction, appearing as an external shield of the basal nuclei. (**b**) Preoperative MRI confirms the insular topography, showing enhancement in a tissue deep to the opercula and external to the basal nuclei; sagital postgadolimium in the left and coronal T2WI in the right (**c**) In the arterial phase, angiography shows an arteriovenous malformation (AVM) irrigated by the M2 segment of the middle cerebral artery. The M1-M2 transition corresponds to the limen insulae, being a good reference for topographic interpretation, confirming a diagnosis of an insular AVM. (**d**) Venous phase demonstrating superficial drainage of the insular AVM, in the right oblique projection and in the left lateral projection

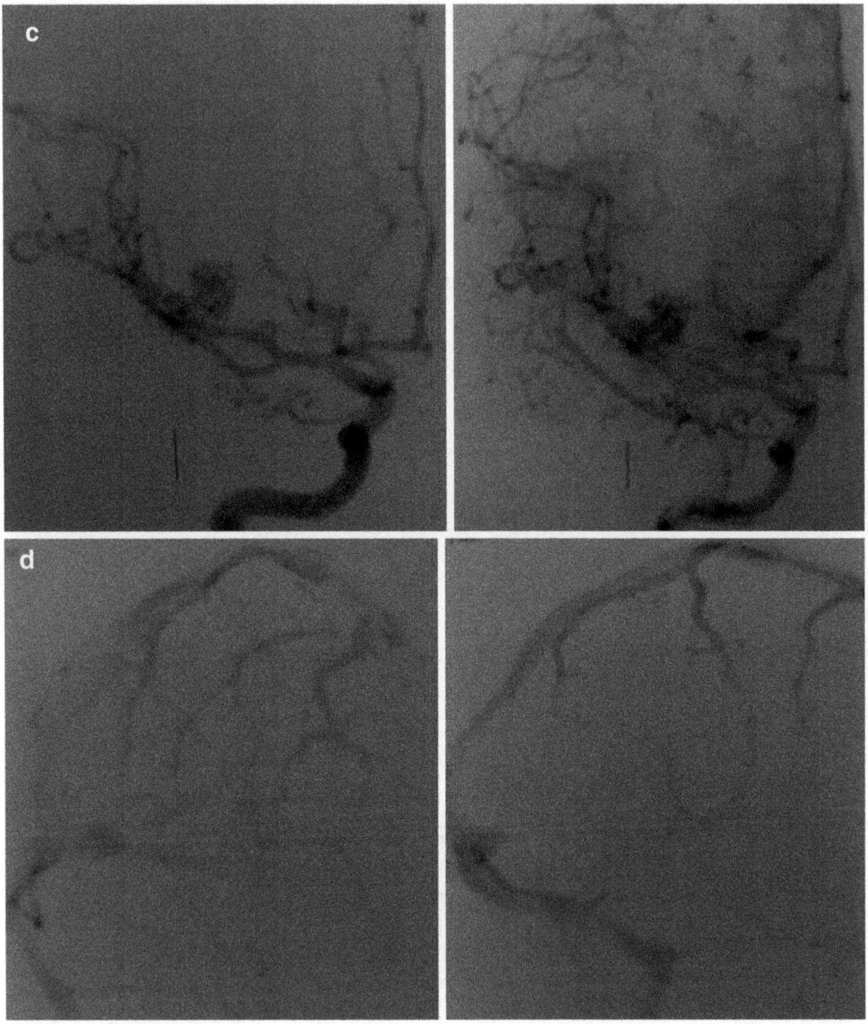

Fig. 5.6 (continued)

fissure allowed exploration of the MCA from the carotid bifurcation and along the entire disposition of M2 over the insular cortex (Fig. 5.8a and b). After extensive exposure, the AVM was identified, in addition to identifying nidal aneurysms (Fig. 5.8c). The entire nidus was devascularized and safely removed (Fig. 5.8d, e and f).

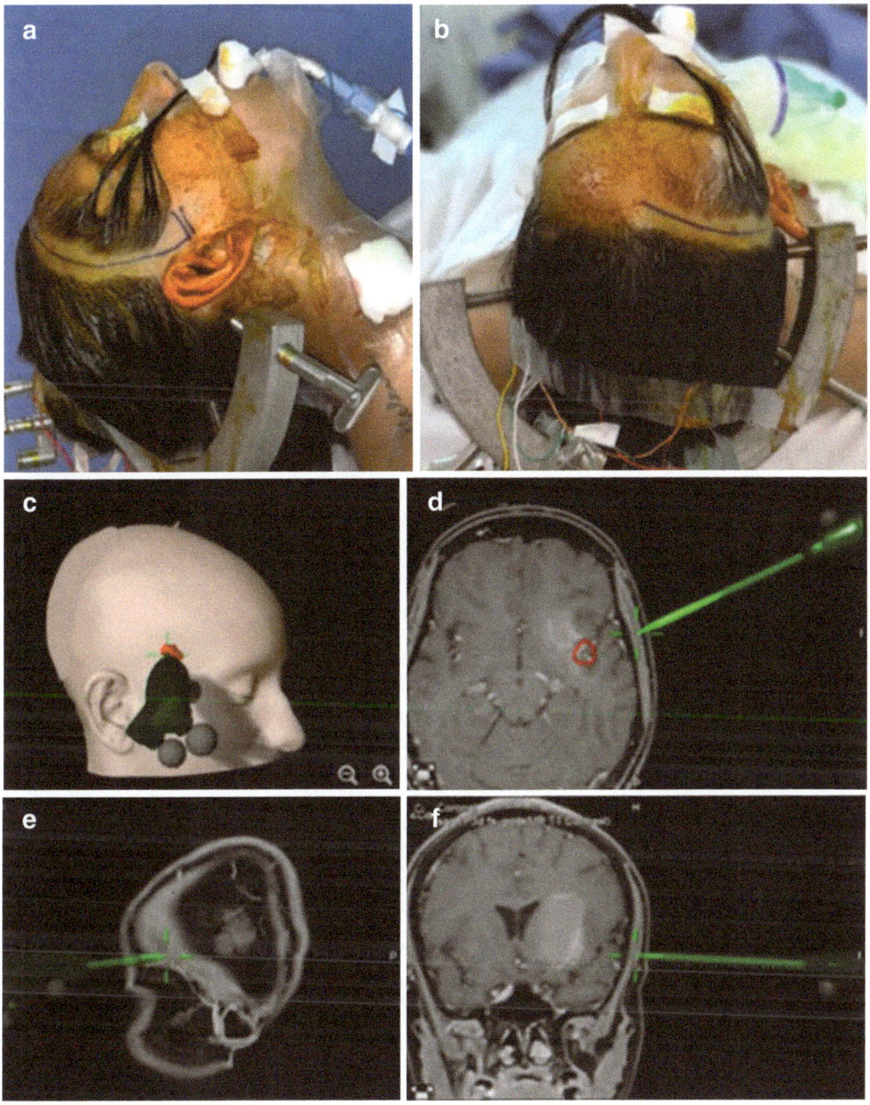

Fig. 5.7 (**a**) Head positioning of the patient in the lateral view, with slight contralateral rotation and deflection of the head, in the superior view (**b**). (**c**) Intraoperative anatomical verification of the lesion using neuronavigation system, (**d**) axial view of the MRI, (**e**) sagittal view, and (**d**) coronal view

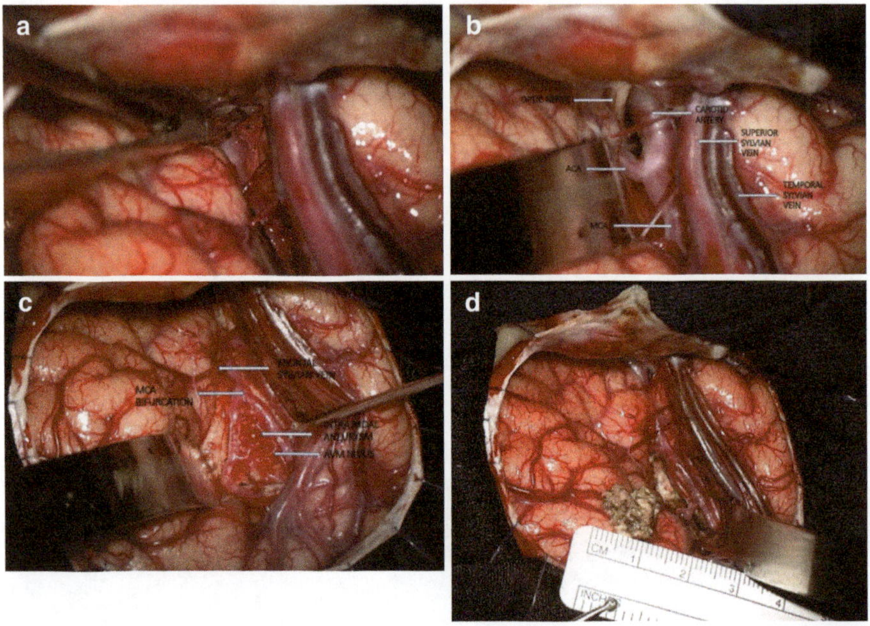

Fig. 5.8 (**a**) Wide dissection of the Sylvian fissure. (**b**) Identification of the internal carotid artery, anterior cerebral artery, middle cerebral artery, and superficial veins. (**c**) Insular arteriovenous malformation (AVM) exposure. (**d**) Complete removal of the insular AVM. MRI scans showing complete removal of the insular AVM in the axial T1WI (**e**) and T2WI (**f**). Postoperative angiography showing complete insular AVM removal and preservation of the arterial circulation, (**g**) postoperative early arterial phase, and (**h**) postoperative late arterial phase

5.3 Extreme Capsule, Claustrum, and External Capsule

The extreme capsule is a strip of subcortical white matter under the insular cortex, composed of U fibers that connect the insular gyri among themselves and with the frontoparietal and temporal opercula. Medially to the extreme capsule lies a fine lamina of gray matter that constitutes the claustrum. The claustrum is a small subcortical nucleus with extensive excitatory connections with many cortical areas. Theoretical and experimental work has suggested that the claustrum is involved in attention, novelty coding, sensorimotor integration, and stress. While the ventral portion of the claustrum is thinner and composed of small islands of gray matter within the white matter, its dorsal portion is thicker and much better defined. The external capsule is a second strip of white matter between the claustrum and the lentiform nucleus. It is composed mainly of fibers originating within the claustrum, and, anteriorly, it is integrated with the uncinate fasciculus and inferior fronto-occipital fasciculus.

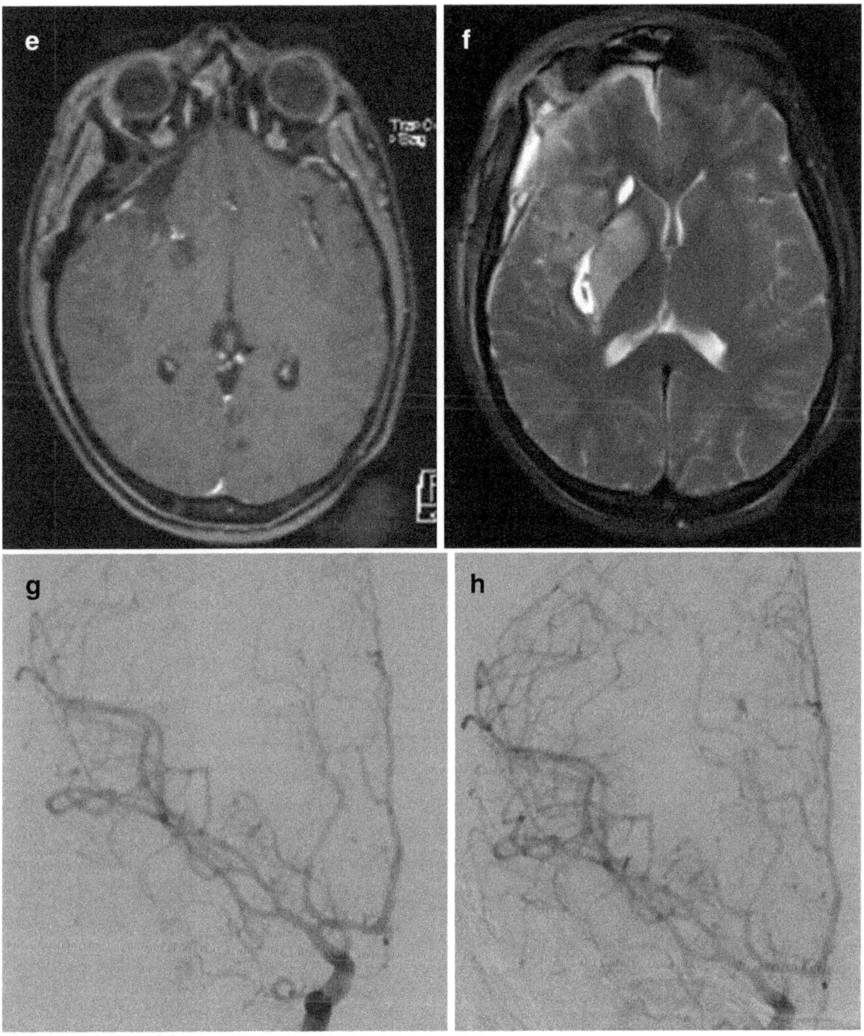

Fig. 5.8 (continued)

The claustrum and external capsule are both divided into ventral and dorsal portions. The dorsal external capsule is formed by claustrocortical projection fibers connecting the claustrum and cortex between the supplementary motor cortex anteriorly and posterior portion of the parietal lobe posteriorly. The ventral external capsule is formed by the fronto-occipital fasciculus dorsally and the uncinate fasciculus ventrally. The dorsal claustrum is located between the extreme and external capsule. The ventral claustrum is formed by islets of gray matter interposed in the ventral external capsule that extends laterally to the amygdaloid nucleus.

5.4 Lentiform Nucleus

The lentiform or lenticular nucleus embraces the putamen and globus pallidus within the basal nuclei (Fig. 5.9). It is a large lens-shaped mass of gray matter between the internal and external capsule. In a coronal section, two medullary laminae are seen dividing the nucleus into three parts: the putamen, internal globus pallidus (GPi), and external globus pallidus (GPe). The putamen has been shown to connect the basal ganglia with regions involved in language and is also involved in learning and motor control. The GPi is one of the output nuclei of the central core, with GABAergic neurons sending axons to the ventral anterior and ventral lateral nuclei of the thalamus and to the pedunculopontine and centromedian complexes. The boundary between the GPe and GPi contains another thin layer of myelinated fibers, namely, the medial medullary laminae.

Vascularization of the lentiform nucleus, as well as of the caudate nucleus, thalamus, and internal capsule, is provided by perforating arteries arising from the anterior, middle, and posterior cerebral arteries and the anterior choroidal arteries. The lenticulostriate arteries are the most important elements for the lentiform nucleus supply. They arise from the middle and anterior cerebral arteries. The recurrent artery of Heubner originates in the transition of the first to the second segment of the anterior cerebral artery (most often in the proximal portion of the second segment to anterior cerebral artery) and supplies some structures, such as the medial portion of the globus pallidus, head of the caudate nucleus, anterior crus of the internal capsule, and anterior hypothalamus. In general, the internal carotid branches supply nuclei of telencephalic origin (caudate and lentiform), while the vertebrobasilar branches supply the diencephalic nuclei (thalamus). Venous drainage of the lentiform nucleus occurs via the striate branches of the internal cerebral vein, which drain into the great cerebral vein (the vein of Galen).

5.5 Caudate Nucleus

The lateral wall of the anterior horn of the lateral ventricle is located in the lateral aspect of the anterior border of the thalamus (Fig. 5.9). The body is located on the upper surface of the thalamus and corresponds to the lateral wall of the ventricular body. The caudate nucleus tail encircles the pulvinar nuclei of the thalamus posteriorly and laterally and then runs along the roof of the inferior horn.

The putamen and caudate nuclei together compose the striatum, an input area of the basal nuclei. Dorsally, they are separated by fibers of the internal capsule, whereas their ventral aspects are continuous underneath the anterior and inferior edge of the internal capsule, constituting the ventral striatum, which corresponds to the nucleus accumbens (*a case example is also demonstrated in the* video 5.1).

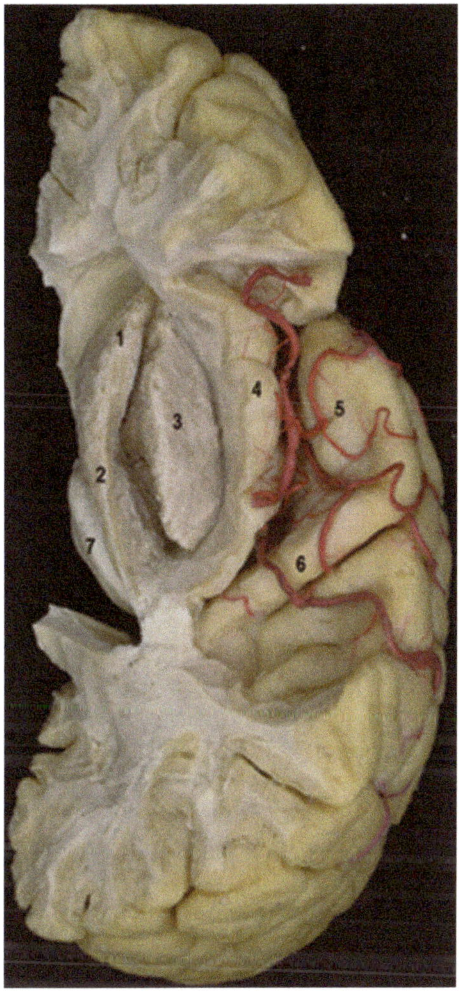

Fig. 5.9 Superior view of the brain after removal of the white matter fibers of the central core. (1) Head of caudate nucleus, (2) body of caudate nucleus, (3) lentiform nucleus, (4) insula, (5) temporal opercula, (6) Heschl's gyrus, (7) thalamus

5.6 Internal Capsule

The internal capsule is a white matter structure that separates the caudate nucleus and thalamus from the lentiform nucleus and is composed of bundles of myelinated projection fibers. Above the superior aspect of the lenticular nucleus, the internal capsule is continuous, forming the centrum semiovale or corona radiata, and inferiorly to the lenticular nucleus, it is continuous and forms the cerebral peduncle. The internal capsule has a V-shaped format, and, in relation to the lenticular nucleus, it is divided into an anterior limb (between the head of the caudate and the putamen in its rostral part and between the head of the caudate and the globus pallidus in its caudal part), a genu (corresponding to the medial apex of the V, adjacent to the interventricular foramen), a posterior limb (between the body of the caudate together with the thalamus and globus pallidus in its rostral part and between the body of the caudate together with the thalamus and putamen in its caudal part), and retrolenticular and sublenticular parts. It is a two-way tract for the transmission of information to and from the cerebral cortex. The central core is attached to the rest of the cerebral hemisphere by isthmi, composed of these different internal capsule parts. Anteriorly and under the anterior limiting insular sulcus lie fibers of the anterior limb of the internal capsule. Superiorly and under the superior limiting sulcus, the remaining anterior limb fibers, as well as the genu and posterior limb fibers that harbor the corticonuclear and corticospinal tracts, are present. Posteriorly and inferiorly to the insular inferior limiting sulcus, retrolenticular and sublenticular internal capsule fibers enclose the auditory and optic radiations.

5.7 Thalamus

The thalamus is an ovoid gray matter structure of the diencephalon, located above the mesencephalon, and has many important roles in human physiology. It is composed of different nuclei with different functions, ranging from relay of sensory and motor signals to regulation of consciousness and alertness. It forms the lateral walls of the third ventricle, while its dorsal surface is part of the floor of the body of the lateral ventricle. Laterally, the thalamus is limited to the posterior arm of the internal capsule. Anterolaterally, it limits the head of the caudate and ventral nucleus, together with the subthalamus and hypothalamus.

The thalamus has a medial, lateral, superior, inferior, anterior, and posterior surface. The lateral surface is convex, adhering to the caudate nucleus superiorly and to the posterior segment of the internal capsule inferiorly. The posterior (pulvinar) surface is divided by the fornix into an ependymal surface called the ventricular surface and into an extraventricular or cisternal surface (pulvinar) related to the quadrigeminal cistern. The medial surface forms the lateral wall of the third ventricle. The upper surface is divided medially by the medullary thalamic stria into two halves: medial and lateral. The lateral half is related to the caudate nucleus and constitutes the lateral body floor of the lateral ventricle, while the medial half is related to the choroidal tissue and fornix. The lower surface lies on the mesencephalon. The posterior border, represented by the pulvinar nucleus, is subdivided by the choroidal fissure in a medial or quadrilateral and a lateral or ventricular segment. This is very important because it allows dividing this sector according to the surgical approach to be used. The anterior border is divided into medial and lateral halves. The lateral half is related to the head of the caudate nucleus, and the medial half forms the posterior limit of the foramen of Monro.

Vascularization of the thalamus is complex and occurs in different arteries. The pre-mamillary artery is defined as the most voluminous perforating branch of the posterior communicating artery. The perforating thalamic arteries arise from the P1 segment of the posterior cerebral artery (PCA) or from the bifurcation of the basilar artery. These arteries penetrate the posterior perforated substance. The superior aspect of the thalamus receives perforators from the terminal branches of the posterior lateral choroidal artery (PLChA), an arterial trunk arising from the PCA. Thalamogeniculate arteries enter around and between the geniculate bodies and arise from the PCA, PLChA, and posterior medial choroidal artery. The venous drainage of the thalamus can be divided into superior (superior and anterior thalamic veins), inferior, lateral, and posterior groups, including a variable number of perforating veins. The thalamostriatal and lateral thalamic veins drain the blood of the superior portion of the thalamus to the internal cerebral vein, and the inferior portion of the thalamus is, particularly, drained by the Rosenthal's basal vein.

Suggested Bibliography

Chaddad-Neto F, Campos Filho JM, Dória-Netto HL, Faria MH, Ribas GC, Oliveira E. The pterional craniotomy: tips and tricks. Arq Neuropsiquiatr. 2012;70(9):727–32.

Chaddad Neto F, de Oliveira E, Paschoal E, Cendes F, Santana Filho M. Características morfológicas do lobo da ínsula em pacientes portadores de epilepsia do lobo temporal medial [Morphological characteristics from the insula's lobe in patients with medial temporal lobe epilepsy]. Arq Neuropsiquiatr. 2006;64(3A):639–44.

Choi CY, Han SR, Yee GT, Lee CH. Central core of the cerebrum. J Neurosurg. 2011;114:463–9.

Costa M, Braga VL, Yağmurlu K, Centeno RS, Cavalheiro S, Chaddad-Neto F. A technical guide for fiber tract dissection of the internal capsule. Turk Neurosurg. 2018;28(6):934–9.

Ribas EC, Yagmurlu K, de Oliveira E, Ribas GC, Rhoton-Jr A. Microsurgical anatomy of the central core of the brain. J Neurosurg. 2018;129:752–69.

Tanriover N, Rhoton-Jr AL, Kawashima M, Ulm AJ, Yasuda A. Microsurgical anatomy of the insula and the Sylvian fissure. J Neurosurg. 2004;100:891–922.

Ture U, Yasargil MG, Al-Mefty O, Yasargil DCH. Arteries of the insula. J Neurosurg. 2000;92:676–87.

Chapter 6
Surgical Anatomy of the Lateral Ventricles

The lateral ventricles are telencephalic cavities arranged around the thalamus, here considered the center of the brain and the head of the brainstem. As the thalamus has an ovoid format, the lateral ventricles naturally acquire a C shape, as do the other structures arranged around it, such as the fornix and caudate nucleus (Fig. 6.1). The lateral ventricle cavities are disposed according to their position in relation to the thalamus, and can be positioned in front, above, behind, and below, forming the frontal horn, body, atrium, posterior horn, and temporal horn, respectively. Each of these cavities is approximately 10 cm^3 and is filled with cerebrospinal fluid.

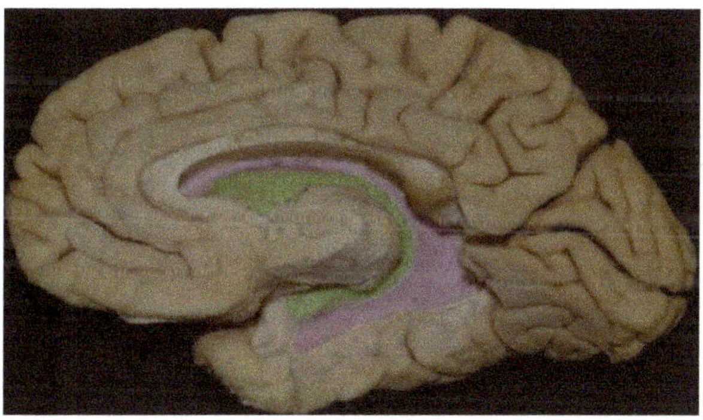

Fig. 6.1 Sagittal section of a right cerebral hemisphere. Note that the lateral ventricles and adjacent structures are arranged around the thalamus, acquiring the shape of a C. In this image, we can see the right lateral ventricle (purple) wrapped around the thalamus and the caudate nucleus (green)

Supplementary Information The online version of this chapter (https://doi.org/10.1007/978-3-030-82747-2_6) contains supplementary material, which is available to authorized users.

F. Chaddad-Neto, M. D. Silva da Costa, *Microneuroanatomy and Surgery*,
https://doi.org/10.1007/978-3-030-82747-2_6

When we explore the lateral ventricles, we divide their related structures into intrinsic or wall components. For example, the head of the caudate nucleus is part of the ventricle walls, and the choroid plexus is one of the intrinsic structures.

The inferior frontal sulcus represents the upper limit of the lateral ventricles on the brain surface and also corresponds to the corpus callosum and to the superior temporal line as a cranial landmark (Fig. 6.2).

Therefore, the middle frontal gyrus and the superior frontal gyrus are positioned above the superior limit of the lateral ventricles. It is notable that the sulci of the lateral surface point to the lateral ventricles, which may explain why many brain pathologies related to these structures acquire a cone shape, such as gliomas and arteriovenous malformations.

The lateral ventricles, as well as the thalamus, which is an oblique structure, extend anteriorly to posteriorly and medially to laterally. Its only straight surface is

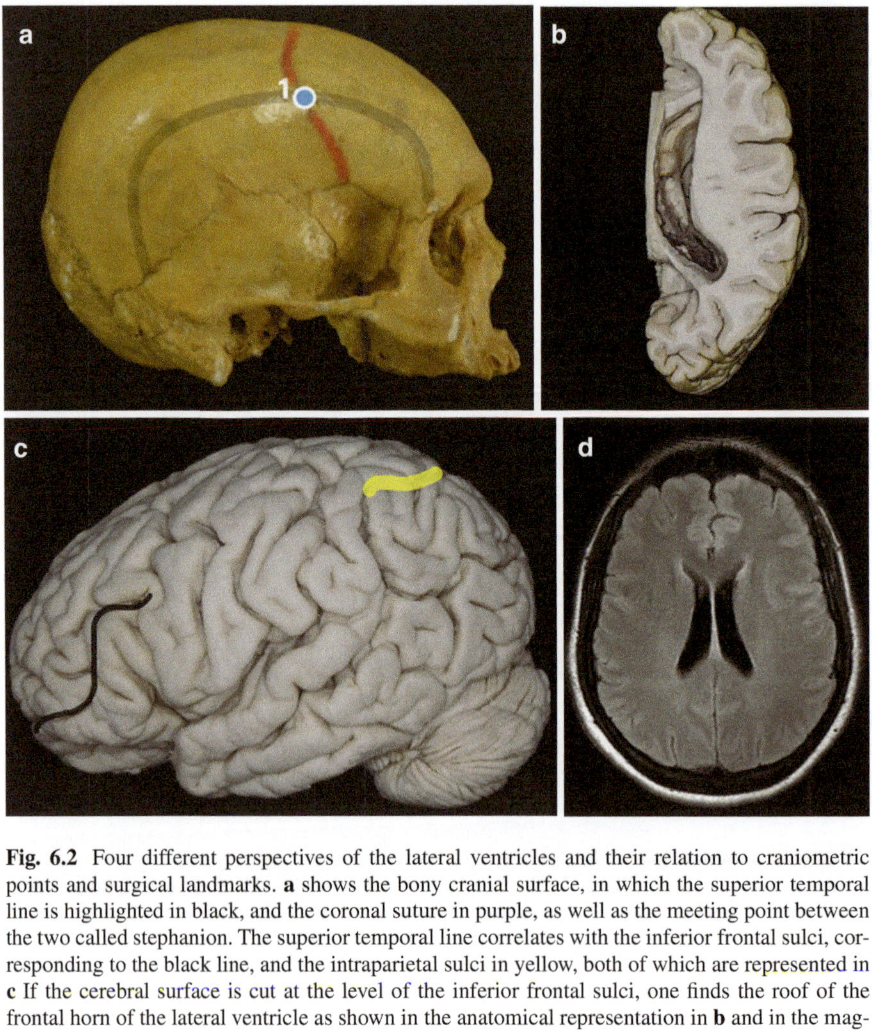

Fig. 6.2 Four different perspectives of the lateral ventricles and their relation to craniometric points and surgical landmarks. **a** shows the bony cranial surface, in which the superior temporal line is highlighted in black, and the coronal suture in purple, as well as the meeting point between the two called stephanion. The superior temporal line correlates with the inferior frontal sulci, corresponding to the black line, and the intraparietal sulci in yellow, both of which are represented in **c** If the cerebral surface is cut at the level of the inferior frontal sulci, one finds the roof of the frontal horn of the lateral ventricle as shown in the anatomical representation in **b** and in the magnetic resonance imaging in **d**

the medial one, which renders the notion of a rectilinear and midline structure. However, the thalamus and the lateral ventricle bend laterally when directed posteriorly. This understanding is important for the ventricular catheterization of the frontal horn or atrium during vp-shunts, as well as for microsurgical ventricular approaches. The frontal horn catheterization is different from the atrial one; these cavities are in different positions and orientations. The frontal horn and the body of the lateral ventricle are midline structures, but the atrium is more lateral, and the temporal horn is more lateral and inferior to the thalamus. When we are performing an approach, it is essential to analyze and consider these differences in the positioning of the different structures of the lateral ventricles. For example, the shortest path to access the frontal horn and the body of the lateral ventricle is through the midline, and the shortest path to access the atrium and temporal horn is through the cerebral cortex. This plays an important role in the surgical decision-making process.

6.1 Fornix

The fornix connects the hippocampus to the mammillary bodies (part of the hypothalamus), connecting the neurovegetative center of the central nervous system to the emotion-processing area and communicating with the two hippocampi through the hippocampal commissure. As the lateral ventricle, the parts of the fornix are classified according to their arrangement around the thalamus. In front lies the column of the fornix, above lies the body, behind lies the crus, and below lies the fimbria of the fornix or fimbria of the hippocampus. The crus of the fornix projects laterally and becomes the fimbria, which is also called the hippocampal fimbria (Fig. 6.3).

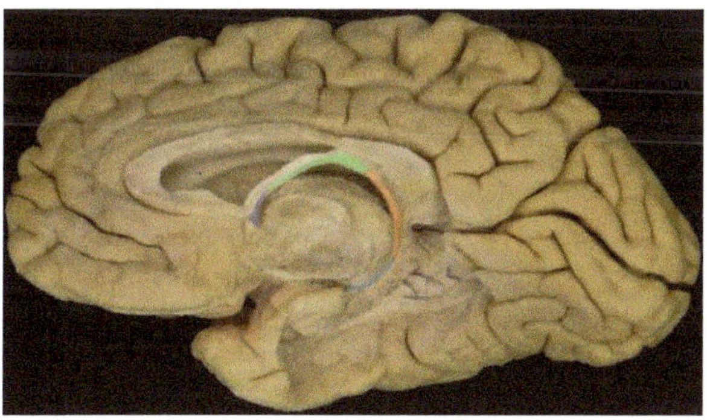

Fig. 6.3 The fornix is divided in column (purple), anterior wall of foramen of Monro, body (green)—in the medial wall of the body of the lateral ventricle, crus (orange), posterior surface of thalamus pulvinar, and fimbris (blue) arising in the temporal horn. Note that the C-format (purple line) of this structure is always wrapped around the thalamus. The fornix starts in the mammillary bodies and runs toward the hippocampus in the temporal horn, connecting these two important neurovegetative points

The fornix has a strict relation with the thalamus. They are bound by membranes, namely the tenia thalamic or choroidea, which connects the thalamus to the choroid plexus, and the tenia fornix, which connects the fornix to the choroid plexus. Therefore, the space formed between the thalamus and fornix is called the choroidal fissure. This relationship is very important to inform surgical strategies because it provides a landmark. Identifying the choroidal fissure means that the location of the thalamus and the fornix can be known. Similarly, identifying the fornix means that it is possible to identify the thalamus and the choroidal fissure.

The fornix forms the medial margin of the choroidal fissure, while the thalamus forms the lateral margin of the choroidal fissure. The choroidal fissure has its superior and medial end in the foramen of Monro or interventricular foramen, while its inferior and lateral end is the inferior choroidal point behind the head of the hippocampus and to the flank of the lateral geniculate body. These are important suggestions for intraoperative localizations (Fig. 6.4).

The fornix is also an important anatomical landmark which separates the pulvinar of the thalamus into two, the medial and lateral parts. The medial part of the

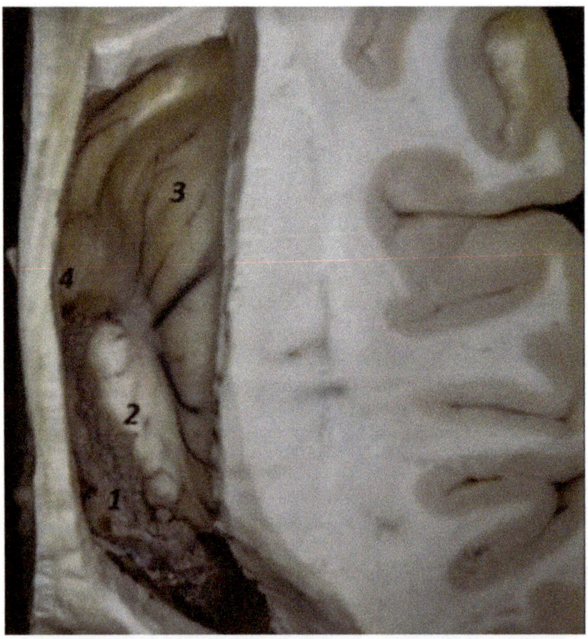

Fig. 6.4 Superior view of the lateral ventricle. As can be seen, the choroid plexus (1) is attached to the thalamus (2) and fornix (3) by the thalamic and fornix tenia, respectively. The lateral caudate nucleus (4) forms part of the lateral wall of the frontal horn. The choroid plexus cannot be seen in the frontal horn because we cannot find the thalamus there. The column of the fornix (5) is the anterior wall of the foramen of Monro

pulvinar of the thalamus is located in the quadrigeminal cistern or pineal region, while the lateral part is located in the atrium of the lateral ventricle.

Surgical Case 1 A 20-year-old man with acute onset of headache and loss of conscience. The investigation revealed a ruptured arteriovenous malformation (AVM) (Figs. 6.5, 6.6, and 6.7). (*This case example is also demonstrated in* Video 6.1.)

The *lateral posterior choroidal artery* arises from the P2 or P3 segment of the posterior cerebral artery or its branches. It runs laterally and enters the choroidal fissure. It supplies not only the atrium but also the peduncle, fornix, pulvinar, and caudate nucleus.

Based on the preoperative images, where is the lesion located? What are the feed arteries and drainage vein? What would be the best way to remove this lesion?

The angiography demonstrates that the lesion is located in the lateral ventricles, the corpus callosum, and choroidal fissure, adjacent to the dorsomedial surface of the right thalamus. The *choroidal fissure* runs from the superior choroidal point

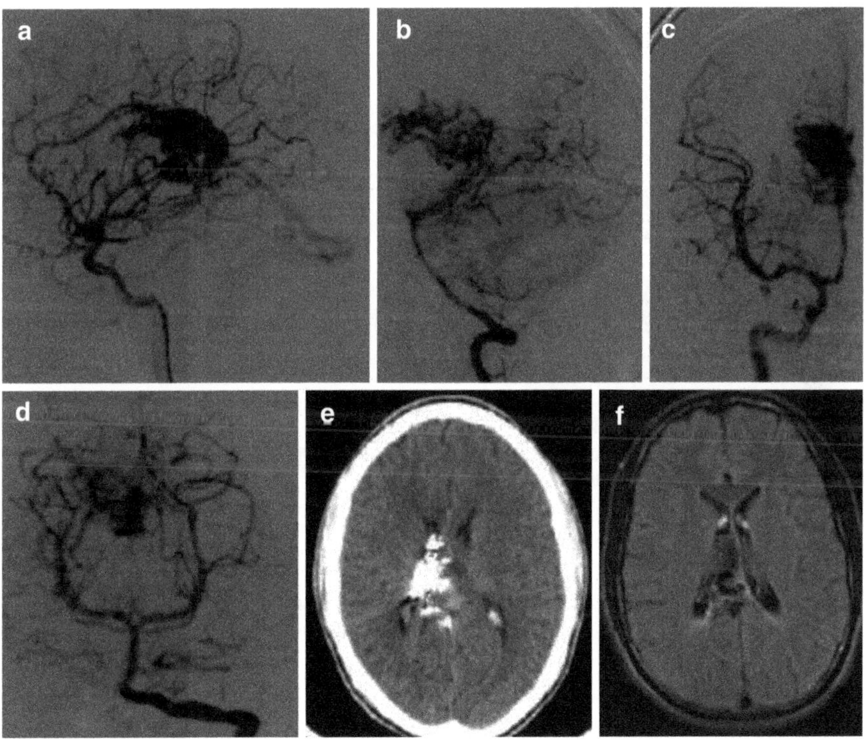

Fig. 6.5 In this panel, we studied the arteriovenous malformation using different platforms. Figure **a** to **d** represents an angiography showing the malformation being nourished by the anterior (**a** and **c**) and posterior (**b** and **d**) circulation in the anteroposterior and lateral views. In addition, the lesion is shown in an axial section tomography (**e**) and MRI (**f**) at the same level

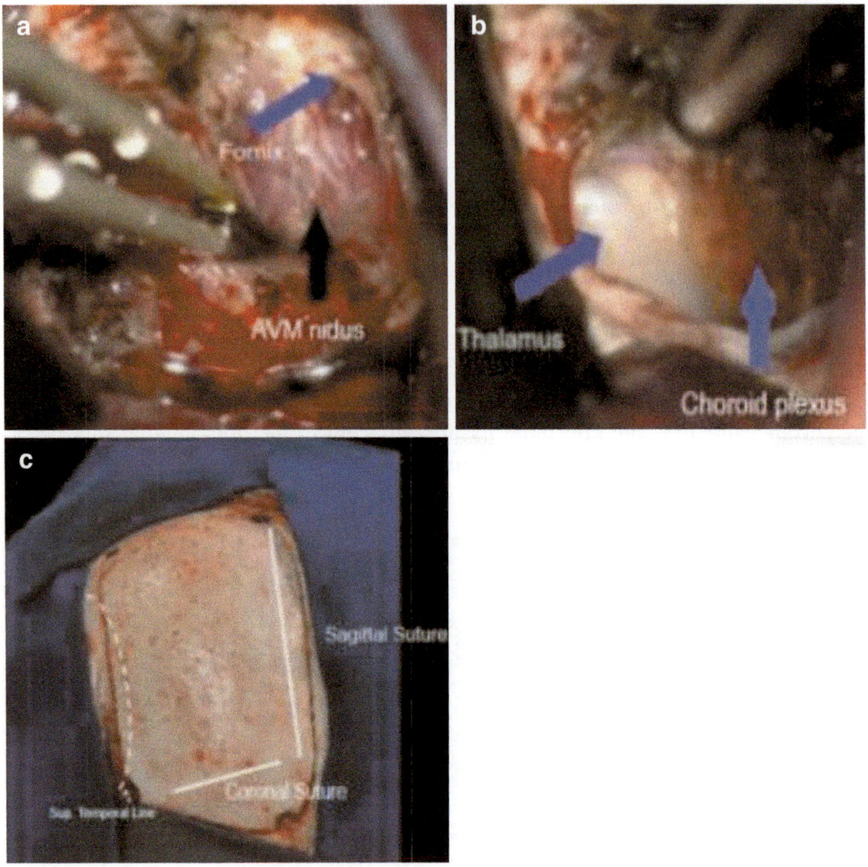

Fig. 6.6 (**a**) The relationship between the fornix and the malformation nidus. (**b**) The dissection of the space between the nidus of the AVM and the dorsomedial surface of the right thalamus is shown. The objective was to find the arterial feeders, start to coagulate them, and cut circumferentially, yielding a surgical view of the thalamus and choroid plexus. (**c**) The craniotomy used in the access to the malformation. The skin incision was made to expose the sagittal suture and the superior temporal line in the right side. Four burr holes were created to expose the border of the superior sagittal sinus after the craniotomy

(foramen of Monro) to the inferior choroidal point, which is immediately behind the uncus and is located at the transition region between the head and body of the hippocampus (Fig. 6.8).

The lesion is mainly fed by branches of the pericallosal artery and the posteromedial and posterolateral choroidal arteries. The posteromedial choroidal artery originates in the proximal portion of the posterior cerebral artery, running parallel and medially to it in the basal cisterns, curving ventrally, and irrigating the choroid plexus in the roof of the third ventricle. The posterolateral choroidal artery travels to the body of the lateral ventricle from the atrial and temporal portions of the choroidal fissure.

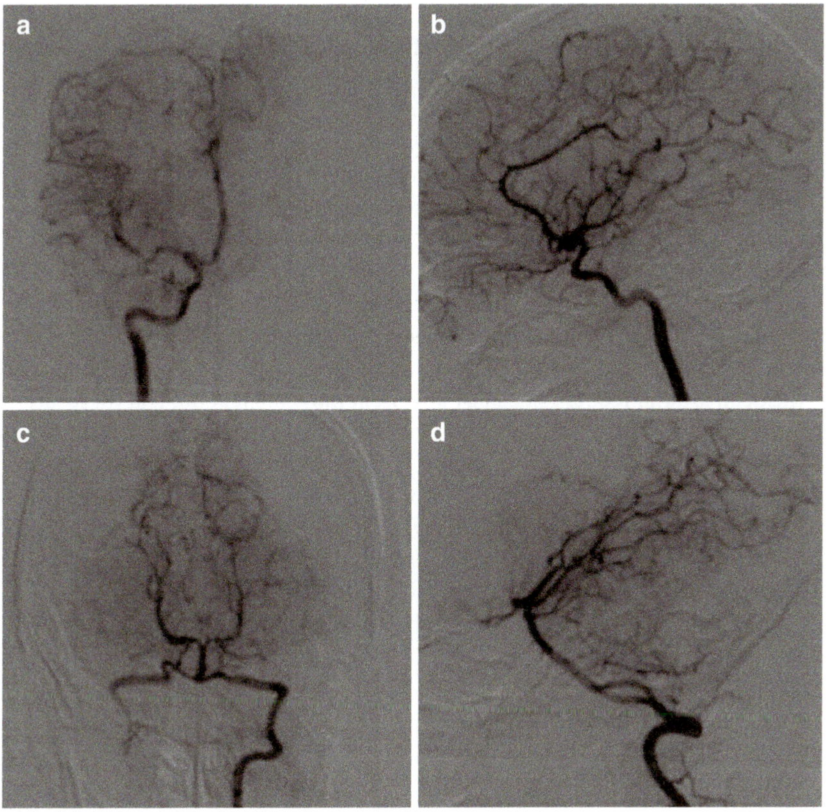

Fig. 6.7 Post-op angiography demonstrating the complete resection of the malformation. We can check the control image according to four different angles, assessing the anterior circulation in the anteroposterior (**a**) and lateral (**b**) views, and the posterior circulation in the anteroposterior (**c**) and lateral (**d**) views

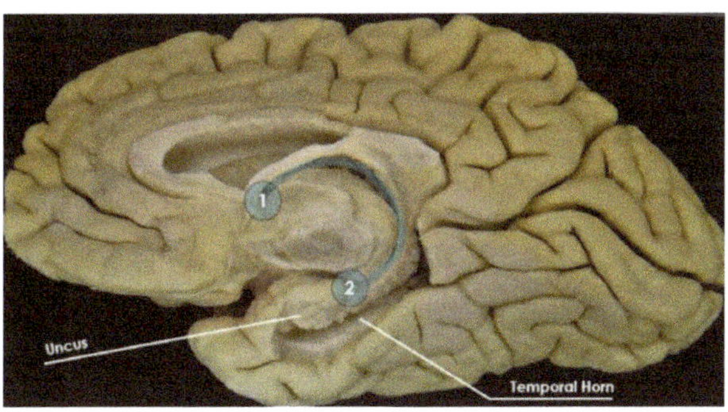

Fig. 6.8 Choroidal fissure running through the superior choroidal point (1) (foramen of Monro) to the inferior choroidal point (2), which is behind the uncus, and the region where the anterior choroidal artery enters the temporal horn and the inferior ventricular vein exits

The choroidal fissure is one of the brain's natural corridors, and some pathologies spread through it. It is subdivided into a body portion and atrial and temporal parts. Using the choroidal fissure, it is possible to communicate the lateral ventricle to the third ventricle according to an interhemispheric transcallosal transchoroidal approach. In the atrium, it connects the cavity with the quadrigeminal cistern. In the temporal horn, it connects with the ambient cistern. The choroidal fissure creates anatomical spaces which we can manage. The choroid plexus runs anterior to posterior in the body and superior to inferior in the atrium of the lateral ventricle.

The treatment strategy here required an important anatomical rationale. The AVM received two major feeders, including one coming from the pericallosal artery and another coming from the posterolateral and posteromedial choroidal arteries, which run deep. As such, for the surgical strategy, it would be interesting to eliminate the deep arterial feeders since any advertent movement could tear these arteries if not embolized. Therefore, the treatment included two stages. The first stage involved the preoperative embolization of the arterial feeders from the posterior circulation, which enabled the reduction of the nidus flow of the AVM. The second involved a microsurgical resection, using an interhemispheric approach, with the patient in a prone position, allowing for access to the anterior circulation feeders and the complete resection of the AVM, without any associated morbidity. Note that in the skin incision was made to expose the sagittal suture and the superior temporal line.

6.2 Corpus Callosum

The corpus callosum is divided into the rostrum, genu, body, splenium, forceps minor, and forceps major (Fig. 6.9). The corpus callosum forms part of the walls of the lateral ventricle. It has fibers in all directions, and its structures are tightly

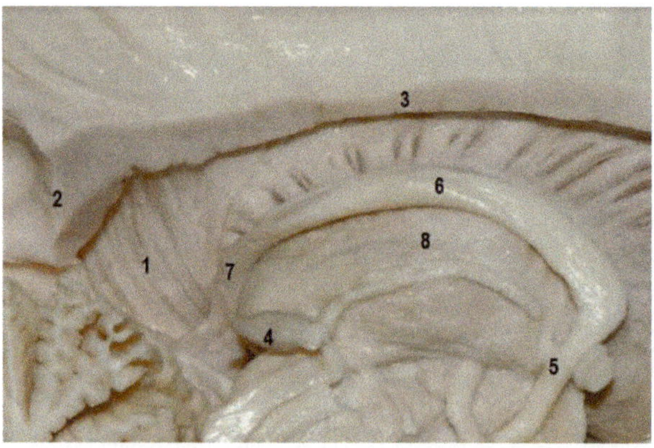

Fig. 6.9 Medial view of the corpus callosum, pineal region, thalamus, and fornix. (1) Tapetum, (2) splenium of the corpus callosum, (3) body of the corpus callosum, (4) pineal gland, (5) column of the fornix, (6) body of the fornix, (7) crura of the fornix, and (8) thalamus

connected to the lateral surface of the brain. The genu has a large bundle of fibers, including a projection called the forceps minor, which forms part of the lateral wall of the frontal horn, which connects both frontal lobes. The frontal horn projection on the lateral surface of the brain is the orbital part of the inferior frontal gyrus.

The forceps minor runs in a U-shape through the genu and connects the two frontal lobes and the posterior forceps, which connects both occipital lobes and crosses the midline via the splenium of the corpus callosum. The structure that connects the corpus callosum to the fornix is called the septum pellucidum.

The splenium of the corpus callosum makes an impression called the callosal bulb in the medial wall of the atrium of the lateral ventricle. When we reach the splenium of the corpus callosum during surgery, we are in the pineal region. If we follow the splenium laterally, we reach the atrium of the lateral ventricle. The splenium is the roof of the pineal region. The vein of Galen is situated below the splenium, which is why it assumes a curved shape.

First described by Johann Christian Reil, the tapetum is a unique part of corpus callosum. It arches from the posterior part of the splenium of corpus callosum forming part of the lateral wall of the atrium and the roof of the inferior horn.

The vast majority of the blood supply to the corpus callosum is provided by the internal carotid artery, mainly by the pericallosal artery (Fig. 6.10). The splenium, as an exception, receives its vascular supply from the vertebrobasilar system. The venous drainage occurs by the callosal and callosal cingular veins, finally draining into the internal cerebral vein.

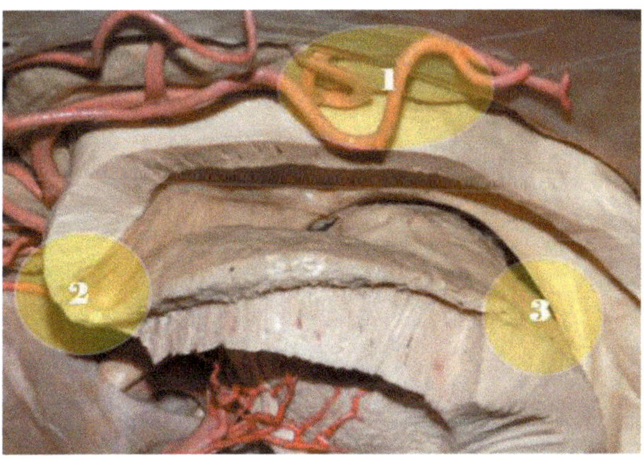

Fig. 6.10 Pericallosal artery (1) running through the callosal sulcus, one of the 100% continuous sulci of the brain. The other three are the collateral sulcus, parietooccipital sulcus, and the Sylvian fissure. The number and the trajectory of this artery can vary across different species. The rostrum of the corpus callosum (2) is a natural pathway to access the communicating anterior complex and the fornix (3), which is wrapped around the thalamus

6.3 Foramen of Monro

Mastering the foramen of Monro and its anatomical relationships is essential to ensuring success in surgical approaches.

The foramen of Monro, which is named the interventricular foramen according to anatomical terminology, connects the lateral ventricles to the third ventricle. Its anterior wall is the column of the fornix, its roof is the body of the fornix, the posterior limit is the anterior thalamic tubercle, and its lateral limit is the genu of the internal capsule. We have to be very careful when manipulating and retracting the lateral wall of this foramen so as to avoid damaging the fibers of the internal capsule; this can happen in colloid cyst surgeries, for example, causing hemiplegia or hemiparesis in such cases.

Surgical Case 2 A 46-year-old woman with a history of progressive headache for 3 months. An investigation revealed a colloid cyst (Fig. 6.11).

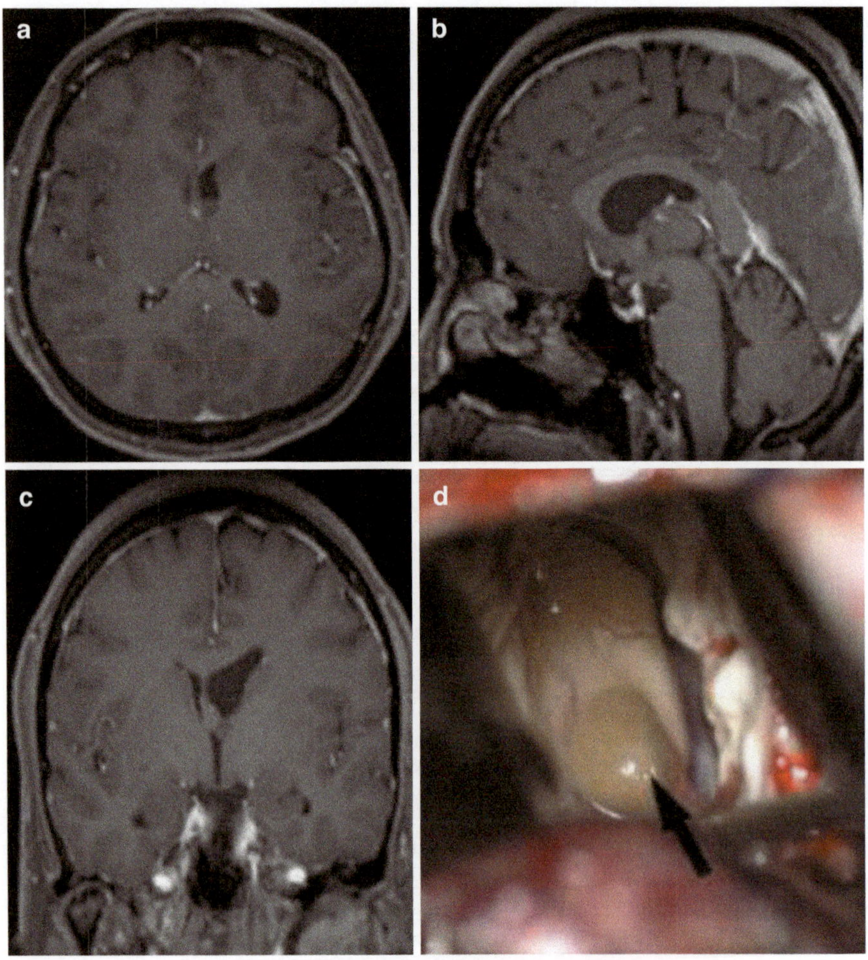

Fig. 6.11 Colloid cyst in axial (**a**), coronal (**b**), and sagittal (**c**) sections. Anatomical view of the lesion. (**d**). The black arrow shows the cyst being visualized in an anatomical view

In the past, the foramen of Monro was identified in angiograms by the venous angle, that is, the union of the anterior septal vein and the thalamostriate vein, which merge to form the internal cerebral vein in the posterior wall of the foramen of Monro. Nowadays, this venous angle remains an important landmark used for intra-operative localization during ventricular endoscopic surgeries.

The cranial landmark for the foramen of Monro is the Bregma, the union of the coronal and sagittal sutures; a perpendicular line traced inferiorly in the projection of the Bregma crosses the foramen of Monro. This foramen separates the frontal horn from the body of the ventricle and divides the internal capsule into anterior and posterior limbs. It is important to note that there is no choroid plexus in front of the foramen of Monro, which means that the choroidal fissure ends superiorly to it.

6.4 Vascularization

The arterial supply for the lateral ventricles comes from the anterior and posterior circulation. Starting from the anterior circulation, we know that the internal carotid artery bifurcation has a tight relation with the frontal horn, as the anterior cerebral arteries pass anteromedially below, sending branches to the lateral wall of the frontal horn and body. The medium cerebral arteries' origin is below the frontal horn, and they send the lenticulostriate to supply the lateral wall of the frontal horn and body of the lateral ventricle.

The anterior choroidal arteries, which are extremely important to the lateral ventricles and choroidal fissure, enter the inferior choroidal point to supply the choroid plexus with the posterior choroidal arteries, as discussed previously in this chapter with regard to surgical case 1. The anterior choroidal arteries normally send branches to the temporal horn and atrium of the lateral ventricles, also supplied in the medial edge by the lateral posterior choroidal arteries. The body of the lateral ventricles and roof of the third ventricle are supplied by the lateral posterior choroidal arteries.

The veins of the lateral ventricle are very important not only due to their intrinsic function but also because they can become a dangerous obstacle during surgery. They are divided into medial and lateral groups based on their relation to the thalamic or forniceal compartments of the choroidal fissure. The frontal horn is drained by both the lateral anterior caudate and medial anterior septal veins. The lateral system of drainage coming from the body consists not only of the thalamostriate vein but also of the thalamocaudate and posterior caudate veins.

It is important to be very careful with the thalamic tenia during surgery, as the drainage of the upper face of the thalamus takes place there. The choroid plexus is mainly drained by the superior and inferior choroidal veins.

The atrium and occipital drainage consists of the medial and lateral atrial veins, and the temporal horn has an interesting drainage system; its roof is drained by the lateral system (inferior ventricular vein) and its floor is drained by the transverse hippocampal veins.

6.5 Frontal Horn

The frontal horn is the ventricular cavity in front of the thalamus. Its anterior limit is the genu of the corpus callosum, its floor is the rostrum of the corpus callosum, its roof is the body of the corpus callosum, and its posterior limit is the foramen of Monro. Its lateral wall consists of the forceps minor of the corpus callosum and the head of the caudate nucleus, which protrudes posteriorly and laterally, as well as the other structures of the lateral ventricles. Its medial wall consists of the septum pellucidum, a structure that connects the fornix to the corpus callosum and separates the lateral ventricles. The frontal horn does not have a choroid plexus, as there is no thalamus or fornix on its wall. The projection of the frontal horn on the lateral surface of the brain is represented by the orbital and triangular parts of the inferior frontal gyrus.

Surgical Case 3 A 54-year-old man with a history of headache and seizure 3 hours before admission. The investigation revealed a giant anterior communicating artery aneurysm (Figs. 6.12 and 6.13).

What is the relationship between the aneurysm shown in the image and the lateral ventricles? What part of the ventricle is related to it?

In order to access the aneurysm, we have to perform an interhemispheric approach. The approach may minimally displace the frontal lobe to make enough space to expose the falx, the anterior cerebral arteries, and the corpus callosum.

The dissection of the interhemispheric fissure can be hampered by the presence of veins that can bleed and become obstacles. This has to be done using a microscope. The falx cerebri are not continuous in the frontal lobe, as they are in the occipital segment. During the procedure, we can find the callosomarginal artery, running above the cingulate gyrus, and the pericallosal artery, which can vary in number, in the callosal sulcus. We can distinguish the white substance from the gray matter by observing the pearly color of the corpus callosum under a light microscope. The incision of the corpus callosum has to be made in a longitudinal direction.

If the floor of the frontal horn is the rostrum of corpus callosum, what is located elow?

Just below the rostrum of the corpus callosum, we have the lamina terminalis cistern and the anterior communicating artery complex. We know we have reached the rostrum of the corpus callosum because it is in front of the column of the fornix and foramen of Monro and because of the absence of choroid plexus. It is possible to remove lesions in the subcallosal region through an interhemispheric transcallosal-transrostral approach. Anterior and inferior routes to the subcallosal region have been described, but they pose a risk of damaging the branches of the anterior cerebral artery (Fig. 6.14).

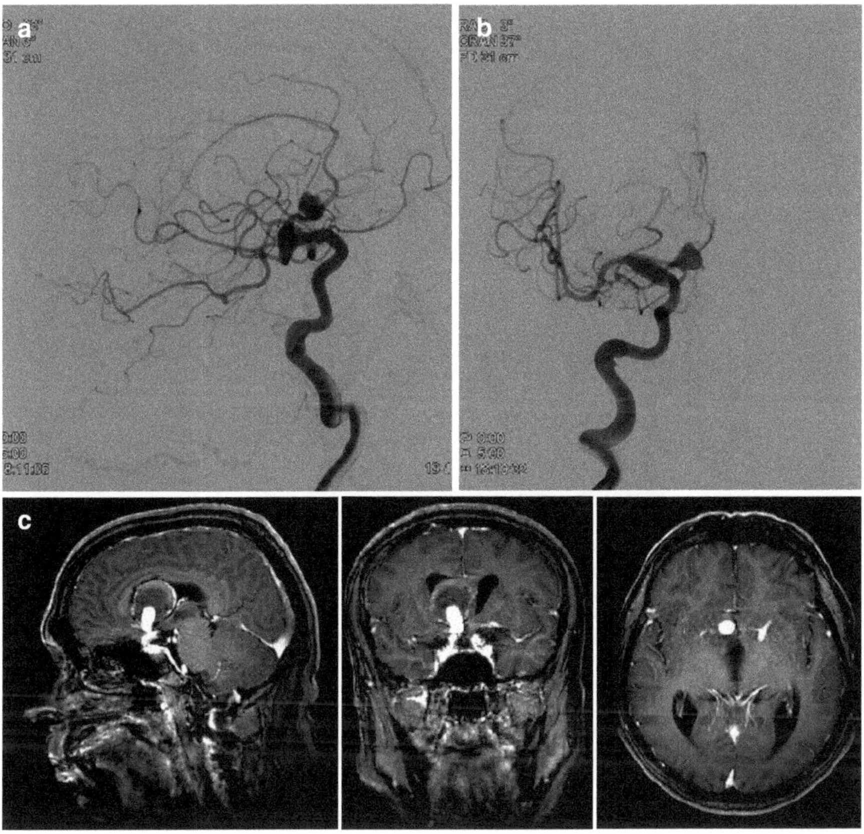

Fig. 6.12 Angiographic study showing a giant communicating artery aneurysm which is located in the frontal horn of the lateral ventricle, projecting superiorly and reaching the roof of the right lateral ventricle. Anteroposterior view (**a**) and lateral view (**b**) of the aneurysm. Figure **c** shows an MRI identifying the giant anterior communicating complex aneurysm, situated in the right frontal horn. It has a superior and posterior guidance. The aneurysm is predominantly thrombosed

6.6 Body of the Lateral Ventricle

Its floor consists of the thalamus and fornix. If we see the thalamus and fornix, we can see the choroidal fissure and choroid plexus. Its lateral limit is the body of the caudate nucleus, which is separated from the thalamus by the thalamostriate sulcus, through which the thalamostriate vein and stria terminalis course. The projection of the body of the lateral ventricle on the lateral surface of the brain is related to the most lateral part of the pre- and post-central gyrus. Its roof is the body of the corpus callosum and the medial wall is the septum pellucidum and the body of the fornix.

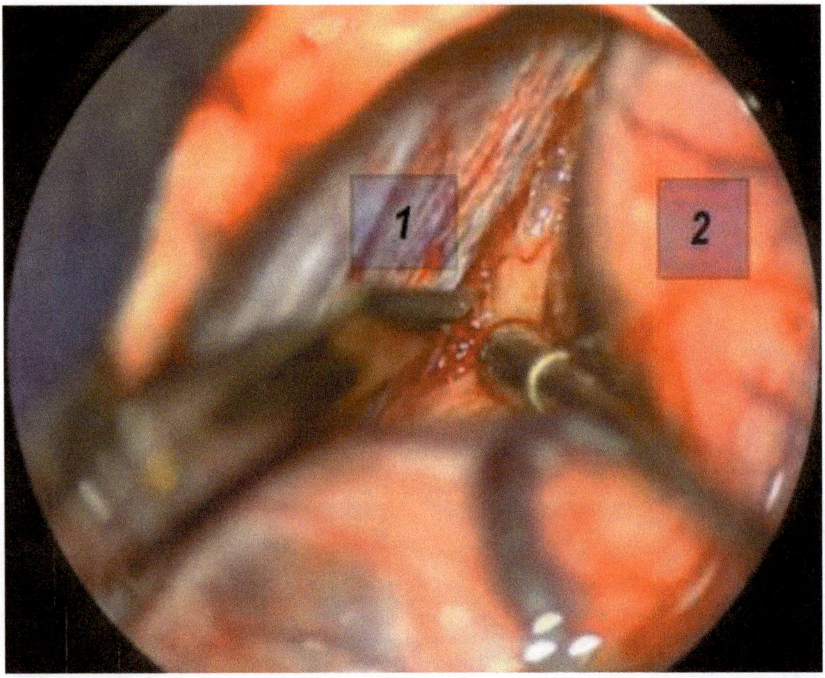

Fig. 6.13 Surgical view of the interhemispheric access used in the surgery. The interhemispheric falx (1) is being retracted from the right frontal lobe (2) to expose the callosal cistern in order to access the corpus callosum

Note that the fornix is both the floor and medial wall of the body of the lateral ventricle. The posterior limit of the body of the lateral ventricle is where the fornix meets the corpus callosum in the medial wall, where there is no longer a pellucid septum. The fornix divides the pulvinar of the thalamus into two parts, namely the extraventricular part, in the cistern of the quadrigeminal plate lamina, and the intraventricular part, in the atrium of the lateral ventricle. The body of lateral ventricle is separated from the third ventricle through the choroidal fissure (Fig. 6.15).

6.7 Atrium and Posterior Horn

The triangular cavity is located behind the thalamus; the anterior limit is the pulvinar of the thalamus (medial), crura of the fornix (lateral), and choroidal fissure. Its lateral wall and roof are formed by the tapetum of the corpus callosum and the floor by the collateral trigone, a triangular area overlying the posterior end of the collateral sulci.

Its medial wall is formed, anteriorly, by the bulb of the corpus callosum and the tail of the hippocampus. Posteriorly, its wall is formed by the callosal bulb and the

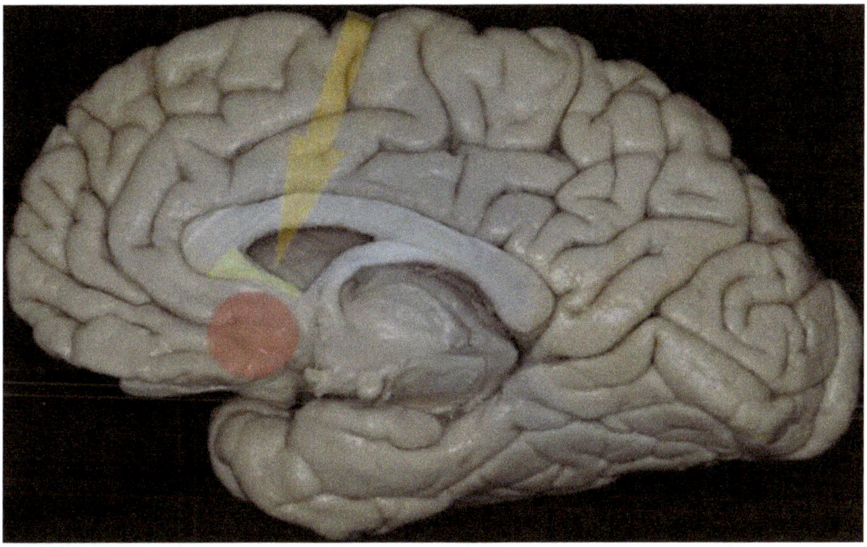

Fig. 6.14 Illustrative figure representing the transcallosal-transrostral access to approach subcallosal lesions. The yellow arrow shows the path taken in the transcallosal-transrostral approach to access the anterior communicating complex (red circle). The rostrum of the corpus callosum is underlined in green

"calcaravis," previously called the minor hippocampus, which is a prominence of the calcarine sulcus. It contains the calcarine artery. The calcarine artery runs along the calcarine sulcus, which is very deep and forms a protuberance in the lower part of the medial wall of the atrium. This sulcus separates the isthmus of the cingulate gyrus from the parahippocampal gyrus and leads the surgeon to the quadrigeminal cistern. It is a landmark in occipital transtentorial approaches.

Surgical Case A 21-year-old man admitted to the emergency room with a history of 6 hours of seizures and right hemiparesis. Tomographic evidence demonstrated an intraparenchymal nucleocapsular hematoma (Fig. 6.16).

The MRI was very clear, showing the cause of the bleeding, which consisted of a tumor in the thalamus pulvinar, the anterior wall of the atrium of the lateral ventricle.

In order to access the lesion and obtain a complete resection, we have to program a left parietal craniotomy and expose the intraparietal and supramarginal sulci, which are the cortex correspondences of the atrium. The lesion is located in the thalamus pulvinar, which is the anterior wall of the atrium of the lateral ventricle.

The gyrus that points to the atrium of the lateral ventricle is the Heschl's gyrus (long anterior transverse temporal gyrus). If the Heschl's gyrus points to the atrium, it means that the insula ends at the pulvinar of the thalamus. Following Heschl's gyrus enables one to reach the atrium of the lateral ventricle (Fig. 6.17). Its medial wall is the isthmus of the cingulate gyrus, one of the surgical corridors or safe zones

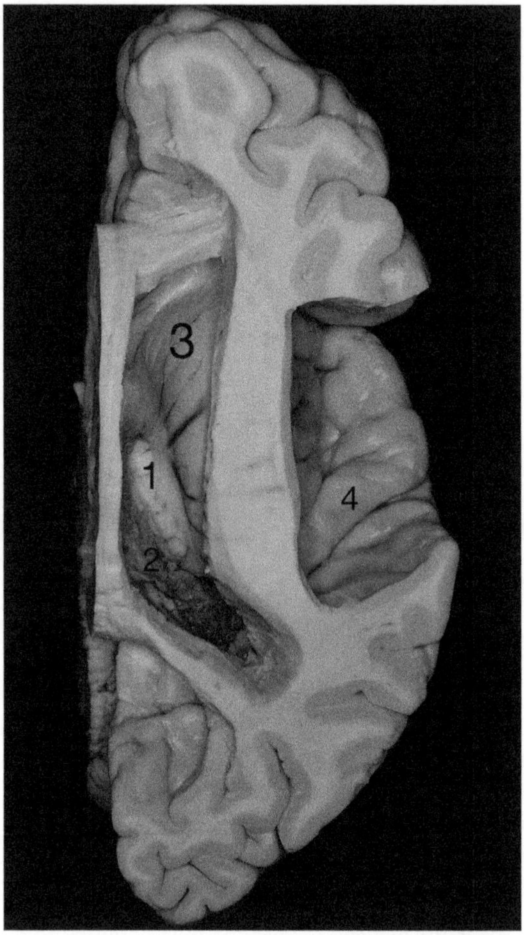

Fig. 6.15 Superior view of the lateral ventricle. The body of the lateral ventricle has a floor which consists of the thalamus (1) and fornix (2). Its lateral wall is the body of the caudate nucleus (3) and its medial limit is the septum pellucidum. Note in the figure that the Hechl's gyrus(4) points to the atrium of the lateral ventricle (5)

of the brain. One of its access reference points is the junction of the upper and posterior limiting sulcus of the insula. We can aspirate the isthmus of the cingulate gyrus to access the medial half of the atrium of the lateral ventricle.

The *occipital* horn runs posteriorly, behind the atrium, and varies in size. We did not find any choroid plexus on its extension. The tapetum forms the roof and lateral wall, separating it from the optic radiations, and the collateral trigone forms the floor. As the atrium, the bulb of the corpus callosum and calcaravis form the medial wall.

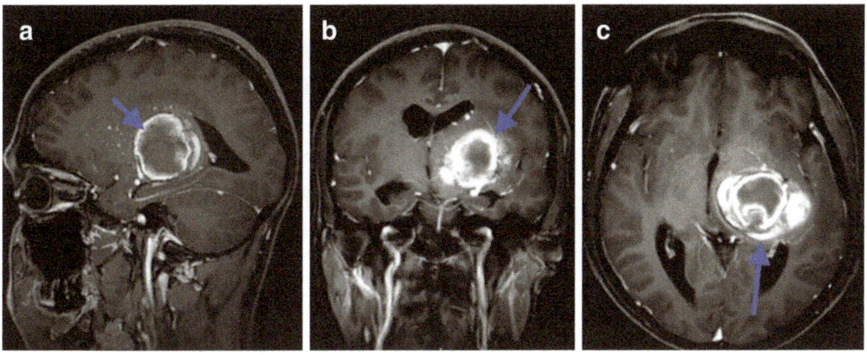

Fig. 6.16 MRI showing an expansive intra-axial formation centered on the left centrencephalic region, heterogeneous in content, presenting hematic intervening foci characterized as a thalamic cavernoma. We can see three different positions of this lesion: lateral (**a**), anteroposterior (**b**), and axial (**c**). The lesion is indicated by the blue arrow

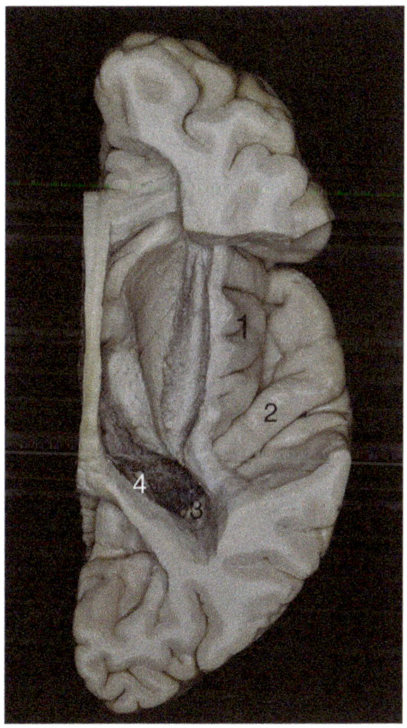

Fig. 6.17 Heschl's gyrus pointing to the atrium of the left lateral ventricle. The thalamus pulvinar is related to the most posterior part of the insula. 1-Insular cortex; 2-Heschl's gyrus; 3-atrium; 4-pulvinar of thalamus

6.8 Temporal Horn

One of the most interesting parts of the lateral ventricle, the temporal horn extends from the atrium to the amygdala. Its anatomical correspondence on the lateral surface is the middle temporal gyrus, and one of the access points is at the opening of the insula's inferior limiting sulci. Its floor is divided into two parts: a medial part, formed by the hippocampal prominence, and a lateral part, formed by the collateral eminence, an impression made by the collateral sulcus, which separates the medial and lateral occipitotemporal sulci (Fig. 6.18).

The roof of the temporal horn is formed by the caudate nucleus, thalamus, and the tapetum of the corpus callosum, which also constitutes the lateral wall, separating it from optic radiations. Its medial wall consists of the fimbria of the fornix. The striothalamic sulci separate the caudate tail and the thalamus. Finally, the medial wall is a narrow structure between the fimbria of the fornix and the inferolateral part of the thalamus.

Accessing the temporal horn is important not only in epilepsy surgeries but also in tumor and vascular surgeries. We can approach the temporal horn by inferior, lateral transcortical, transsulcal, and transsylvian approaches. It is important to remember that disturbances in memory, vision, and language can occur after manipulation. The lateral approach is used by entering the middle temporal gyrus, at a distance of approximately 2.5 cm from the temporal pole. The temporal horn communicates with the ambient cistern through the choroidal fissure.

The lateral ventricles pose particular difficulties and allow strategies that make them objects of interest of many researchers through history, ever since Aristotle and Andreas Vesalius. New imaging modalities have already enabled the development of new perspectives regarding this structure, and there is yet much more to discover.

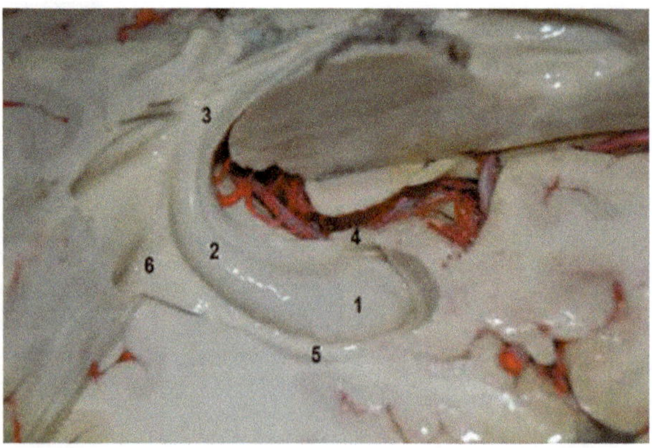

Fig. 6.18 The temporal horn in a lateral view of the right hemisphere. (1) Head of hippocampus, (2) fimbria of the fornix, (3) crura of the fornix, (4) inferior choroidal point, (5) collateral eminence, and (6) collateral trigone

Suggested Bibliography

Chaddad-Neto F, Devanir Silva da Costa M, Bozkurt B, Leonardo Doria-Netto H, de Araujo Paz D, da Silva Centeno R, et al. Contralateral anterior interhemispheric-transcallosal-transrostral approach to the subcallosal region: a novel surgical technique. J Neurosurg. 2018;129(2):508–14. https://doi.org/10.3171/2017.4.jns16951.

da Costa MDS, Lopes RRS, Serrato-Avila JL, Cavalheiro S, Chaddad-Neto F. Endoscopic brain-wash after clipping a ruptured aneurysm of the communicating segment of the intracranial carotid artery. Surg Neurol Int. 2020;11:396.

Da Costa MDS, Santos BFO, Bouchabki de Almeida Guardini F, Chaddad-Neto F. Microsurgical treatment for arteriovenous malformation of the corpus callosum and choroidal fissure. Neurosurg Focus. 2017;43(VideoSuppl1):V12. https://doi.org/10.3171/2017.7.FocusVid.1733.

Le Gars D, Lejeune JP, Peltier J. Surgical anatomy and surgical approaches to the lateral ventricles. In: Advances and technical standards in neurosurgery, Advances and technical standards in neurosurgery, vol. 34. Vienna: Springer; 2009.

Rhoton AL Jr. Rhoton's cranial anatomy and surgical approaches. Oxford University Press; 2019.

Sarikcioglu L, Ozsoy U, Unver G. Tapetum Corporis Callosi: carpet of the brain. J Hist Neurosci. 2007;16(4):434–6. https://doi.org/10.1080/09647040600719013.

Scelsi CL, Rahim TA, Morris JA, Kramer GJ, Gilbert BC, Forseen SE. The lateral ventricles: a detailed review of anatomy, development, and anatomic variations. AJNR Am J Neuroradiol. 2020;41(4):566–72. https://doi.org/10.3174/ajnr.A6456.

Chapter 7
Surgical Anatomy of the Third Ventricle

All four cavities that compose the ventricular system are located in the midline (Fig. 7.1). The third ventricle is a unique brain cavity that is very difficult to access surgically. Also called the diencephalic cavity, it is located between the thalami and below the fornix. It has six walls: the roof, the floor, two lateral walls, and the anterior and posterior walls. Cerebrospinal fluid drains from the lateral ventricles into the third ventricle through the foramen of Monro, then draining into the fourth ventricle through the Sylvian (mesencephalic) aqueduct. It is surrounded by critical structures such as the thalamus, hypothalamus, and pituitary and pineal glands.

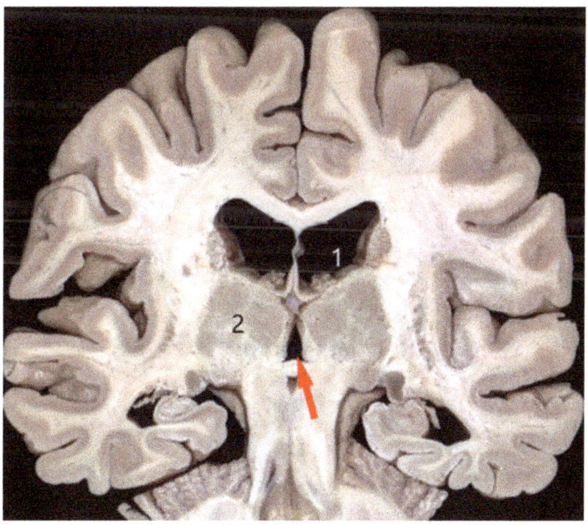

Fig. 7.1 Coronal anatomical section showing the third ventricle (red arrow). 1, Lateral ventricles; 2, thalamus

Supplementary Information The online version of this chapter (https://doi.org/10.1007/978-3-030-82747-2_7) contains supplementary material, which is available to authorized users.

F. Chaddad-Neto, M. D. Silva da Costa, *Microneuroanatomy and Surgery*,
https://doi.org/10.1007/978-3-030-82747-2_7

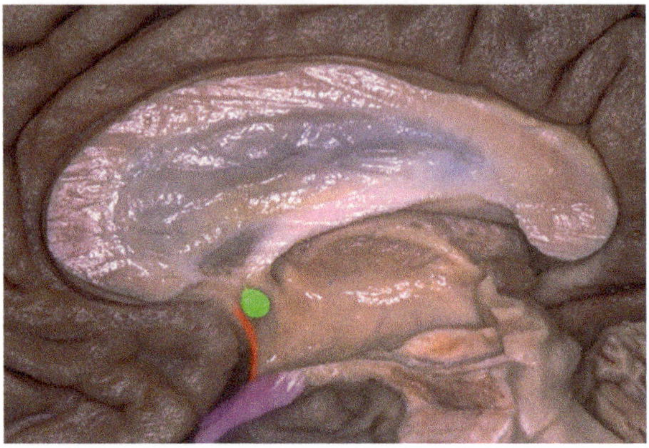

Fig. 7.2 Sagittal view of the third ventricle. The anterior commissure is marked in green. Lamina terminalis, red; optic chiasm, purple

7.1 Anterior Wall

The anterior wall of the third ventricle extends from the foramen of Monro to the optic chiasm. It is composed mainly of the lamina terminalis, a thin membrane bound superiorly by the anterior commissure and inferiorly by the superior aspect of the optic chiasm, which forms not only part of the anterior wall but also part of the floor of the third ventricle (Fig. 7.2). The anterior commissure is the meeting point between the anterior wall and the roof of the third ventricle. The anterior wall structures are best identified in the sagittal plane.

Many different craniotomies can be used to access the anterior wall of the third ventricle. Subfrontal access is the main route to reach the cistern of the lamina terminalis, which contains the A1 segments of the anterior cerebral arteries, the anterior communicating complex, the recurrent arteries of Heubner, the fronto-orbital arteries, subcallosal arteries, hypothalamic arteries, and the anterior communicating veins.

The lamina terminalis separates the third ventricle from the lamina terminalis cistern. Therefore, opening the lamina terminalis connects the third ventricle to the basal cisterns. Also called the anterior third ventriculostomy, this procedure drains cerebrospinal fluid to obtain the best conditions for neurosurgery. Aneurysms of the anterior communicating artery facing inferiorly are situated close to the optic chiasm and can cause temporal visual loss (Fig. 7.3).

7.2 Floor

The floor of the third ventricle extends from the optic chiasm to the Sylvian aqueduct and is medial to the uncus and anterior perforated substance. It is composed of subthalamic and hypothalamic structures. In order of anterior to posterior, these include the optic chiasm, infundibulum of the hypothalamus with the pituitary stalk, tuber cinereum (region used in the inferior third ventriculostomy), mammillary bodies, posterior perforated substance (where the perforating arteries penetrate), and the subthalamus. The third ventricle has two expansions called recesses. The infundibular recess, also called the pars cava infundibuli, lies in the anterior portion of the third ventricle floor, which continues as the pituitary stalk. The chiasmatic recess, associated with the optic chiasm, lies between the junction of the anterior wall and the floor of the third ventricle (*case example is also demonstrated in* Video 7.1).

The third ventriculostomy is an interesting procedure used in the treatment of obstructive hydrocephalus. Its classical indications are aqueductal stenosis, posterior third ventricular, and quadrigeminal plate tumors, and different etiologies that cause fourth ventricular outlet obstruction (Fig. 7.4). The main goal is to create a

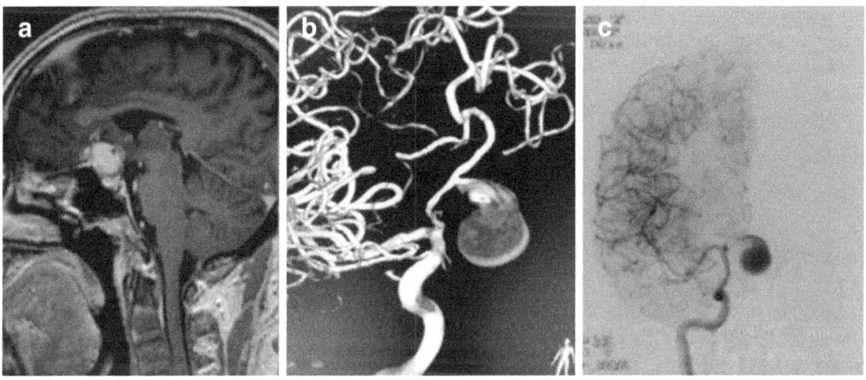

Fig. 7.3 Giant inferiorly directed aneurysm of the anterior communicating complex compressing the optic chiasm. The patient presented with progressive visual loss as the primary symptom. (**a**) Sagittal magnetic resonance imaging view; (**b** and **c**) angiographic study of the aneurysm

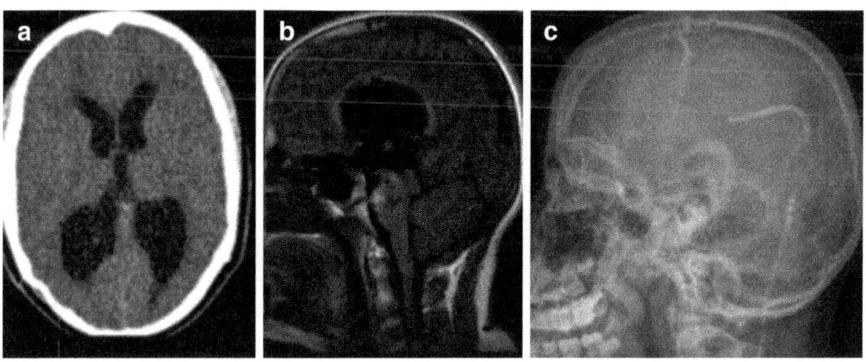

Fig. 7.4 Surgical case: A 12-year-old boy presented with refractory headache and intense episodes of nausea and vomiting. He had previously undergone ventriculoperitoneal shunting due to hydrocephalus. The initial investigation showed recurrence of hydrocephalus and shunt malfunction, visualized as enlarged ventricles on computerized tomography (**a**) and magnetic resonance imaging (**b**). Sagittal radiography of the head (**c**) shows the "beaten-silver" sign, correlated with chronic intracranial hypertension. A third ventriculostomy was performed to treat the hydrocephalus

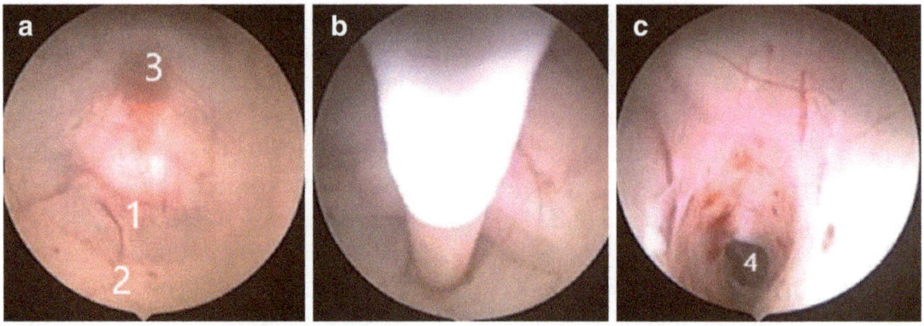

Fig. 7.5 Endoscopic view of the third ventricle floor in a third ventriculostomy. (**a**) After entering the third ventricle through the foramen of Monro, we can identify the tuber cinereum (1), where the ventriculostomy is formed, the mammillary bodies (2), and the infundibular recess (3). (**b**) Perforating the floor with the Fogarty catheter. (**c**) Final image after the procedure. It is important that all the membranes have been opened (4) without injuring the adjacent structures—mainly the basilar artery

natural shunt between the ventricles and the arachnoid by perforating the tuber cinereum, which forms part of the third ventricle floor (Fig. 7.5). It is also important for the neurosurgeon to identify and avoid injuring the basilar and superior cerebellar arteries and the pituitary gland and stalk.

7.3 Roof

The roof of the third ventricle is an arch-like structure composed of five layers, extending from the foramen of Monro anteriorly to the suprapineal recess posteriorly (Fig. 7.6). The first layer is the body of the fornix. If we visualize the fornix, we can also visualize the choroidal fissure. We can only visualize the choroid plexus in the third ventricle roof, as it is the only place where a choroidal fissure is present. The second layer is the superior membrane of the tela choroidea. The third layer is composed of vessels, not only the internal cerebral vein, but also the posteromedial choroidal artery. The fourth layer is the inferior membrane of the tela choroidea, and the fifth layer is the choroid plexus (*a case example is also demonstrated in* Video 7.2).

The velum interpositum is a triangular space between the two layers of the tela choroidea (Fig. 7.7) and hosts the vascular components of the third ventricle roof. It is an anterior extension of the quadrigeminal plate cistern, and a dilation of its potential space is known as the cavum of the velum interpositum. The second layer of the tela choroidea houses the stria medullaris, which extends from the foramen of Monro to the region of the habenular commissure.

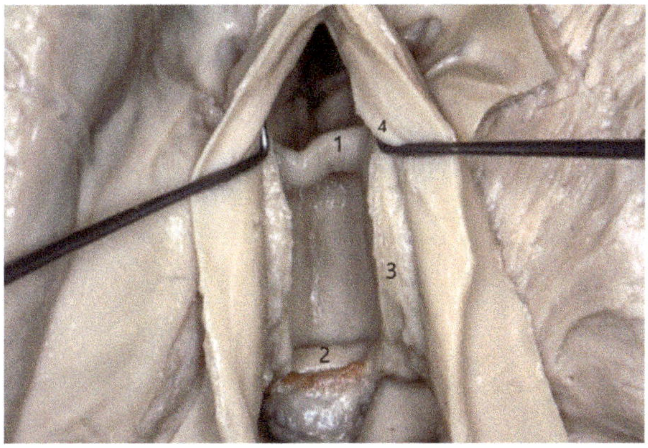

Fig. 7.6 Superior view of the third ventricle. 1, Anterior commissure; 2, posterior commissure; 3, thalamus; 4, fornix

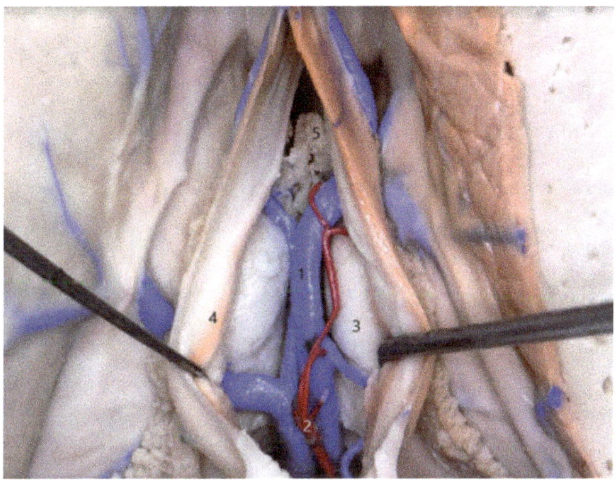

Fig. 7.7 Roof of the third ventricle and its layers. 1, Internal cerebral vein, a component of the third or vascular layer; 2, posteromedial choroidal artery; 3, thalamus; 4, fornix; 5, choroid plexus

7.4 Posterior Wall

The posterior wall extends from the suprapineal recess to the Sylvian aqueduct. It is composed mainly of the epithalamus and has at its center the pineal gland, located below the splenium of the corpus callosum (the roof of the pineal region).

The posterior wall of the lateral ventricle is compounded by two commissures (habenular and epithalamic) and two recesses (suprapineal and pineal) (Fig. 7.8). The pineal recess is the deepest structure of the posterior third ventricle wall. The suprapineal recess is a thin, posterior extension of the third ventricle and adheres to the stria medullaris and the lower layer of the tela choroidea.

The center of the pineal region is the pineal gland (Fig. 7.9), which is surrounded by many critical structures such as the complex cerebral vein of Galen. Accessing this region is a challenge for neurosurgeons. Above the pineal gland, we visualize the habenular commissure, and the epithalamus is confirmed below, adjacent to the tectum of the midbrain and the quadrigeminal cistern. The calcarine sulcus serves

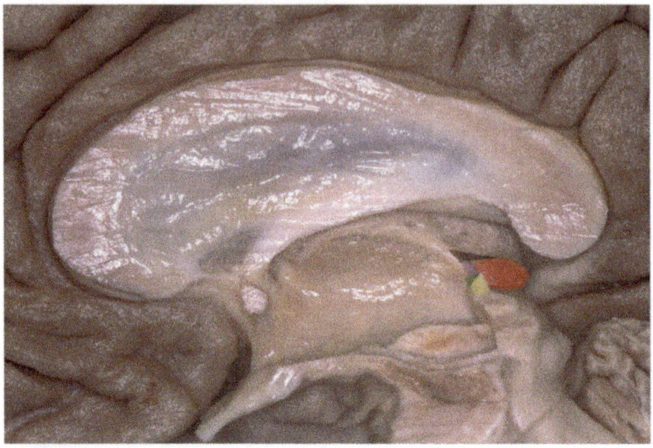

Fig. 7.8 Sagittal view of the third ventricle. Red, pineal gland; yellow, posterior commissure; green, pineal recess; purple, habenular commissure; black, suprapineal recess

Fig. 7.9 Posterolateral view of the pineal region and the posterior cranial fossa landmarks. 1, Suprapineal recess; 2, pineal recess; 3, pineal gland; 4, superior colliculus; 5, inferior colliculus; 6, superior medullary velum; 7, superior cerebellar peduncle; 8, medial cerebellar peduncle; 9, fourth nerve; 10, splenium of corpus callosum (the roof of the pineal region). The fourth nerve is the main landmark of the posterior cranial fossa

as a guide to the quadrigeminal cistern. The pineal gland is mainly supplied by the posteromedial choroidal arteries and branches of the posterior cerebral artery.

Lesions in the posterior third ventricle can be approached in several ways, depending on the location, but the main approaches are the infratentorial supracerebellar and the occipito-transtentorial approaches.

7.5 Lateral Wall

The lateral wall of the third ventricle is composed of the thalamus, and in its inferior portion, by the subthalamus and hypothalamus (Fig. 7.10). The lateral wall is separated by the hypothalamic sulcus, which extends from the foramen of Monro to the

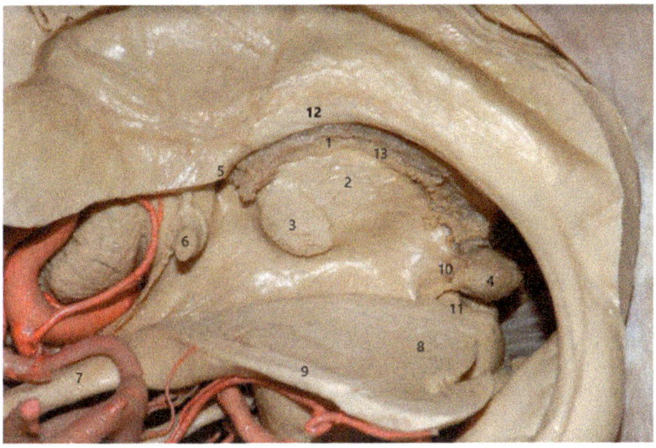

Fig. 7.10 Structures of the anterior wall, lateral wall, roof, and posterior wall of the third ventricle. 1, Vascular layer of the roof containing the internal cerebral vein and the posteromedial choroidal arteries; 2, thalamus, a component of the lateral wall; 3, interthalamic adhesion; 4, pineal gland, a component of the posterior wall; 5, foramen of Monro; 6, anterior commissure; 7, optic chiasm; 8, posterior perforated substance; 9, optic tract; 10, suprapineal recess; 11, pineal recess; 12, body of the fornix; 13, choroid plexus

aqueduct of Sylvius posteriorly. Below the hypothalamic sulcus, the lateral wall of the third ventricle is composed of the hypothalamus anteriorly and the subthalamus posteriorly. The massa intermedia is a band of gray matter in the upper medial surface of the thalamus that crosses the third ventricle. It is not considered eloquent, as no fibers are contained in this structure (*a case example is also demonstrated in* Video 7.3).

7.6 Vascularization

The third ventricular drainage system is associated with the deep venous system of the brain. It can be divided into veins related to the roof (internal cerebral vein) and the floor (basal vein of Rosenthal) (Fig. 7.11). The internal cerebral vein is situated in the third layer of the roof, in the velum interpositum-subarachnoid space between the two layers of the tela choroidea. It is formed by the junction of the anterior septal and thalamostriate veins, running posteriorly to the vein of Galen. The thalamostriate vein drains the posterior frontal and anterior parietal lobes, caudate nucleus, and internal capsule. It is a subependymal structure becoming ependymal in the foramen of Monro, acquiring a double-curved shape.

The vein of Galen has a circular configuration, the shape of the splenium of the corpus callosum. It receives blood from the internal cerebral vein, basal vein, and inferior sagittal sinus, draining into the straight sinus. The basal vein is related to the floor of the third ventricle, originating in the surface of the anterior perforated substance and passing through the ambient and crural cisterns. It joins the vein of Galen with the internal cerebral vein in the quadrigeminal cistern (Fig. 7.11).

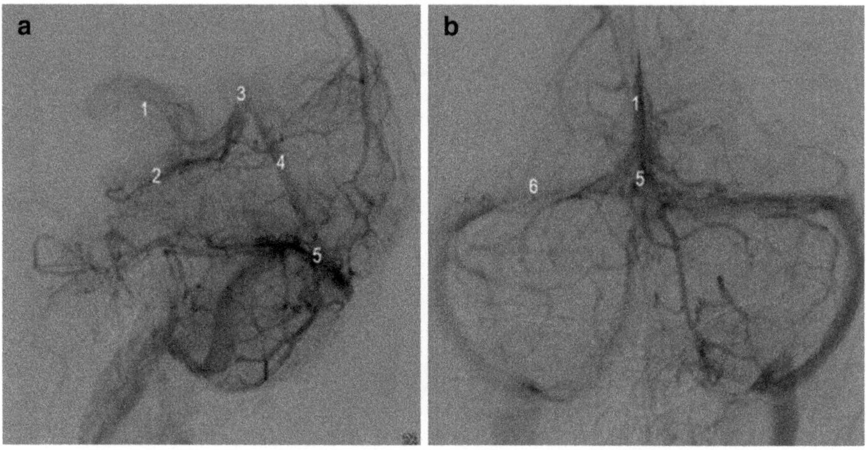

Fig. 7.11 Lateral (A) and anteroposterior (B) views of a venous angiography. We can identify the veins that drain the third ventricle: 1, internal cerebral vein; 2, basal vein of Rosenthal; 3, vein of Galen; 4, straight sinus; 5, torcular (confluence of sinuses); 6, transverse sinus

The thalamic veins are divided into superior, anterior, and inferior. The superior veins drain into the vein of Galen, the anterior into the thalamostriate, and the inferior into the inferior portion of the thalamus.

The medial posterior choroidal arteries are branches of the peduncular segment of the posterior cerebral arteries, running parallel and turning forward in the quadrigeminal plate cistern. There they enter the velum interpositum, comprising part of the third layer of the third ventricle roof with the internal cerebral vein. They supply not only the pineal gland but also the roof of the third ventricle and superior and inferior colliculi.

The lateral posterior choroidal arteries arise from the posterior cerebral artery in the ambient and quadrigeminal cistern, entering behind the anterior choroidal arteries to course outside the lateral ventricles.

Suggest Bibliography

Cimşit NC, Türe U, Ekinci G, Necmettin Pamir M, Erzen C. Venous variations in the region of the third ventricle: the role of MR venography. Neuroradiology. 2003;45(12):900–4. https://doi.org/10.1007/s00234-003-1103-0.

Kiliç T, Akakin A. Anatomy of cerebral veins and sinuses. Front Neurol Neurosci. 2008;23:4–15. https://doi.org/10.1159/000111256.

Rhoton A. Rhoton's cranial anatomy and surgical approaches. Oxford University Press Inc. USA; 2019.

Serrato-Avila JL, Da Costa MDS, Frudit ME, Carrasco-Hernandez JP, Alejandro SA, Chaddad-Neto F. Interhemispheric transcallosal transforaminal approach for decompression of a giant superior cerebellar artery thrombosed aneurysm: Three-dimensional operative video. Surg Neurol Int. 2020;11:84.

Simon E, Afif A, M'Baye M, Mertens P. Anatomy of the pineal region applied to its surgical approach. Neurochirurgie. 2015;61(2–3):70–6. https://doi.org/10.1016/j.neuchi.2013.11.008.

Tsutsumi S, Ishii H, Ono H, Yasumoto Y. The third ventricle roof: an anatomical study using constructive interference in steady-state magnetic resonance imaging. Surg Radiol Anat. 2018;40(2):123–8. https://doi.org/10.1007/s00276-017-1905-0.

Tubbs RS, Nguyen HS, Loukas M, Cohen-Gadol AA. Anatomic study of the lamina terminalis: neurosurgical relevance in approaching lesions within and around the third ventricle. Childs Nerv Syst. 2012;28(8):1149–56. https://doi.org/10.1007/s00381-012-1831-8.

Yamamoto I, Albert L, Rhoton AL Jr, Peace DA. Microsurgery of the third ventricle: part 1: microsurgical anatomy. Neurosurgery. 1981;8(3):334–56. https://doi.org/10.1227/00006123-198103000-00006.

Chapter 8
Surgical Anatomy of the Cerebellum and the Fourth Ventricle

The posterior cranial fossa communicates with the supratentorial space and the spinal canal through the tentorial incisura and the foramen magnum, respectively. It encases the cerebellum and the brainstem, constituting important pathways to vital signs and consciousness (Fig. 8.1).

The vertebral and basilar arteries as well as the cranial nerves are intricately associated with the cerebellum and the brainstem, which influence surgical approach strategies to this region. The three main vessels of the posterior fossa, namely, superior cerebellar artery (SCA), anterior inferior cerebellar artery (AICA), and posterior inferior cerebellar artery (PICA), create three neurovascular complexes with the nerves and fissures (Fig. 8.2). The SCA enters the cerebellomesencephalic fissure and becomes associated with cranial nerves III, IV, and V. The AICA is related to the cerebellopontine fissure and to cranial nerves VI, VII, and VIII. Finally, the PICA runs along the cerebellomedullary fissure and becomes associated with the lower cranial nerves.

Surgical Case 1 A 32-year-old man presented to the emergency department with complaints of sudden headache after physical activity. Physical examination revealed that he had dysmetria and dysdiadochokinesia on the right side. He underwent brain magnetic resonance imaging (MRI), as indicated below (Fig. 8.3) (*another case example is also demonstrated in the* video 8.1).

The arteriovenous malformation (AVM)) is related to the lower surface of the tentorium and with one of the three surfaces of the cerebellum: the tentorial surface. The other two surfaces are the petrosal and suboccipital surfaces.

The tentorial surface of the cerebellum is directed down from the apex and encircles the posterior part of the midbrain, which also limits posteriorly the

Supplementary Information The online version of this chapter (https://doi.org/10.1007/978-3-030-82747-2_8) contains supplementary material, which is available to authorized users.

F. Chaddad-Neto, M. D. Silva da Costa, *Microneuroanatomy and Surgery*,
https://doi.org/10.1007/978-3-030-82747-2_8

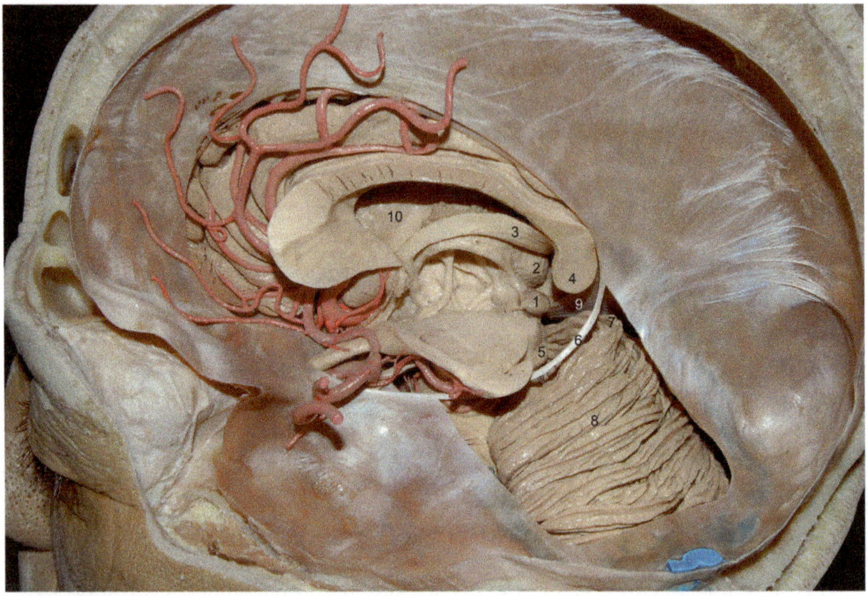

Fig. 8.1 Image of the entire intracranial space. The tentorial fold (6) separates the supratentorial and the infratentorial structures. (1) Pineal gland. (2) Pulvinar nuclei of the thalamus. (3) Fornix. (4) Splenium of the corpus callosum. (5) Quadrigeminal lamina. (7) Vermis of the cerebellum. (8) Hemisphere of the cerebellum. (9) Cerebellomesencephalic fissure. (10) Lateral ventricle

cerebellomesencephalic fissure. The vermian portions of this surface are culmen, declive, and folium. Each vermian portion corresponds to a lobule that compounds the hemispheric part of this surface: quadrangular, simple, and superior semilunar. The most prominent division is the primary fissure, also known as tentorial fissure, which separates the culmen and quadrangular lobules from the declive and simple lobules.

The supracerebellar infratentorial approach deals with the pineal region and the posterior surface of the midbrain. It reaches the corridor between the tentorial surface of the cerebellum and the tentorial fold. After craniotomy, it is easy to identify a vermian segment that, in some cases, can make access difficult to the mentioned operated structures: this is the culmen (Fig. 8.4).

Which is the most important artery for this surface? Digital angiography was performed in this case (Fig. 8.5).

The SCA (Fig. 8.6) is the dominant artery of the cerebellomesencephalic fissure and the tentorial surface. It is divided into four segments according to its relationship with the brainstem and the cerebellum: anterior pontomesencephalic, lateral pontomesencephalic, cerebellomesencephalic, and cortical.

The petrosal surface is seen in the petrous part of the temporal bone. It is divided by the cerebellopontine fissure, which exhibits a V shape and is the point of origin of some cranial nerves. The superior limb communicates with the cerebellomesencephalic fissure and the inferior limb, to the cerebellomedullary fissure. It is also possible to identify the flocculus and the choroid plexus above the inferior limb of

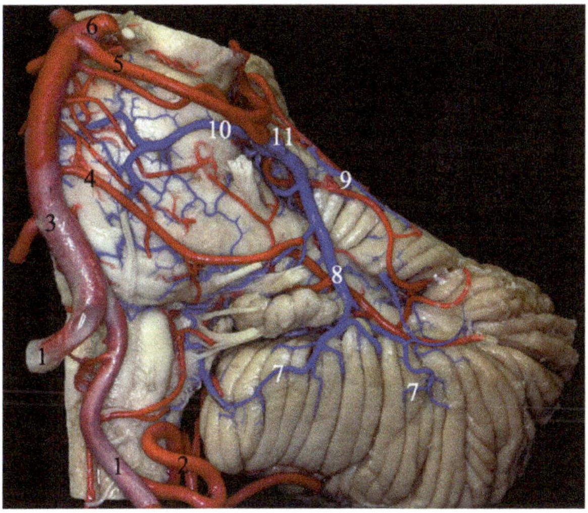

Fig. 8.2 Note that the main vessels of the posterior fossa create three neurovascular complexes with the nerves and the fissures. The vertebral arteries (1) enter the posterior fossa through the atlantooccipital membrane. In the majority of the cases, these arteries give rise to the PICA (2). They meet at the bulbopontine sulcus, forming the basilar artery (3). The basilar artery gives rise to the perforating and paramedian arteries while passing above the sulcus to the basilar artery. The AICA (4) also originates from the basilar artery. The basilar artery bifurcates and gives rise to the SCA (5) and again bifurcates into the posterior cerebral artery (6). Notably, the SCA has a double origin in this specimen. Note that each one of the main artery is associated with three (or, in the case of the PICA, four) cranial nerves. The venous drainage is also seen in this specimen. The anterior hemispheric veins (7) anastomose to form the vein of the cerebellopontine fissure (8), which ascends above the middle cerebellar peduncle. The vein of the cerebellopontine fissure receives another anterior hemispheric vein (9) and the pontine transverse vein (10), giving rise to the superior petrosal vein (11)

the cerebellopontine fissure after it exits through the foramen of Luschka (Fig. 8.7). The fifth nerve is superior to the flocculus, but the ninth nerve is inferior to the flocculus. Above the foramen of Luschka and at the lateral portion of the pontomedullary sulcus, the seventh and eighth cranial nerves are seen.

The vermian portions that are superior to the fourth ventricle are the lingual lobule, central lobule, and culmen. The first one does not have a hemispheric correspondent, and the other is related to the wing of the central lobule and the quadrangular lobule. Caudal to the fourth ventricle, the nodule and the uvula are related to the flocculi and tonsils, respectively. The simple, biventral, and semilunar lobules also form the hemispheric portion of the petrosal surface.

The AICA (Fig. 8.7) is associated with the petrosal surface and the cerebellopontine fissure. It is divided into four segments: anterior pontine, lateral pontine, flocculonodular, and cortical.

The suboccipital surface is exposed at the suboccipital craniotomies (Fig. 8.8). The vermis is hidden in the depth of the posterior cerebellar notch, where the dura mater folds to form the falx cerebelli. On the other hand, each vermian portion

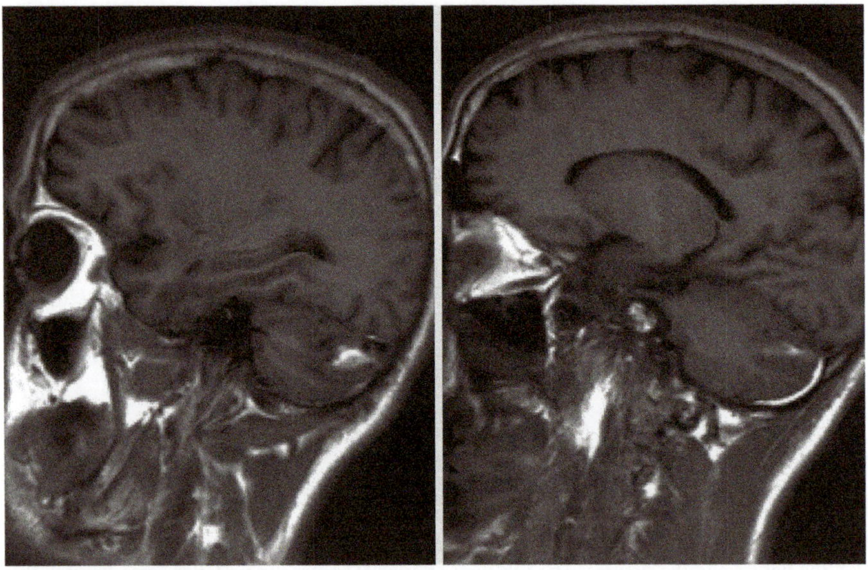

Fig. 8.3 Brain magnetic resonance imaging of arteriovenous malformation at the right cerebellar hemisphere associated with intraparenchymal hemorrhage

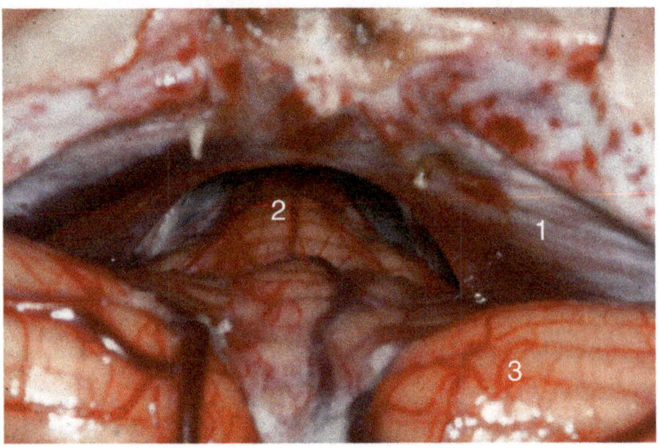

Fig. 8.4 Surgical view during a supracerebellar infratentorial approach, wherein the corridor between the tentorial fold (1) and the tentorial surface of the cerebellum is exposed. The most prominent part of the vermis, in this view, is the culmen (2). This approach also exposes part of the suboccipital surface of the cerebellum (3)

corresponds to a hemispheric part: folium and superior semilunar lobules, tuber and inferior semilunar lobules, pyramid and biventral lobules, uvula, and tonsils. The suboccipital surface is divided by the suboccipital fissure (junction of the prepyramidal and prebiventral fissures) and the petrosal fissure, which is formed at the petrosal surface of the cerebellum and extends to the suboccipital surface.

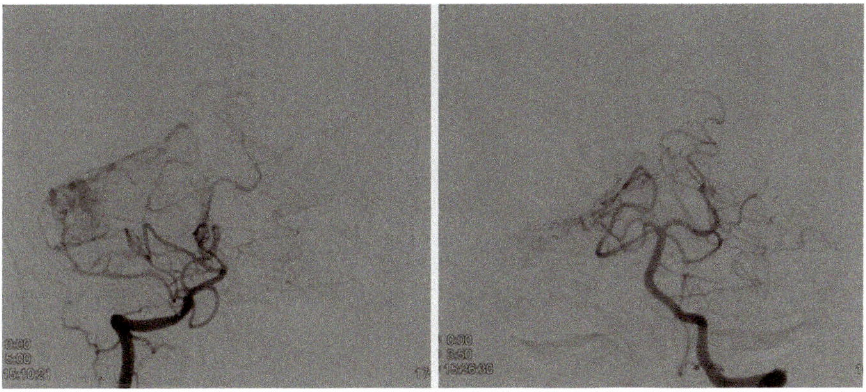

Fig. 8.5 Digital angiography performed in the arterial phase demonstrating the AVM supplied by the superior cerebellar, anterior inferior cerebellar, and posterior inferior cerebellar arteries

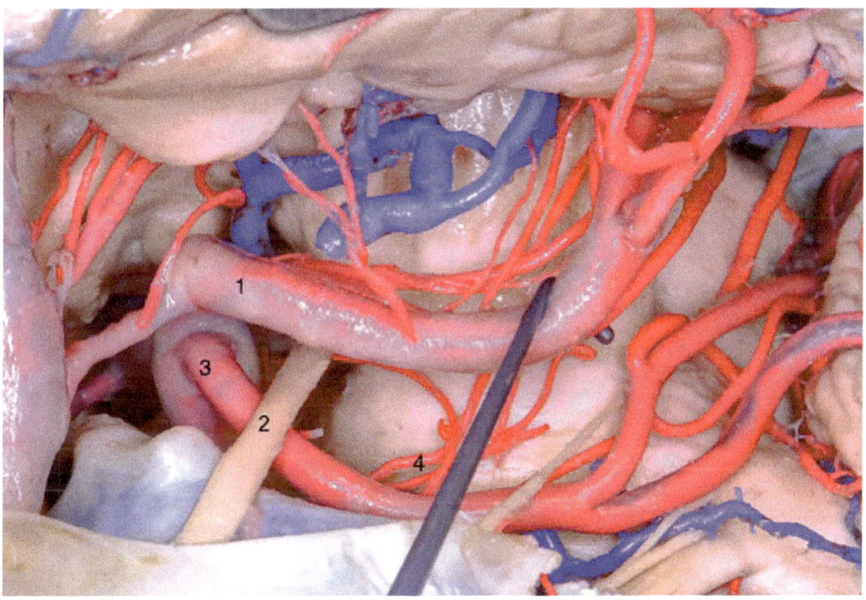

Fig. 8.6 Image of the posterior cerebral artery (1) above the oculomotor nerve (2) and the SCA (3) below this nerve. The SCA outlines the pontomesencephalic junction; at this point, the vessel bifurcates as illustrated in this specimen. Note the perforating branches (4) ending at the cerebellar peduncles. The artery continues through the cerebellomesencephalic fissure and bifurcates into hemispheric branches

The PICA outlines the cerebellomedullary fissure and irrigates the suboccipital surface. It is divided into five segments, some of which are just identified after the tonsils are retracted.

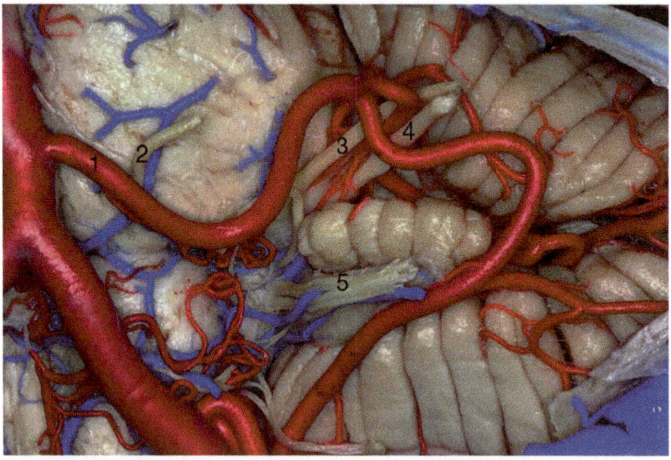

Fig. 8.7 Note in this specimen the AICA (1) running adjacent to the bulbopontine sulcus and below cranial nerve VI (2). This neurovascular complex also involves the cranial nerves VII (3) and VIII (4). In the portion of the flocculonodular segment of the AICA, the vessel is divided into two rami; the rostral runs to the cerebellomedullary fissure, while the caudal directs toward the cerebellopontine fissure within the petrous fissure. Also note the lower cranial nerves (5 – glossopharyngeal) below the flocculus. The cerebellomedullary fissure has a V shape, as demonstrated in this specimen

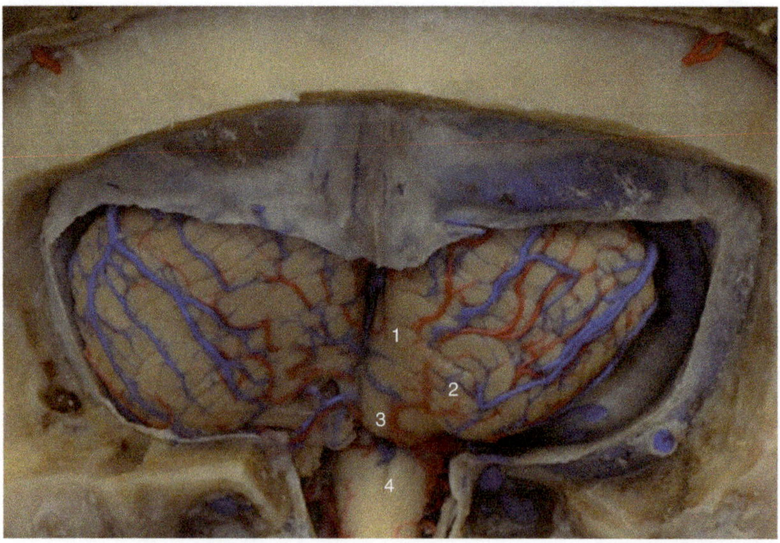

Fig. 8.8 Suboccipital surface exposed after a suboccipital craniotomy. Its superior and lateral limits are the transverse and sigmoid sinuses, respectively. The vermis is hidden between the two cerebellar hemispheres, and these hemispheres are separated by the falx cerebelli. The suboccipital fissure separates the inferior semilunar lobule (1) from the biventral lobule (2) at the hemisphere and the tuber from the pyramid at the vermis. The tonsils (3) are components of the cerebellar hemispheres. They are separated from the medulla (4) by the cerebellomedullary fissure

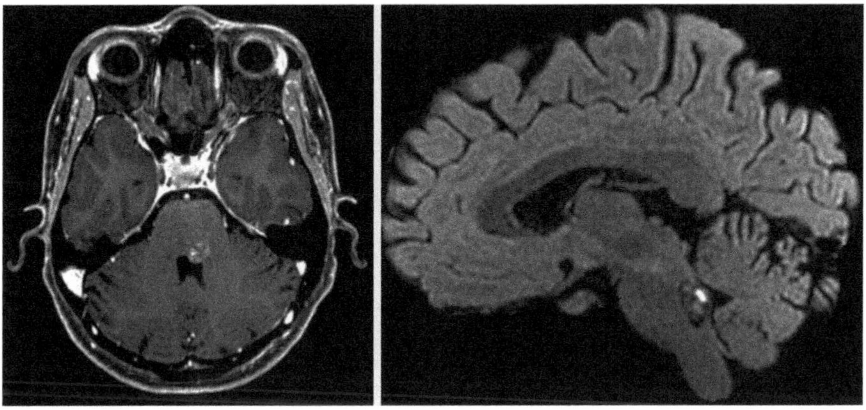

Fig. 8.9 Brainstem cavernous malformation

Surgical Case 2 Female, 39 years old, presented in the past 6 months several transient episodes of left hemifacial paresthesia and numbness of the right superior limb. Her brain MRI demonstrated a cavernous malformation, as demonstrated in Fig. 8.9.

Looking carefully to the MRI, it is possible to identify that this malformation is directly related posteriorly to the fourth ventricle. Then, it is located on the floor. The fourth ventricle is a cavity between the cerebellum and the brainstem that communicates with the space encased between the tonsils, called vallecula, through the foramen of Magendie. Superiorly, it is continuous with the cerebral aqueduct. The lateral recesses communicate to the cerebellopontine angles through the foramen of Luschka.

The floor of the fourth ventricle (Fig. 8.10) is posterior to the pons and the medulla, and its cranial and caudal limits are, respectively, the cerebral aqueduct and the obex. It is divided by the median sulcus of the fourth ventricle, called the median, which is an unpaired structure. On one side of this sulcus, there is a medial eminence on each side, limited laterally by the sulcus limitans. The most cranial structure of the medial eminence is the facial colliculus, which is formed by fibers of the seventh nerve (internal genu) that outlines the sixth nerve nucleus. Lateral to the medial eminence, there is the vestibular area, which extends to the lateral recess and overlies the vestibular nuclei as well as the auditory tubercle. Inferiorly to the facial colliculus, there are three trigonal structures, namely, hypoglossal and vagal triangles, which are related to the nucleus of the hypoglossal nerve and the dorsal nucleus of the vagal nerve and the area postrema, respectively.

An important intraoperative reference for the facial colliculus is the stria medullaris of the fourth ventricle, originating at the median sulcus and directed toward the lateral recess. The stria medullaris occurs below the facial colliculus (Fig. 8.11).

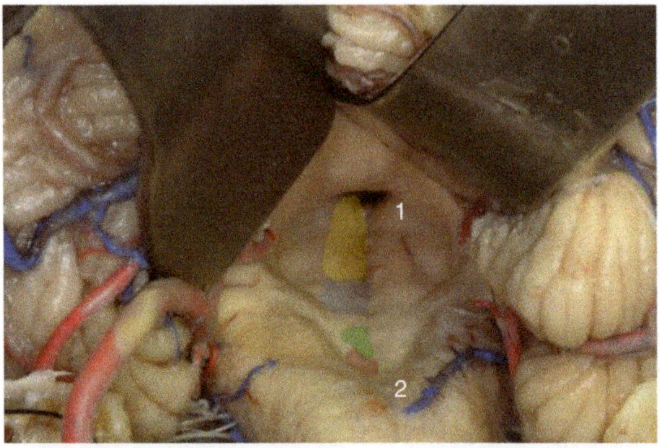

Fig. 8.10 The floor of the fourth ventricle is exposed after the retraction of the tonsils and the removal of the roof. It is connected to the third ventricle through the cerebral aqueduct (1) and is inferiorly limited by the obex (2). The medial eminence (yellow) is separated from each other by the median sulcus. The inferior portion of the medial eminence is the facial colliculus (blue). The inferior segment of the floor contains the hypoglossal (green) and the vagal (red) triangles

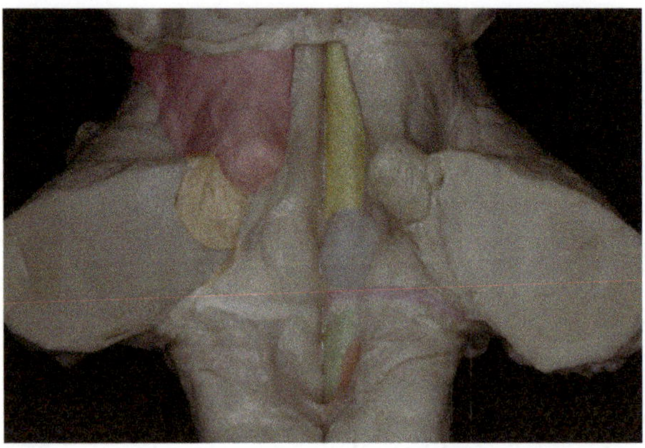

Fig. 8.11 Image of the stria medullaris of the fourth ventricle (purple) that crosses the floor and extends from the median sulcus toward the lateral recess, a lateral extension of the vestibular area (white). The superior (pink) and the inferior (orange) cerebellar peduncles are directly associated with the fourth ventricle, and then the middle (black) peduncle is the only one not related to it. Notice the inferior segment of the floor, which contains the hypoglossal (green) and the vagal (red) triangles, separated from the area postrema (brown) by the funiculus separans. The fourth cranial nerve is the most cranial structure and is an important landmark to identify the quadrigeminal plate

Surgical Case 2 Which surgical approach is reasonable to treat this cavernous malformation?

A patient underwent suboccipital craniotomy, and the suboccipital surface of the cerebellum was exposed. The tonsils cover the inferior segment of the roof of the fourth ventricle as well as part of the inferior posterior cerebellar artery (Fig. 8.12).

To expose the fourth ventricle, telovelar access was proposed. The tonsils was retracted and the tela choroidea opened. If sufficient exposition is not achieved with this maneuver, then the inferior medullary velum can also be incised (Fig. 8.13).

The roof of the fourth ventricle (Fig. 8.14) is divided into the superior and inferior portions that meet each other in the apex, named fastigium. The superior part of

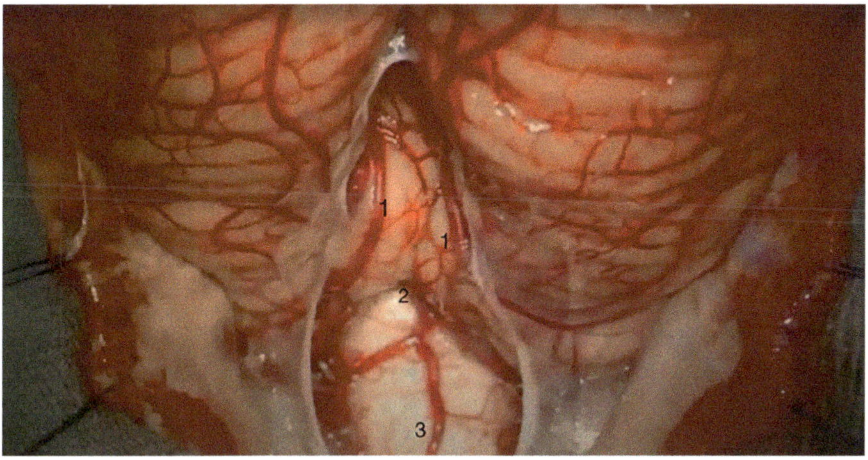

Fig. 8.12 Image of the suboccipital surface of the cerebellum that is being exposed. Note that it is necessary to transpose the tonsils (1) to reach the lesion situated at the floor of the fourth ventricle. (2) Right cerebellar hemisphere. (3) Inferior medulla

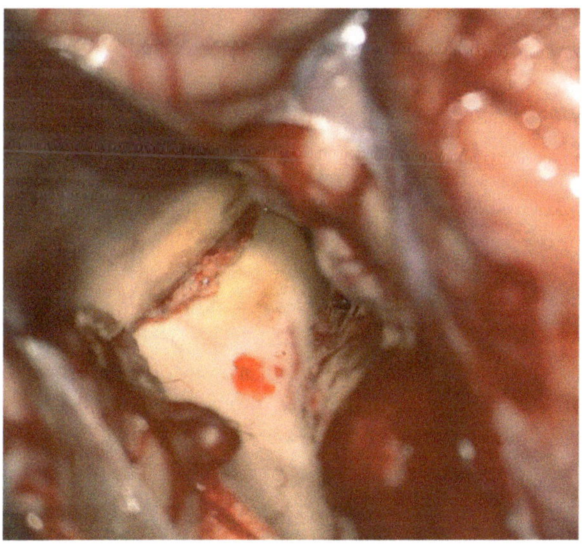

Fig. 8.13 Telovelar access exposes the floor of the fourth ventricle after the tonsil retraction and the opening of the tela choroidea, in the telovelar access

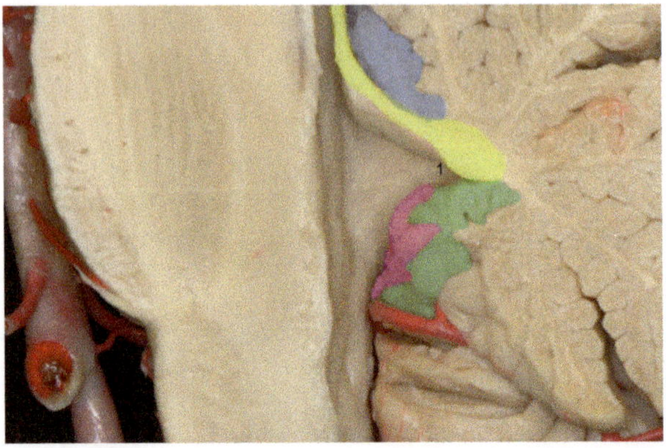

Fig. 8.14 The roof of the fourth ventricle has a tent shape, and the most prominent point of the tent is the fastigium (1). In the median sagittal view, as shown in this figure, the superior part of the tent is composed of the superior medullary velum (yellow), covered at the cisternal surface by the lingula (blue). The ventricle part of the lower half of the roof is composed of the inferior medullary velum, which is covered externally by the nodule (green). The choroid plexus (purple) extends at the roof as a T-shaped structure

the roof is composed of both superior cerebellar peduncles connected by the superior medullary velum. The cisternal surface of the superior medullary velum is covered by an extension of the vermis, the lingula (Fig. 8.8). The inferior part of the roof is composed of the nodule; the link between this and the flocculus, which is named the inferior medullary velum; and the tela choroidea. The choroid plexus is a T-shaped structure located at the roof of the fourth ventricle. Both medial segments extend to the foramen of Magendie, and the transversal portions extend to the lateral recess and the foramen of Luschka.

The dentate nucleus is located above the posterolateral portion of the roof of the fourth ventricle and encompasses the superior pole of the tonsil, which is separated by the inferior medullary velum. This relationship should be well understood to avoid lesion occurrence in this important cerebellar nucleus.

The veins of the posterior fossa are named according to the associated structures. Their final portions terminate as bridging veins that drain into one of the drainage groups: galenic, petrosal, or torcular/tentorial. The galenic system drains the tentorial surface of the cerebellum and the midbrain to into the vein of Galen (Fig. 8.15). The torcular/tentorial group drains the suboccipital (Fig. 8.16) surface of the cerebellum and the fourth ventricle into the tentorial sinuses. The petrosal system drains the petrosal surface of the cerebellum, the inferior segment of the pons, and the medulla into the superior and inferior petrosal sinuses.

In addition to this functional division, the posterior fossa vein system is divided into superficial, deep, and brainstem. The superficial veins are divided according to the drainage into the cerebellar surface. The deep vein courses in the fissures (cerebellomesencephalic, cerebellopontine, and cerebellomedullary) and drain into the cerebellar peduncles. The brainstem veins drain the midbrain, pons, or medulla.

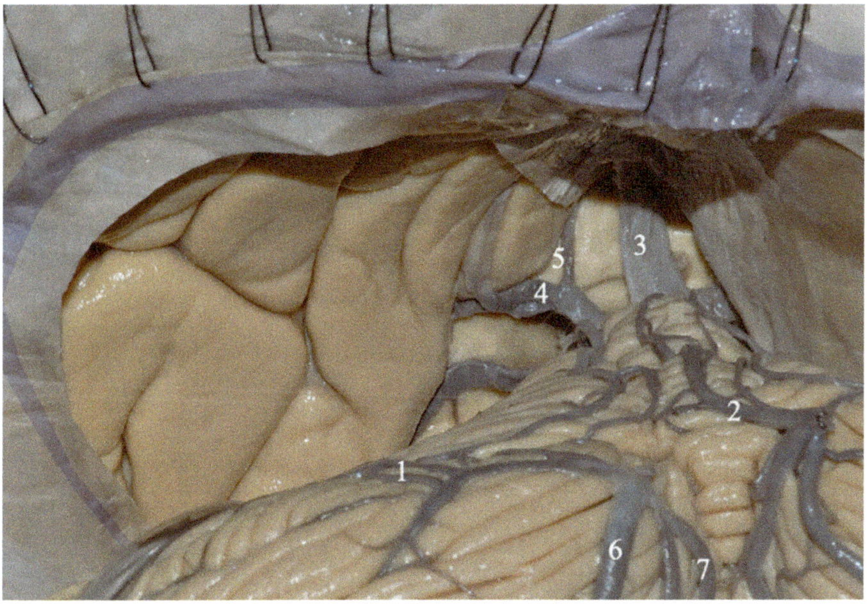

Fig. 8.15 Image of the tentorium that was cut at the right side. The tentorial surface of the cerebellum and the superior part of the suboccipital surface were exposed. The superior cerebellar (1) and vermian (2) veins drain the tentorial surface of the cerebellum into the vein of Galen (3). Notice the anastomosis of the basal vein (4) and the internal cerebral (5) vein at the right side. The inferior cerebellar (6) and vermian veins (7) emerge from the suboccipital surface toward the tentorial surface to drain into the tentorial sinuses

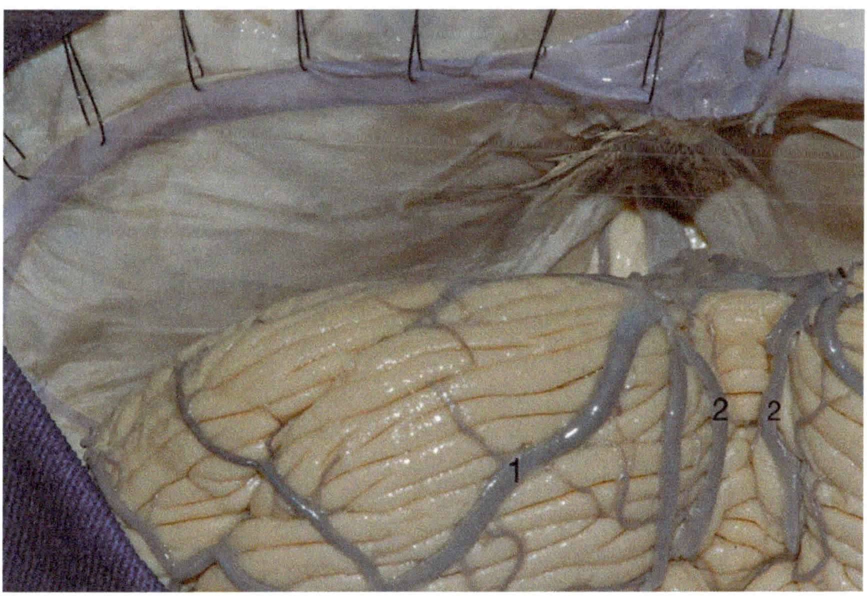

Fig. 8.16 Inferior cerebellar (1) and vermian (2) veins drain the suboccipital surface of the cerebellum. They cross the tentorial surface of the cerebellum and are tributaries of the tentorial bridging veins

Suggested Bibliography

Deshmukh VR, Figueiredo EG, Deshmukh P, et al. Quantification and comparison of telovelar and transvermian approaches to the fourth ventricle. Neurosurgery. 2006;58:ONS-202–6.

Lessa SS, Paz-Archilla JA, Amorim BL, Filho JMC, de Siqueira Campos CM, de Deus SL, Choque RGP, da Costa MDS, Chaddad-Neto F. Microsurgical treatment for cerebellomesencephalic fissure arteriovenous malformations after multiple sessions of endovascular treatment. Surg Neurol Int. 2021;12:214.

Operative cranial neurosurgery anatomy. New York: Thieme, 2019.

Ramos A, Chaddad-Neto F, Dória-Netto HL, Campos-Filho JM, Oliveira E. Cerebellar anatomy as applied to cerebellar microsurgical resections. Arq Neuropsiquiatr. 2012;70(6):441–6.

Rhoton. Cranial anatomy and surgical approaches. Philadelphia: Lippincott Williams & Wilkins; 2003.

Wainberg RC, da Costa MDS, Hernández YAU, Caramanti RL, Alves Filho CAF, Palmiero H, Saick RP, Chaddad-Neto F. Microsurgical clipping of ruptured distal posterior inferior cerebellar artery aneurysm: 3-dimensional operative video. Oper Neurosurg (Hagerstown). 2019;16(2):48–9.

Wen DY, Heros RC. Surgical approaches to the brain stem. Neurosurg Clin North Am. 1993;4:457–68.

Chapter 9
Surgical Anatomy of the Midbrain

9.1 Introduction

The midbrain, or mesencephalon, is the smallest and most rostral portion of the brainstem, located between the pons caudally and the diencephalon rostrally. Embryologically, the midbrain is derived from the second of the three primary neural tube vesicles and evolves in a straightforward manner from this vesicle, with no further subdivisions during the reminder of neural development, unlike the forebrain and hindbrain. Nevertheless, the midbrain plays an important role in mediating reflexes and the transmission of sensory and motor information through a collection of neurons aggregated as nuclei and fibers associated with tracts, fasciculi, and lemnisci. It contains a complex network of ascending and descending pathways that connect the forebrain and hindbrain and has a role in many important functions, including receiving and integrating sensory information, particularly visual and auditory information; eye movements; motor coordination; and conscious level.

The midbrain has an oblique ventral and rostral direction and occupies most of the incisural space. It is approximately 16 mm long and 13 mm and 19 mm wide at its caudal and rostral parts, respectively, with an anteroposterior diameter of 21 mm. Externally, the midbrain can be divided into four faces and two extremities: an anterior face, which is related to the cerebral peduncles and posterior perforated substance; a posterior face, which is related to the quadrigeminal colliculi; two lateral faces, which lie between the cerebral peduncles and the quadrigeminal colliculi; and the superior and inferior extremities, which are related to the diencephalon rostrally and the pons caudally, respectively. Internally, it is divided into an anterior and posterior portion by an imaginary line that crosses the cerebral aqueduct; the cerebral

---

Supplementary Information The online version of this chapter (https://doi.org/10.1007/978-3-030-82747-2_9) contains supplementary material, which is available to authorized users.

F. Chaddad-Neto, M. D. Silva da Costa, *Microneuroanatomy and Surgery*,
https://doi.org/10.1007/978-3-030-82747-2_9

peduncles lie anterior to this line, and the tectum lies posterior to this line. Each cerebral peduncle is further subdivided into the crus cerebri and the tegmentum by the substantia nigra.

The aim of this chapter is to review the anatomy of the midbrain and to discuss this in the context of clinical cases. Safe surgical entry zones to the midbrain are also discussed.

Clinical Case 1 A 51-year-old man with left hemiparesis, hypoesthesia, right palpebral ptosis, and anisocoria (Figs. 9.1 and 9.4).

Based on the previous magnetic resonance images, the following questions arose: In exactly which part of the midbrain is the lesion located? How can this lesion be approached surgically? What are the superficial landmarks for the surgical approach? Which anatomical structures should be borne in mind?

The midbrain has an internal and external morphology. Internally, the midbrain is divided by an imaginary transverse plane across the aqueduct in the posterior part, the tectum, and the anterior part, the cerebral peduncle. Consequently, the cerebral peduncle is divided by the substantia nigra into the crus cerebri anteriorly and the tegmentum posteriorly. Each of these three components of the internal morphology of the midbrain is represented by one of the different surfaces (anterior, lateral, or posterior).

Externally, the midbrain has superior and inferior limits: the optic tract and the pontomesencephalic sulcus, respectively (Fig. 9.2). The anterior surface is composed of two cerebral peduncles, which diverge in an inferior to superior direction; this creates a triangular space with the superior foramen cecum as the apex, the medial sulcus of the cerebral peduncle and oculomotor nerve forming the lateral edges, and the mammillary bodies forming the base. This triangular space is known as the interpeduncular fossa and contains the posterior perforated substance (Fig. 9.3).

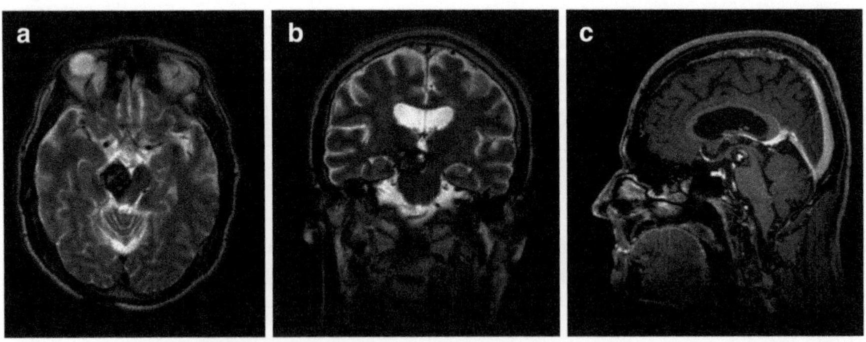

Fig. 9.1 Initial magnetic resonance imaging (MRI) scan. (**a**) Axial and (**b**) coronal T2-weighted MRI scan showing a lesion at the level of the midbrain with a central core of mixed signal intensity, surrounded by a rim of signal hypointensity. (**c**) Sagittal T1-weighted post-contrast MRI scan showing the heterogeneous lesion located at the anterior portion of the midbrain

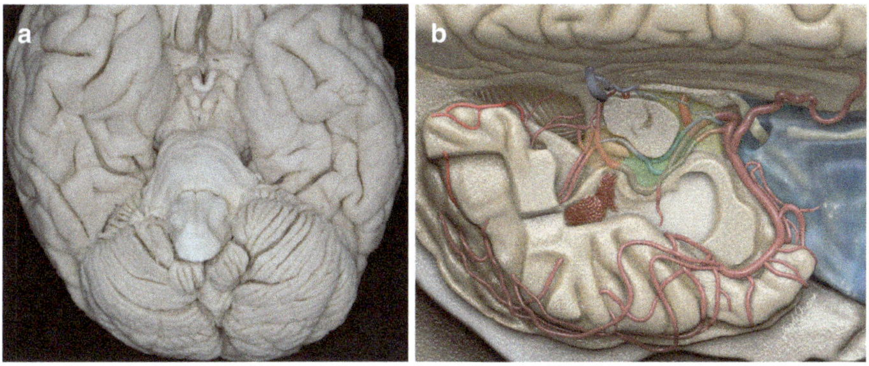

Fig. 9.2 Extrinsic anatomy of the midbrain, and the cisternal as well as the vascular relationships. (**a**) The relationship of the midbrain, located at the top of the brainstem, to the medial surface of the temporal lobe. (**b**) The cisternal and vascular relationships of the midbrain. The interpeduncular cistern, with the precommunicating segment of the posterior cerebral artery (P1) and the pontomesencephalic segment of the superior cerebellar artery, is shaded in yellow. The crural cistern, related to the crural segment of the posterior communicating artery, the cisternal segment of the anterior choroidal artery, and the basal vein, is shaded in green. The ambient cistern, related to the ambient segment of the posterior communicating artery and the basal vein, is shaded in orange

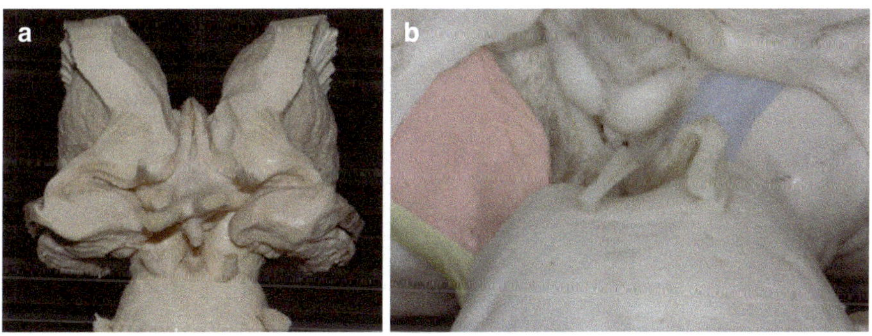

Fig. 9.3 Anterior surface of the midbrain. (**a**) The relationship of the midbrain to the pons, diencephalon, and basal ganglia. (**b**) The anterior surface of the midbrain, showing the frontopontine fibers of the crus cerebri (shaded blue), the corticospinal tract (shaded red), and the temporoparietooccipitopontine fibers (shaded yellow)

The anterior surface is hidden behind the uncus of the temporal lobe, and, as for all of the surfaces of the midbrain, it is related to multiple structures, with cisternal, vascular, and cranial nerve relationships.

9.2 Cisternal Relationships

The anterior surface of the midbrain is related to the middle portion of the interpeduncular cistern, corresponding to the level of the interpeduncular fossa, and with the crural cistern laterally, in relation to the cerebral peduncle.

9.3 Vascular Relationships

At the interpeduncular cistern, the major arteries encountered are the tip of the basilar artery, precommunicating segment of the posterior cerebral artery (P1), and the pontomesencephalic segment of the superior cerebellar artery. At this level, many branches originate from these arteries, principally central perforating arteries to the diencephalon and midbrain, which are divided into direct and circumflex perforating branches. From the tip of the basilar artery and the posterior and superior walls of the P1 segment of the posterior cerebral artery arise multiple direct perforating arteries directed to the peduncle and the posterior perforated substance. The most important of these is the thalamoperforating artery, arising from the P1 segment of the posterior cerebral artery; multiple short and long circumflex arteries arise from the P1 segment, extending to the geniculate bodies and the colliculi.

At the crural cistern, the major arteries are the posterior communicating artery, the cisternal segment of the anterior choroidal artery giving branches to the cerebral peduncle, and the crural segment of the posterior cerebral artery (P2A). From the P2A segment of the posterior cerebral artery arise direct peduncular perforating arteries, short and long circumflex arteries, the thalamogeniculate artery, the medial posterior choroidal arteries directed to the third ventricle, and the hippocampal arteries.

The related venous structures are principally the peduncular and the basal veins, as well the vein of the pontomesencephalic sulcus, tributaries to the basal vein of Rosenthal from the ventricular group, the anterior incisural space, and anterior part of the middle incisural space.

9.4 Cranial Nerve Relationships

The oculomotor nerve emerges from the medial sulcus of the cerebral peduncle. It crosses the interpeduncular cistern between the precommunicating segment of the posterior cerebral artery and the pontomesencephalic segment of the superior cerebellar artery toward the oculomotor trigone in the roof of the cavernous sinus.

The trochlear nerve emerges from the posterior surface of the midbrain below the inferior colliculi and surrounds the midbrain, passing through the cisterns around it in close relation to the posterior cerebral and superior cerebellar arteries, between which it passes at the level of the ambient cistern. The trochlear nerve then continues anteriorly to form the posterolateral edge of the oculomotor triangle.

At this point, the relevant superficial anatomy has been determined; however, what internal structures are related to the anterior surface, and what would be the safe entry zone or zones to this region?

As previously mentioned, each surface is related to each part of the internal morphology of the midbrain. In this case, the anterior surface is related to the crus cerebri. The crus cerebri is limited posteriorly by the substantia nigra and has a semilunar shape on axial section. It is the continuation of the fibers coming from the internal capsule and is therefore composed of longitudinal fibers, making it possible to divide the crus into three portions related to the fibers that comprise each portion. The first portion is the internal segment, representing one-fifth of the crus, which is composed of frontopontine fibers and corticonuclear fibers originating from the precentral gyrus in its inferior portion. The second portion, comprising three-fifths of the crus, comprise the intermediate segment, which is formed by the corticospinal pathway. The final one-fifth is the external segment, which is formed by temporopontine, parietopontine, and occipitopontine fibers.

Given the internal and external structures related to the anterior surface of the midbrain, it is possible to analyze how to approach a lesion located on this surface and the crus cerebri. It is important to emphasize that all lesions of the brainstem must be approached from the point closest to the surface whenever the lesion is near a surface. On the anterior surface, lesions related to the anterior midbrain can be approached through an orbitozygomatic or pretemporal approach and the perioculomotor safe entry zone, limited medially by the oculomotor nerve, laterally by the corticospinal tract, superiorly by the precommunicating segment of the posterior cerebral artery, and inferiorly by the anterior pontomesencephalic segment of the superior cerebellar artery (Fig. 9.4).

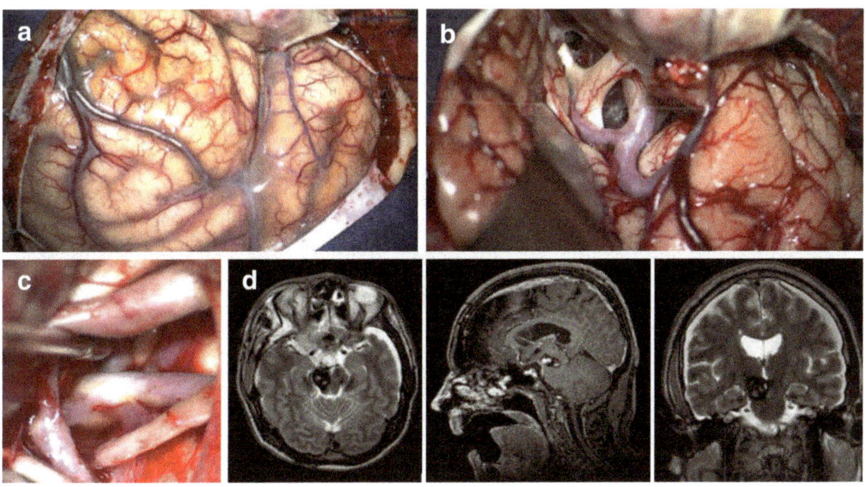

Fig. 9.4 Intraoperative images. The patient was diagnosed with a cavernoma at the level of the crus cerebri on the right side. (**a**, **b**, and **c**) With the patient in the supine position, a right orbitozygomatic approach was performed in order to obtain a view from "below to above" and expose the anterior surface of the midbrain. (**d**) Postoperative magnetic resonance imaging, showing a complete resection of the cavernoma

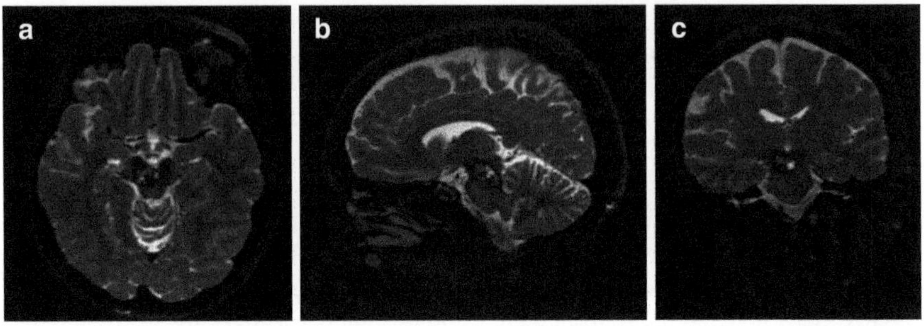

Fig. 9.5 Preoperative magnetic resonance imaging. (**a**, **b**, and **c**) T2-weighted images showing an area of mixed signal intensity surrounded by a rim of hypointensity at the level of the middle portion of the midbrain

Clinical Case 2 A 45-year-old woman with a left hemiparesis, dysarthria, diplopia, and right palpebral ptosis (Figs. 9.5, 9.8, and 9.9) (*this case example is also demonstrated in the* video 9.1).

Based on the previous magnetic resonance images: At what portion of the internal configuration of the midbrain is the lesion located? How can this lesion be approached? What are the superficial landmarks for the surgical approach? Which anatomical structures should be borne in mind?

The lateral surface of the midbrain is located posteromedial to the uncus and the parahippocampal gyrus of the temporal lobe and is divided by the lateral mesencephalic sulcus into an anterior part that is continuous with the cerebral peduncle and a posterior part that is called the lemniscal trigone (lateral lemniscus), which is limited by the posterior conjunctival brachium, the superior cerebellar peduncle, and the lateral mesencephalic sulcus (Fig. 9.6).

9.5 Cisternal Relationships

The lateral surface of the midbrain is related to the ambient cistern, which extends from the posterior edge of the cerebral peduncle communicating with the crural cistern anteriorly, and terminates at the posterior edge of the lateral surface where it meets the lateral portion of the quadrigeminal cistern.

9.6 Vascular Relationships

The lateral surface of the midbrain is related to one major artery, the ambient segment of the posterior cerebral artery (P2P), which runs in the hippocampal sulcus of the medial surface of the temporal lobe. The P2P segments give rise to peduncular

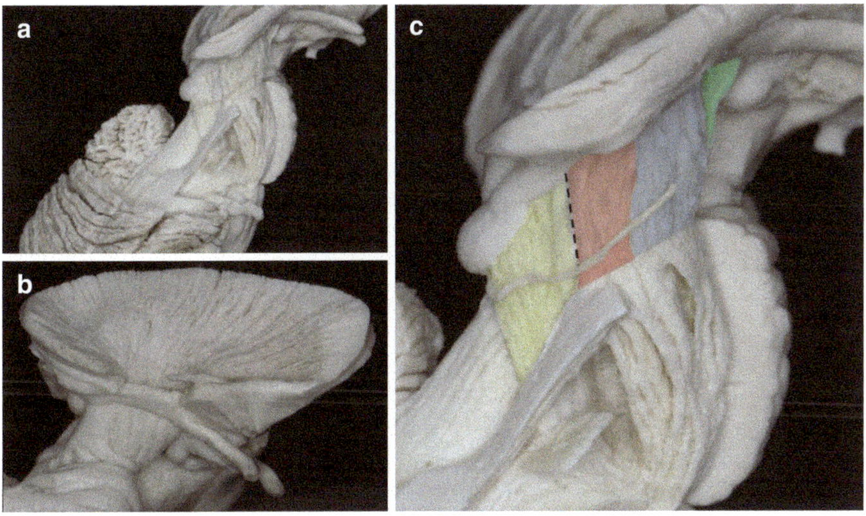

Fig. 9.6 Lateral surface of the midbrain and fiber dissection. (**a**) Fiber dissection of the lateral surface of the midbrain and the relationships with the pons, medulla, and cerebellum. (**b**) Lateral surface of the midbrain and the relationship to the fibers of the internal capsule, showing the concept of the crus cerebri as the continuation of the fibers of the internal capsule. (**c**) Magnified picture of the lateral surface, showing the constitution of the crus cerebri: the frontopontine fibers (shaded green), the corticospinal tract (shaded blue), and the temporoparietooccipitopontine fibers (shaded red). The dotted line shows the trajectory of the lateral mesencephalic sulcus. In the dissection, it is possible to observe the relationship between the lateral lemniscus (shaded yellow) and the superior cerebellar peduncle, the inferior colliculus, and the posterior conjunctival brachium

perforating arteries, short and long circumflex arteries directed to variable distances around the midbrain, the thalamogeniculate artery (which may arise from either the P2A or P2P segments), the lateral posterior choroidal arteries, and the middle and posterior temporal arteries.

In this portion of the midbrain, venous relationships comprise of the basal vein of Rosenthal on the roof of the ambient cistern and with its tributaries at this level, which are principally the lateral mesencephalic vein and the medial temporal vein.

9.7 Cranial Nerve Relationships

The trochlear nerve is the only cranial nerve related to the lateral surface of the midbrain. As previously mentioned, this nerve emerges from the inferior border of the inferior colliculi and surrounds the midbrain while running through the ambient cistern anteriorly until it reaches the crural cistern, where it joins the lower margin of the free edge of the tentorium and continues toward the posterolateral edge of the oculomotor trigone.

The related superficial anatomy and structures have been elucidated; nevertheless, it is important to answer the following questions: what is the internal morphology of the midbrain related to the lateral surface, and what would be the safe entry zone or zones to this region?

The lateral surface of the midbrain is related to the tegmentum of the midbrain. The tegmentum is limited anteriorly by the crus cerebri and posteriorly by the imaginary plane across the cerebral aqueduct. The tegmentum is a complex area composed of white matter (ascending, descending, and transverse fibers), gray matter homologous to the spinal cord (third and fourth nerve nuclei), the midbrain's own gray matter (the substantia nigra and the red nucleus), and the reticular formation. The substantia nigra is located at the anterior limit of the tegmentum just behind the crus cerebri and extends from the medial sulcus of the cerebral peduncle to the lateral mesencephalic sulcus, creating a characteristic dark semilunar shape with an anterior convexity in axial sections. The red nucleus is located in the center of the tegmentum, and from it, the spatial correlation of the main structures found in the tegmentum is easier to understand. First, the reticular formation, a network of fibers and nuclei present throughout the brainstem, is located behind the red nucleus, lateral to the central gray matter and anterior to the posterior commissure. The trochlear nucleus is located posterior to the red nucleus and anterior to the periaqueductal gray matter at the level of the inferior colliculus. The three nuclei that comprise the oculomotor nucleus (the oculomotor nucleus, Perlia's nucleus, and the accessory oculomotor nucleus) form a column at the level of the superior colliculus that is located anteromedially to the periaqueductal gray matter, posteromedial to the medial longitudinal fasciculus, and posterior to the red nucleus. The principal white matter fibers located at the tegmentum are the fibers of the dentatorubrothalamic pathway, running from the superior cerebellar peduncle to the inferior surface of the red nucleus and extending from the superior portion of the red nucleus to the thalamus; the rubrospinal fibers, which run inferiorly from the red nucleus to the anterior horn of the spinal cord; the medial lemniscus, which ascends lateral to the red nucleus and posteriorly to the substantia nigra; the lateral lemniscus, which ascends lateral to the medial lemniscus over the lateral surface of the midbrain toward the inferior colliculus; the trigeminal lemniscus, which ascends posteriorly to the medial lemniscus in the lateral part of the tegmentum; the spinal lemniscus, which runs posteriorly to the trigeminal lemniscus; and the medial longitudinal fasciculus, which is located anterior to the oculomotor nucleus adjacent to the midline (Fig. 9.7).

After understanding the main anatomical relationships of the lateral surface of the midbrain and the tegmentum, the reasoning about how to surgically approach lesions located in this area begins. The safe entry zone for this area is the lateral mesencephalic sulcus, which is hidden under the lateral mesencephalic vein between the substantia nigra anteriorly and the medial lemniscus posteriorly. The lateral mesencephalic sulcus can be approached in two ways: through the subtemporal approach and the supracerebellar infratentorial approach.

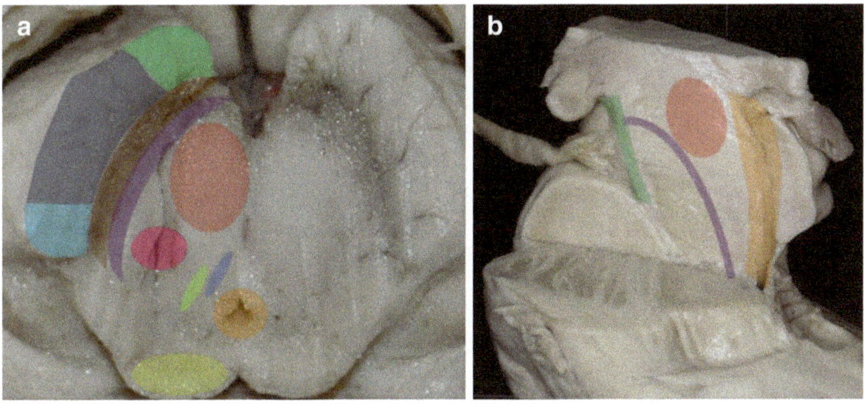

Fig. 9.7 Internal morphology of the midbrain. (**a**) An axial slice of the midbrain, showing on the right side, the crus cerebri comprising, from anterior to posterior, the frontopontine fibers (shaded dark green), corticospinal tract (shaded dark blue), and the temporoparietooccipitopontine fibers (shaded turquoise). The substantia nigra can be identified posteriorly (shaded brown), alongside the medial lemniscus (shaded purple). At the level of the tegmentum, it is possible to identify the red nucleus (shaded red), central tegmental tract (shaded pink), third cranial nerve nucleus (shaded light blue), medial longitudinal fasciculus (shaded light green) and, posteromedially, the periaqueductal gray matter (shaded orange). At this level, in the tectum, the superior colliculus is visible (shaded yellow). (**b**) A sagittal slice of the midbrain showing some of the structures mentioned above, represented by the same colors

Clinical Case 3 A 62-year-old woman with symptoms of ataxia, tremor, nystagmus, and left hypoacusis (Figs. 9.10, 9.12, and 9.13).

Based on the previous magnetic resonance images: In which part of the internal configuration of the midbrain is the lesion located? How can this lesion be approached surgically? What are the superficial landmarks for the surgical approach? Which anatomical structures should be borne in mind?

The posterior surface of the midbrain is characterized by the presence of the quadrigeminal plate, a structure composed of two superior oval and two inferior round protuberances, and the quadrigeminal tubercle or colliculi, separated by the cruciform sulcus. The superior colliculi connect through the anterior conjunctival brachium with the lateral geniculate body and are related to the optic pathway. The inferior colliculi connect through the posterior conjunctival brachium with the medial geniculate body and are involved in the acoustic pathway. From the inferior border of the inferior colliculus, the trochlear nerve courses laterally and then anteriorly around the midbrain (Fig. 9.11).

Similar to the other surfaces discussed above, the posterior surface has cisternal, vascular, and cranial nerve relationships; however, the cerebellomesencephalic fissure deserves a special mention as the most superior of the spaces between the cerebellum and the brainstem, created by the developmental fold of the cerebellum around the brainstem.

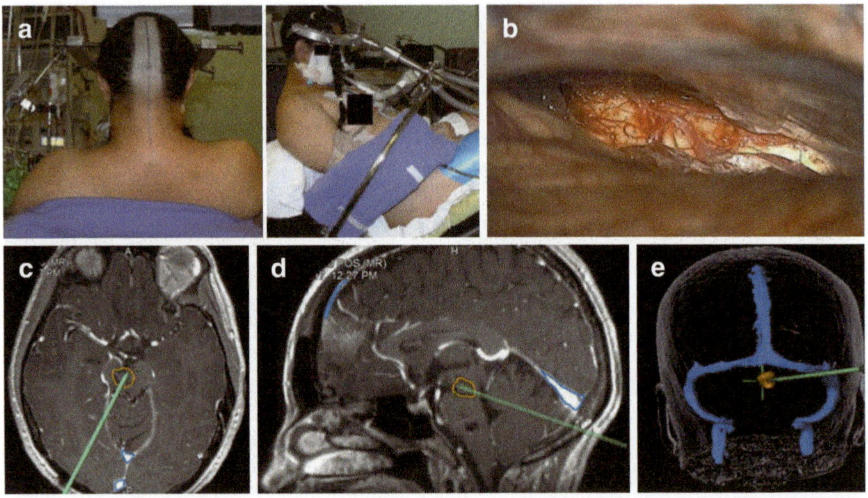

Fig. 9.8 Intraoperative images. The patient was diagnosed with a right tegmental cavernoma. (**a**) With the patient in the semisitting position, a right supracerebellar infratentorial approach was performed. (**b**) The lateral mesencephalic vein and sulcus were identified as the safe entry zone for resection of the lesion. (**c**, **d**, and **e**) Neuronavigational assistance was used in order to increase the accuracy of the procedure and minimize the risk of secondary neurological deficits

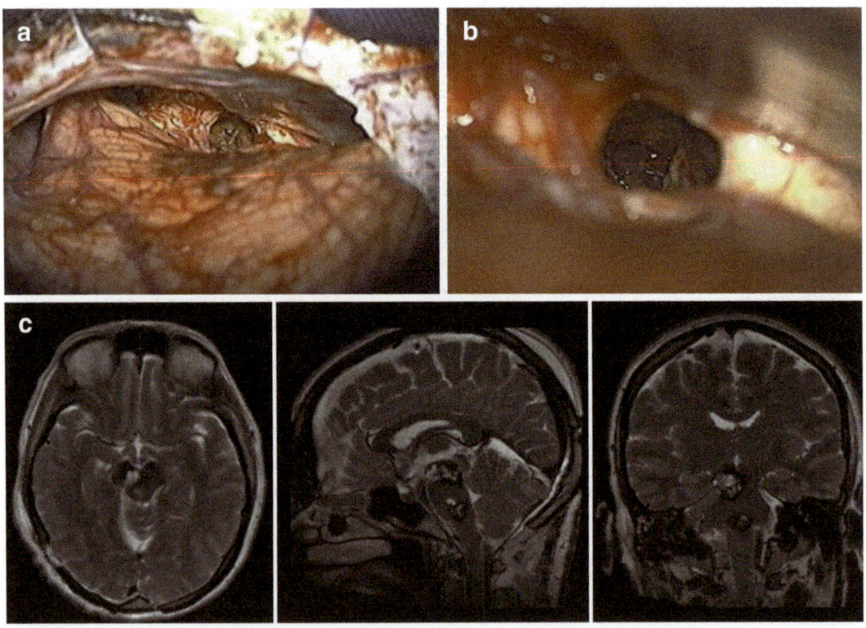

Fig. 9.9 Intraoperative and postoperative images. (**a**) An intraoperative image showing the surgical corridor through the supracerebellar infratentorial approach. (**b**) The resection cavity through the lateral mesencephalic sulcus safe entry zone. (**c**) Postoperative magnetic resonance images. T2-weighted images showing a complete resection of the cavernoma at the level of the midbrain, with communication between the lateral mesencephalic sulcus and the medial sulcus of the cerebral peduncle

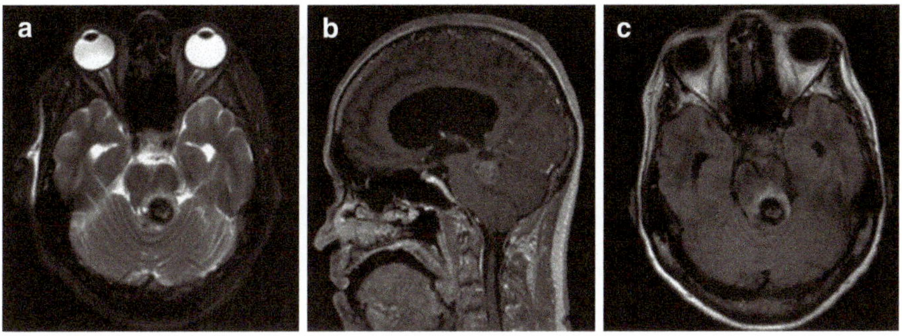

Fig. 9.10 Preoperative magnetic resonance imaging. (**a**) An axial T2-weighted image showing a heterogeneous lesion, with mixed zones of hyper- and hypointensity within the posterior portion of the midbrain. (**b**) A sagittal T1-weighted image with contrast, showing that the lesion is compromising the flow of cerebrospinal fluid. (**c**) Axial fluid-attenuated inversion recovery (FLAIR) image showing the lesion with mixed intensities and perilesional edema

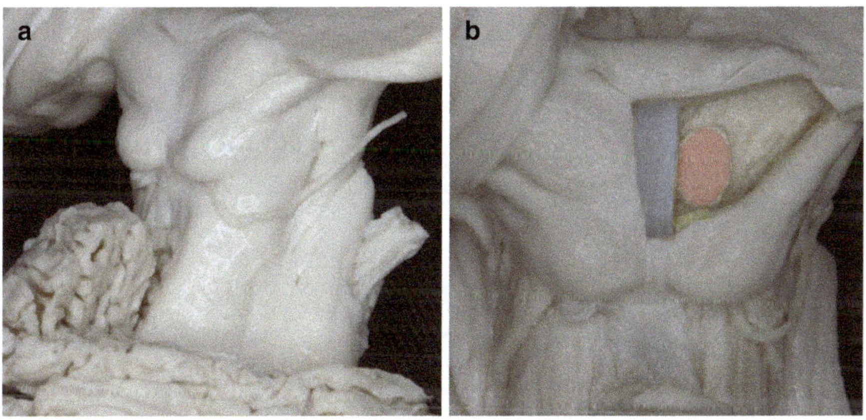

Fig. 9.11 Posterior surface of the midbrain. (**a**) The posterior surface of the midbrain is formed by the quadrigeminal tubercles or colliculi, two located superiorly and two located inferiorly. They are all separated by the cruciform sulcus. (**b**) At this level the internal morphology of the midbrain corresponds to the tectum. It is possible to observe the relationship between the colliculi and the aqueduct (anterior wall shaded in blue), the third nerve nucleus (shaded in red), and the fourth nerve nucleus (shaded in yellow)

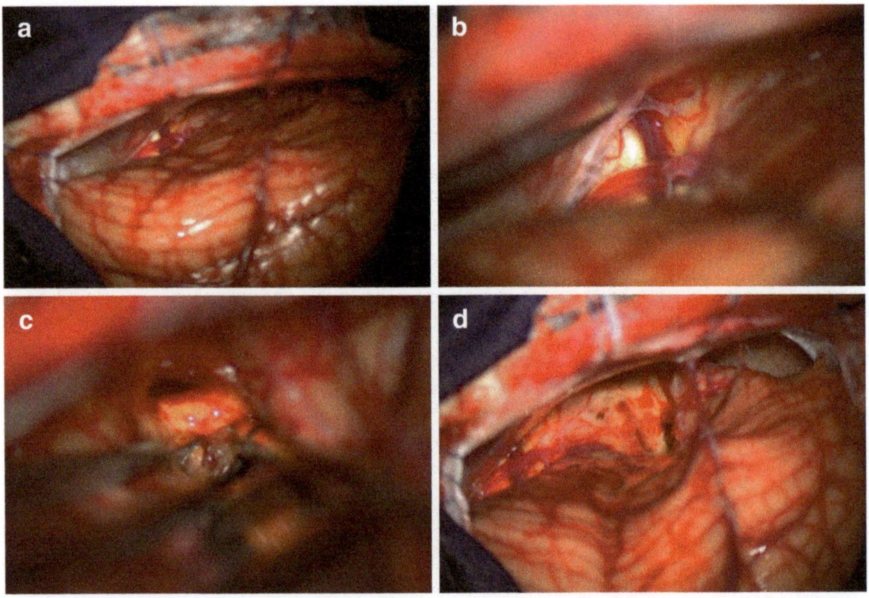

Fig. 9.12 Intraoperative pictures. (**a**) The patient was diagnosed with a cavernoma of the tectum. With the patient in the semisitting position, a left supracerebellar infratentorial approach was performed. (**b**) The lateral mesencephalic sulcus was identified. It is possible to identify a yellowish area over the surface of the midbrain corresponding to the area over the cavernoma. (**c**) The cavernoma was approached over the yellow area (the closest point of the lesion to the surface of the midbrain). (**d**) The resection cavity in the depths of the surgical corridor

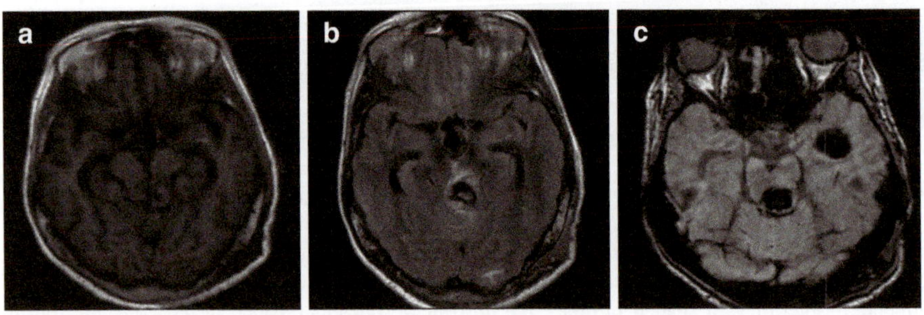

Fig. 9.13 Postoperative magnetic resonance imaging. (**a**) Axial T1-weighted image showing postoperative changes in the midbrain. (**b** and **c**) Axial fluid-attenuated inversion recovery and susceptibility-weighted images showing the resection cavity and postoperative edema

9.8 Cisternal Relationships

The posterior surface of the midbrain is related to the quadrigeminal cistern, which communicates at the level of the posterior border of the lateral surface with the ambient cistern, inferiorly to the cerebellomesencephalic fissure, which can communicate with the velum interpositum cistern.

9.9 Vascular Relationships

At this surface, the principal artery encountered is the quadrigeminal segment of the posterior cerebral artery (P3), which extends posteromedially from the lateral limit of the quadrigeminal cistern to the anterior limit of the calcarine sulcus. An important concept here is the collicular point, the closest site where the two posterior cerebral arteries meet each other. This point can be formed by a principal trunk or, if the bifurcation appears early in relation to the collicular point, by the calcarine artery. The principal branches at this level are the proximal segments of the terminal branches of the posterior cerebral artery, the calcarine and parieto-occipital arteries, branches directed to the superior colliculi, and the dorsal pericallosal arteries.

The anterior incisural space contains the most important arterial relationships, and the posterior incisural space, which is related to the posterior surface, houses the major intracranial veins. The venous structures at this level are the most complex and crucial to be aware of during microsurgical procedures in this region. The most important veins at this level are the posterior portion of the basal veins of Rosenthal, the internal veins, and the vein of Galen. This group of veins meet at the superior portion of the quadrigeminal cistern, below the splenium of the corpus callosum, where the basal veins coming from each side and the internal veins coming out of the velum interpositum join to create the vein of Galen, with the respective tributaries of each one, in the quadrigeminal cistern coming from the ventricular system, the temporal and occipital medial, and inferior surfaces.

9.10 Cerebellomesencephalic Fissure

The cerebellomesencephalic fissure is the most superior space between the cerebellum and midbrain, formed by the developmental fold of the cerebellum around the brainstem. This fissure has an anterior and posterior wall. The anterior wall is formed medially by the vermian lingula and laterally by the superior surfaces of the superior and middle cerebellar peduncles. The posterior wall is formed in the midline by the culmen and central lobule and laterally by the central lobule wings and quadrangular lobules.

The cerebellomesencephalic fissure opens superiorly into the quadrigeminal cistern, which is related to the posterior surface of the midbrain. Multiple structures pass across this fissure, including the cerebellomesencephalic segment of the superior cerebellar artery, supplying the inferior colliculi; the lateral mesencephalic, superior cerebellar peduncle, and cerebellomesencephalic veins; and the trochlear nerve, which arises below the inferior colliculi and runs in an anterior and lateral direction toward the ambient cistern.

After studying the surface anatomy of the posterior wall, it is important to understand the internal morphology that facilitates an answer to the following question: What would be the safe entry zone or zones to this region?

At the tectum, the nuclei of the superior and inferior colliculi correspond to the level of the superior and inferior colliculi, respectively. The nucleus of the superior colliculus is associated with behavioral responses to external stimuli, while the inferior colliculus nucleus is involved in the acoustic pathway.

When treating lesions located in the quadrigeminal plate, it is possible to access them using a supracerebellar infratentorial or a suboccipital approach. On the posterior surface of the midbrain, three possible safe entry zones are described: the supracollicular zone, with a transverse incision immediately above the superior edge of the superior colliculi; the infracollicular zone, with a transverse incision below the inferior edge of the inferior colliculi between the trochlear nerve output; and finally the intercollicular zone, with an incision along the vertical component of the cruciform sulcus. It is wise not to advance in depth to beyond the level of the aqueduct in order to avoid lesions to important tracts running anterior to it.

Suggested Bibliography

Cavalcanti DD, Preul MC, Kalani MYS, Spetzler RF. Microsurgical anatomy of safe entry zones to the brainstem. J Neurosurg. 2015;124:1359–76.

Cavalheiro S, Serrato-Avila JL, Párraga RG, et al. Interpeduncular sulcus approach to the posterolateral Pons. World Neurosurg. 2020;138:e795–805.

Cavalheiro S, Yagmurlu K, da Costa MD, et al. Surgical approaches for brainstem tumors in pediatric patients. Childs Nerv Syst. 2015;31(10):1815–40. https://doi.org/10.1007/s00381-015-2799-y.

Párraga RG, Possatti LL, Alves RV, Ribas GC, Türe U, de Oliveira E. Microsurgical anatomy and internal architecture of the brainstem in 3D images: surgical considerations. J Neurosurg. 2015;124:1377–95.

Yagmurlu K, Rhoton AL, Tanriover N, Bennett JA. Three-dimensional microsurgical anatomy and the safe entry zones of the brainstem. Neurosurgery. 2014;10:602–19.

Yang Y, van Niftrik B, Ma X, et al. Analysis of safe entry zones into the brainstem. Neurosurg Rev. 2019;42:721–9.

Chapter 10
Surgical Anatomy of the Pons

10.1 Introduction

Surgery of the pons is a challenge, even for most experienced neurosurgeons. It demands detailed anatomical knowledge of the intrinsic and extrinsic anatomies, a well-developed microsurgical technique, mastery of skull base and complex surgical approaches, and a complete understanding of the underlying pathology.

The pons is a small but highly eloquent region located in the middle of the brainstem, between the midbrain and medulla oblongata. It is derived from the most caudal division of the three primary neural embryological vesicles: the rhombencephalon. This primary structure in the fourth week of life subdivides into two secondary vesicles, the metencephalon and myelencephalon, with the metencephalon finally giving rise to the pons and cerebellum.

The pons has several vital functions related to conscious level, ocular and facial movements, auditory function, and maintenance of balance. Moreover, it provides a route for the ascending and descending projection tracts, carrying sensation, proprioception, pain, and auditory as well as motor information.

The pons is oriented towards a rostroventral direction, forming an oblique line together with the brainstem. It is spherical in shape and continues posterolaterally with the cerebellum through the middle cerebellar peduncles. The main dimensions of the pons are 27 mm, 38 mm, and 25 mm in the rostrocaudal, laterolateral, and dorsoventral axes, respectively (Fig. 10.1).

The pons is continuous rostrally with the base of the cerebral peduncles and inferiorly with the medulla oblongata. On the ventrolateral surface, the superior limit of the pons is the pontomesencephalic sulcus, and the inferior limit is the pontomedullary sulcus. On the dorsal surface, the superior limit is the fourth cranial

Supplementary Information The online version of this chapter (https://doi.org/10.1007/978-3-030-82747-2_10) contains supplementary material, which is available to authorized users.

F. Chaddad-Neto, M. D. Silva da Costa, *Microneuroanatomy and Surgery*,
https://doi.org/10.1007/978-3-030-82747-2_10

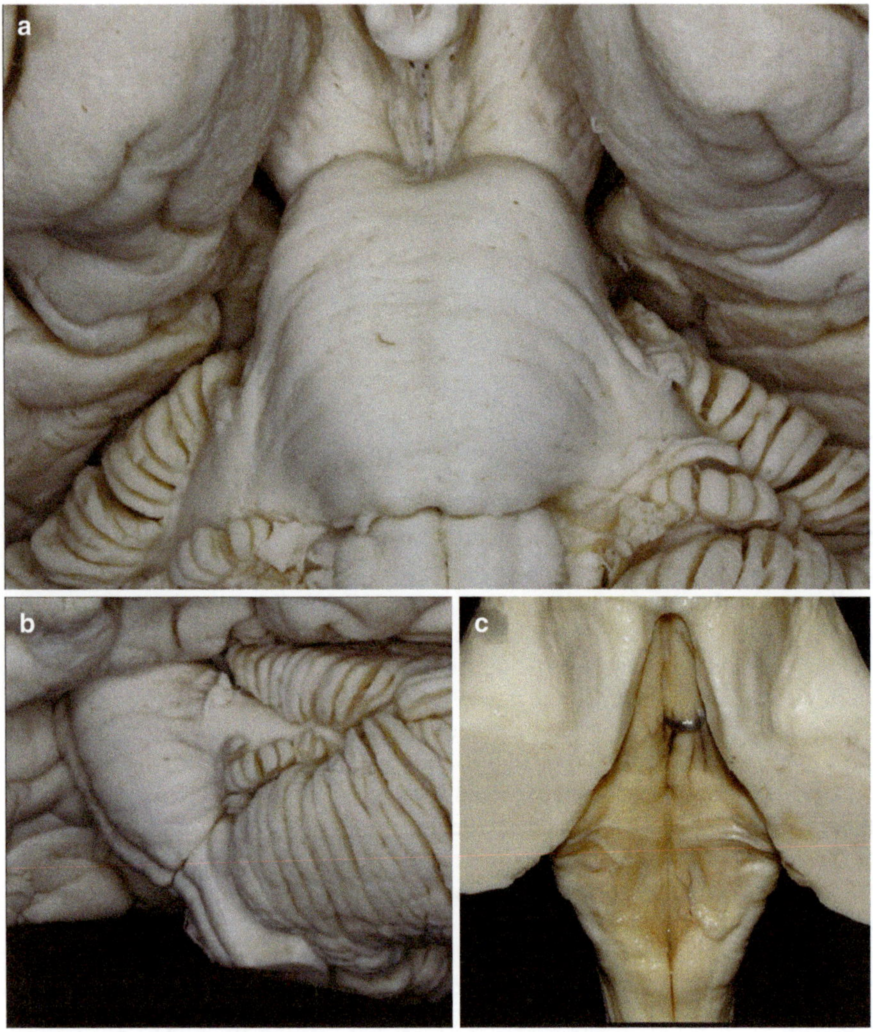

Fig. 10.1 Extrinsic anatomy of the pons. (**a**) The anterior surface. The temporal lobe surrounds the superior aspect of the pons. The pons continues directly with the base of the crus cerebri superiorly and inferiorly with the medulla. (**b**) The lateral surface, showing the ventrolateral limits of the pons. (**c**) The posterior or ventricular surface. The cerebellum has been removed to show the posterior limits, which comprise of the fourth nerve and the lateral recess of the fourth ventricle

nerve, which corresponds with the inferior border of the inferior colliculi, and the inferior limit is the lateral recess of the fourth ventricle.

The aim of this chapter is to review the anatomy of the pons and to discuss this in the context of clinical cases.

In an axial image (Fig. 10.2), we can divide the pons by the medial lemniscus into a ventral part or base and a dorsal part or tegmentum. The lateral mesencephalic

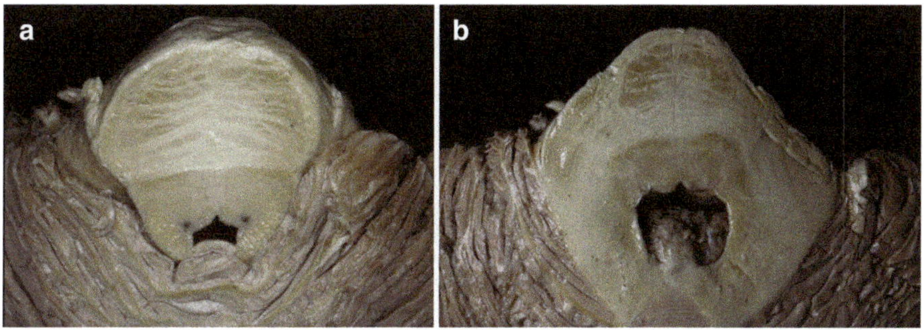

Fig. 10.2 Axial cut of the pons. (**a**) The upper pons, showing the base located anterior to the medial lemniscus. The tegmentum corresponds with the ventral and lateral surfaces. (**b**) Lower pons at the level of the nodule of the cerebellum. Note the close relation of the nodule with the floor of the fourth ventricle. The floor of the fourth ventricle is the dorsal part of the pons and corresponds with the most dorsal aspect of the tegmentum

sulcus is another important landmark because anteriorly lies the base and posteriorly lies the pons tegmentum.

10.2 Illustrative Case 1

A 45-year-old woman with a clinical history of systemic arterial hypertension presented with sudden mental confusion and amnesia. The patient underwent investigation, and a magnetic resonance imaging (MRI) scan was performed (Fig. 10.3).

To continue the patient's workup, a vascular etiology was suspected and investigated. Therefore, a computed tomography angiography (CTA) scan was performed (Fig. 10.4), which confirmed the suspected diagnosis.

This case should trigger several questions, and the neurosurgeon must answer them completely before the intervention is performed in order to offer the best treatment to the patient.

Once the diagnosis has been confirmed and the best treatment option has been decided as surgery by a multidisciplinary team, such as in this case, the neurosurgeon should start planning the procedure by asking the following questions: *Where is the lesion located anatomically, and what important structures are nearby?* To answer this question and to plan the surgery, a formal digital subtraction angiogram was performed after the MRI scan (Fig. 10.5).

In this case, even though the lesions causing the clinical status of the patient were in the distal vascular territory, the underlying cause must be treated. The aneurysm was located posterior and inferior to the dorsum sellae. It was possible to deduce, without reviewing the MRI scan, that the aneurysm was located adjacent to the anterior and superior surfaces of the base of the pons and in the prepontine cistern due to the low basilar bifurcation.

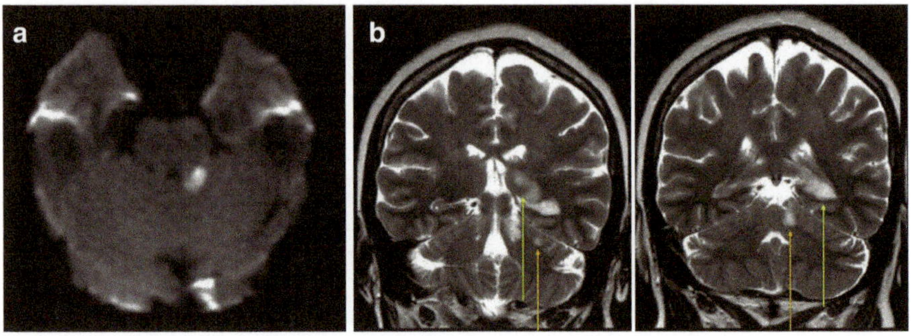

Fig. 10.3 Initial brain magnetic resonance imaging. (**a**) An axial diffusion-weighted image showing an area of diffusion restriction located in the left superior cerebellar peduncle. (**b**) A coronal T2-weighted image at the level of the fourth ventricle showing multiple hyperintense areas in the territory of the posterior cerebral artery (light yellow arrows) and the superior cerebellar artery (dark yellow arrows), suggestive of a vascular etiology

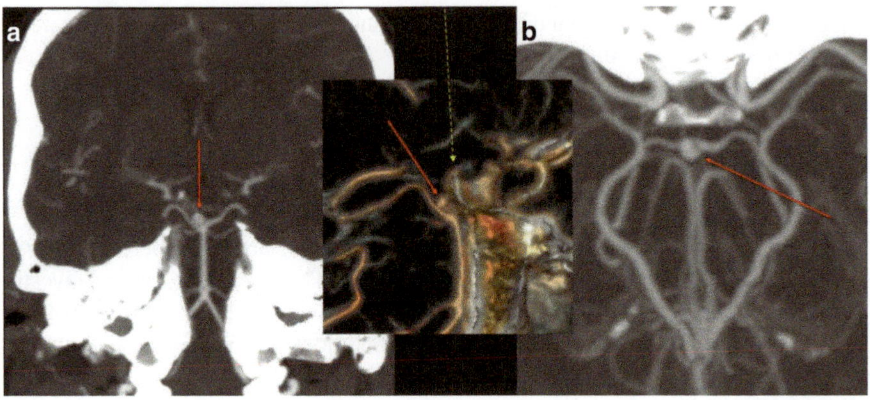

Fig. 10.4 Computed tomography angiography with arterial 3D reconstruction showing a low-riding basilar bifurcation aneurysm (red arrow). The aneurysm was located 1 cm below the dorsum sellae (yellow arrow). Note the upward trajectory of the posterior cerebral arteries, which indicates that the basilar artery has a low bifurcation

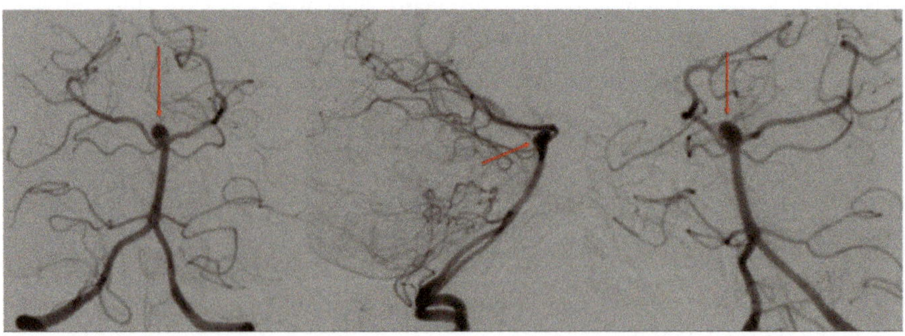

Fig. 10.5 Digital subtraction angiography of the vertebral artery. Anteroposterior, lateral, and oblique projections, respectively, showing an anterior projecting aneurysm originating at the basilar bifurcation, with the dome oriented to the right side

The base of the pons has two surfaces: an anterior and a lateral surface. Each surface is related to one cranial base cistern and may contain arteries and cranial nerves.

10.2.1 Anterior Surface: Cisternal, Vascular, and Cranial Nerve Relationships

The anterior surface is in direct contact with the prepontine cistern, which contains the basilar artery, proximal aspects of the superior cerebellar artery, the anterior cerebellar artery, long- and short-perforating branches, anteromedial pontomesencephalic vein, and transverse pontine veins. Superiorly it continues with the interpeduncular cistern, inferiorly with the premedullary cistern, and laterally with the cerebellopontine cistern.

10.2.2 Lateral Surface: Cisternal, Vascular, and Cranial Nerve Relationships

The cerebellum pontine cistern is in direct contact with the lateral surface of the base of the pons. It contains the trigeminal nerve in its superolateral part, the abducens nerve in its inferomedial part, and the facial and vestibulocochlear nerves in the inferolateral part of the cistern. Another neural structure that is an important surgical landmark is the flocculus of the cerebellum. It is located immediately behind the VII and VIII cranial nerves, and it helps to locate these nerves, guiding the surgeon to the internal acoustic meatus during surgery using a retrosigmoid approach. The arteries located in this cistern are the superior cerebellar artery, the anterior inferior cerebellar artery, and the perforating branches of the basilar artery. The superior cerebellar artery can bifurcate into its superior and inferior trunks, either in the prepontine cistern or in the pontomesencephalic cistern. It courses below the trochlear nerve and above the trigeminal nerve before entering the cerebellomesencephalic fissure. Long circumflex arteries that enter the cerebellomesencephalic fissure can arise from the superior cerebellar artery. The anterior inferior cerebellar artery usually bifurcates into the superior and inferior trunks in the cerebellomesencephalic cistern to supply the petrosal surface of the cerebellum. The main veins related to this cistern are the anterior hemispheric veins, the vein of the cerebellopontine fissure, the vein of the middle cerebellar peduncle, the transverse pontine veins, the lateral pontomesencephalic vein, the vein of the pontomesencephalic sulcus, and the pontotrigeminal vein. These veins usually join to form the superior petrosal vein; in order to widen the surgical corridor during a supracerebellar infratentorial approach, this vein and the precentrocerebellar vein can be coagulated and divided if collateral venous drainage is present.

Now that we know where to access, the question is *how are we going to get there safely and achieve our surgical goals*? To answer this question, we need to be familiar with the different approaches to the prepontine cistern and the anterior surface of the pons. Several approaches have been described to access the superior aspect of the pons. From an anterior perspective, we can access this region using a trans-Sylvian or a temporopolar approach through a pretemporal or an orbitozygomatic craniotomy. If a lateral approach is required, a subtemporal approach can be performed through a temporal or pretemporal craniotomy. To access the inferior aspect of the pons from an anteromedial perspective between the trigeminal nerves, the approach that offers the best view is the presigmoid. If the pathology is located lateral to the trigeminal nerves, the retrosigmoid approach offers the best view.

In this case, we selected a pretemporal transcavernous approach (Fig. 10.6) in order to have a wide exposure of the low basilar bifurcation and related vessels. By selecting this approach, we had the oculomotor nerve in the center of the surgical field. Therefore, we were able to work around the nerve and open the roof of the cavernous sinus to widen the surgical window (Fig. 10.10). It is especially important to maintain a clear surgical field, a wide exposure of the surrounding surgical anatomy, and to maintain a methodical microsurgical technique throughout the entire surgery (Figs. 10.6, 10.7, 10.8, 10.9, 10.10, 10.11, and 10.12).

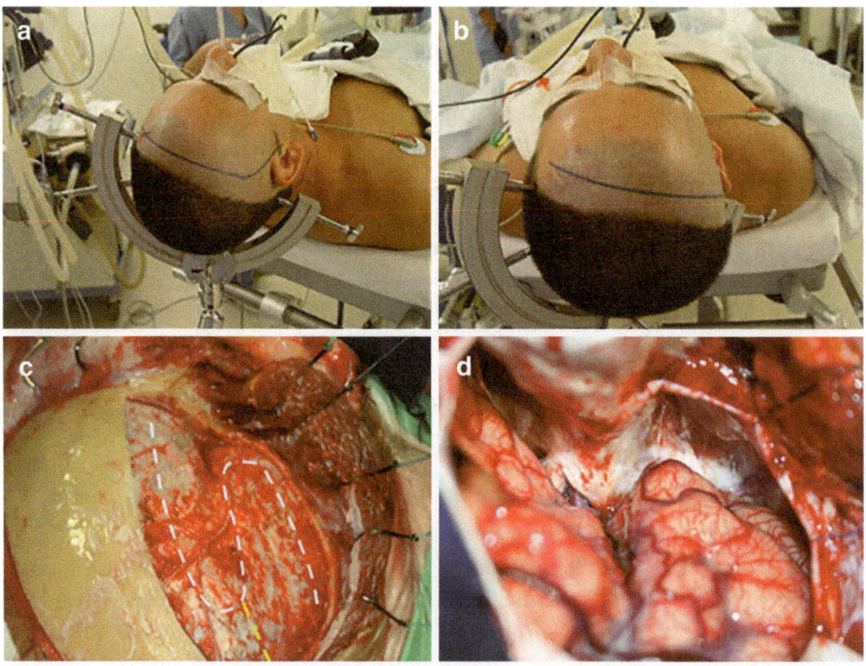

Fig. 10.6 Intraoperative photographs of the pretemporal approach. The patient positioning and the planned skin incision marked with a pen, (**a**) lateral view and (**b**) superior view. A pretemporal craniotomy and dural incision were performed. Note the drilling of the lateral wall and roof of the orbit and the exposure of the floor of the middle fossa, before opening of the dura (**c**) and the wide exposure to the anterior portion of the temporal lobe (**d**)

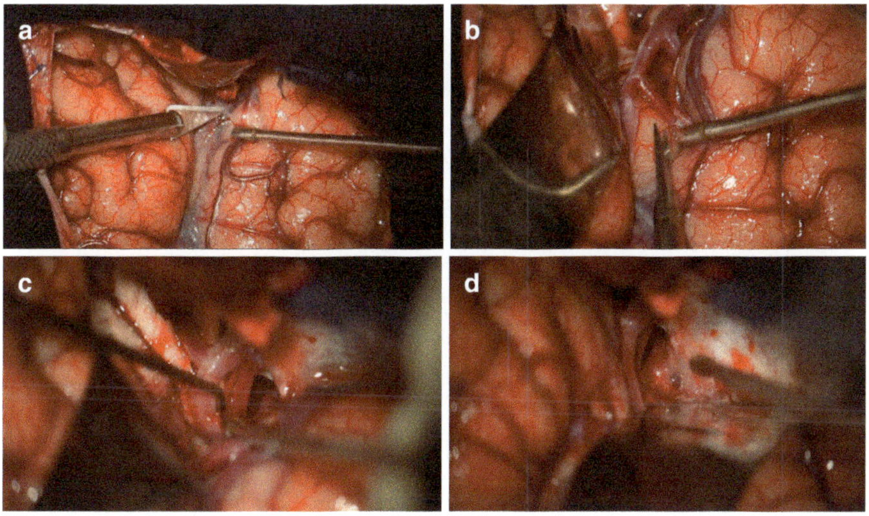

Fig. 10.7 Intraoperative photographs during microsurgical dissection. (**a**) The Sylvian fissure was opened, starting from the pars triangularis. (**b**, **c**) Complete dissection of the Sylvian fissure and opening of the basal cisterns was started from the optic cistern and then the carotid cistern. The anterior choroidal artery and the posterior communicating artery were exposed. (**d**) The Liliequist membrane attachment to the oculomotor nerve and tentorial free edge was exposed well

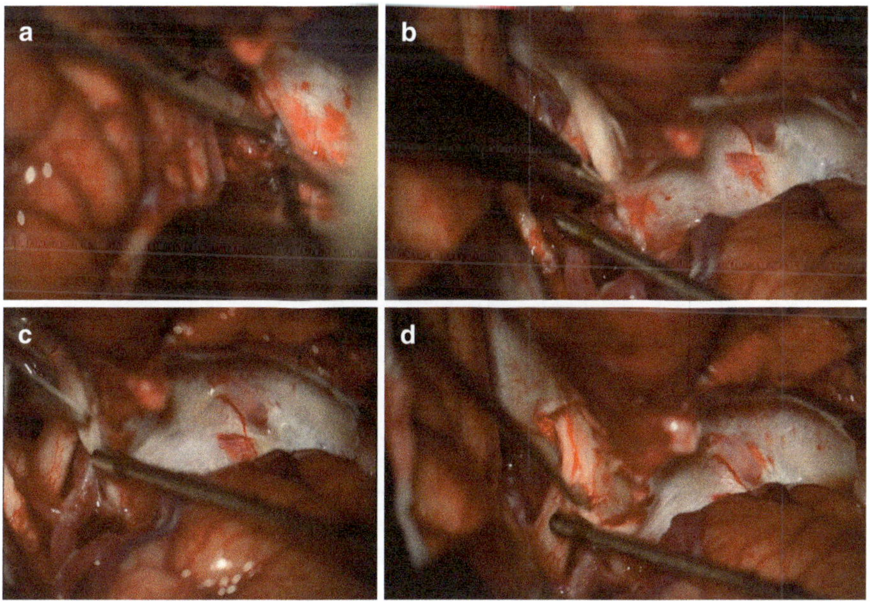

Fig. 10.8 Intraoperative photographs during microsurgical dissection. The Liliequist membrane was cut to access the prepontine cistern (**a**). Intradural clinoidectomy was performed: first, the dura was coagulated (**b**); second, the dura was cut parallel, perpendicular, and below the optic nerve (**c**); third, the dura was released using a round micro-dissector (**d**)

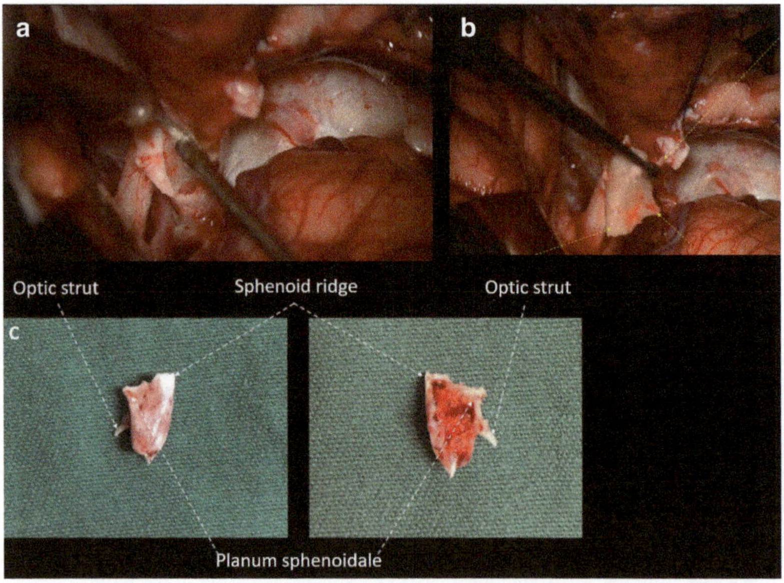

Fig. 10.9 Intraoperative photographs during anterior clinoidectomy. (**a**, **b**) The clinoid was detached by drilling its three points of attachment (yellow arrows): the sphenoid ridge, the planum sphenoidale, and the optic strut. (**c**) Panoramic view of the anterior clinoid process, resected in one piece

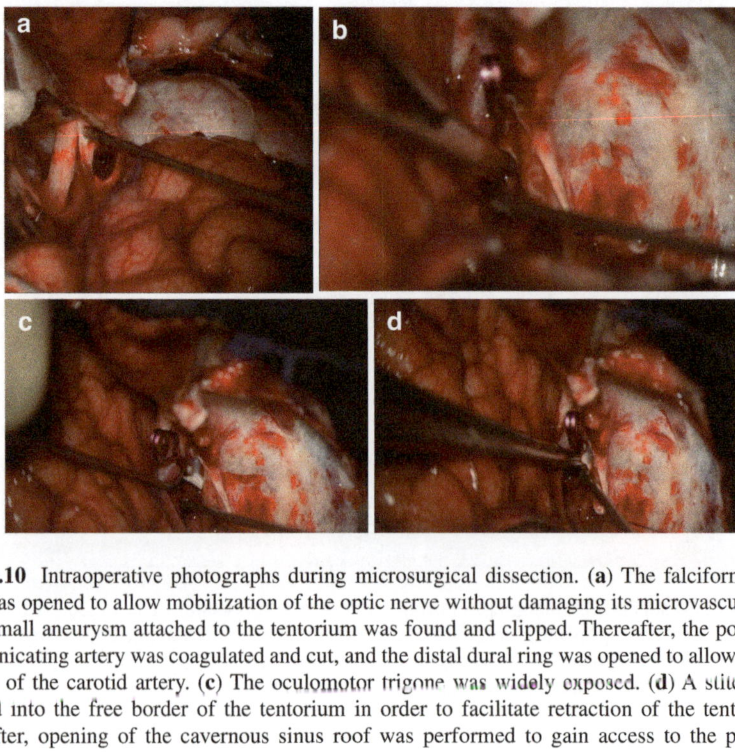

Fig. 10.10 Intraoperative photographs during microsurgical dissection. (**a**) The falciform ligament was opened to allow mobilization of the optic nerve without damaging its microvasculature. (**b**) A small aneurysm attached to the tentorium was found and clipped. Thereafter, the posterior communicating artery was coagulated and cut, and the distal dural ring was opened to allow mobilization of the carotid artery. (**c**) The oculomotor trigone was widely exposed. (**d**) A stitch was inserted into the free border of the tentorium in order to facilitate retraction of the tentorium. Thereafter, opening of the cavernous sinus roof was performed to gain access to the prepontine cistern

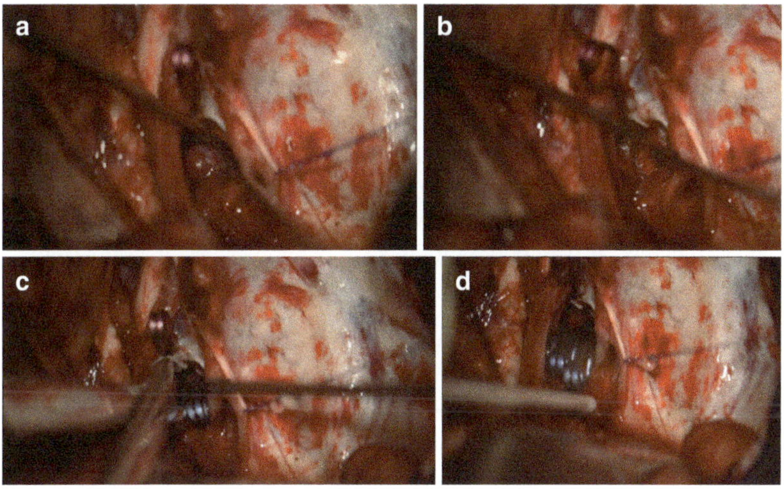

Fig. 10.11 Intraoperative photographs during the aneurysm clipping. (**a**) The aneurysm dome was exposed adjacent to the ipsilateral posterior cerebral artery and the superior cerebellar artery. (**b**) The thalamoperforating arteries were exposed and dissected from the neck of the aneurysm. (**c**) The aneurysm was successfully clipped. (**d**) The flow of the distal vessels was corroborated with Doppler ultrasound

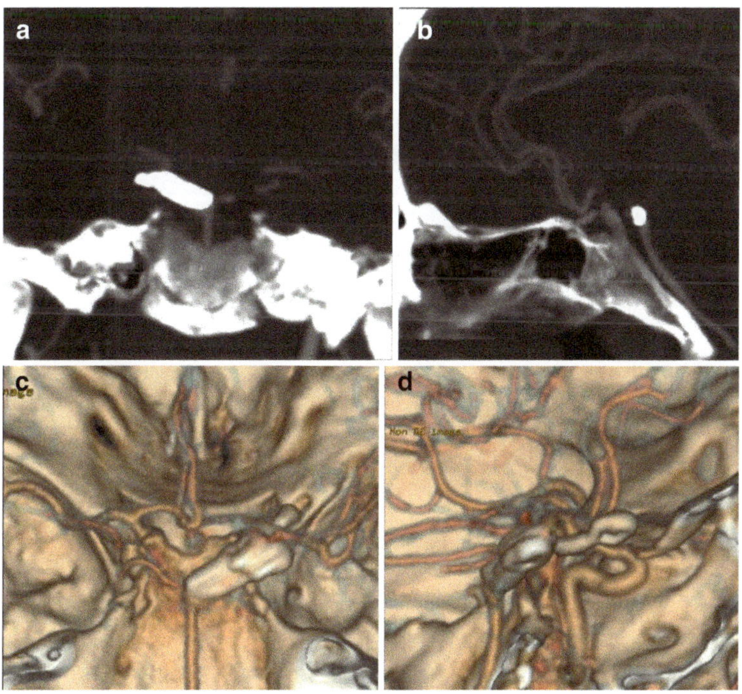

Fig. 10.12 Postoperative computed tomography angiography scan. (**a**, **b**) Coronal and sagittal projections showing adequate placement of the surgical clips and patency of the distal branches. (**c**, **d**) Arterial 3D reconstruction showing complete occlusion of the aneurysm

10.3 Illustrative Case 2

A 46-year-old female patient presented with symptoms of headache for the last 3 days and dizziness, without any other deficit. The patient underwent investigation, and an MRI scan was performed (Fig. 10.13).

First, we need to be sure *where the lesion is located?* At this point, we have studied the cisternal, vascular, and neural relations with the base of the pons, but not the internal architecture. It is important to now study the intrinsic anatomy of the pons (Fig. 10.14).

The base of the pons is composed of longitudinal and transverse fibers. The longitudinal fibers comprise of, from anterior to posterior, the corticospinal, corticopontine, and corticonuclear tracts. The most representative of these tracts is the corticospinal tract, which carries motor information from the precentral cortex to the spinal cord. The transverse fibers of the base are composed of pontocerebellar fibers that originate in the pontine nuclei to form the middle cerebellar peduncle and end in the cerebellar cortex.

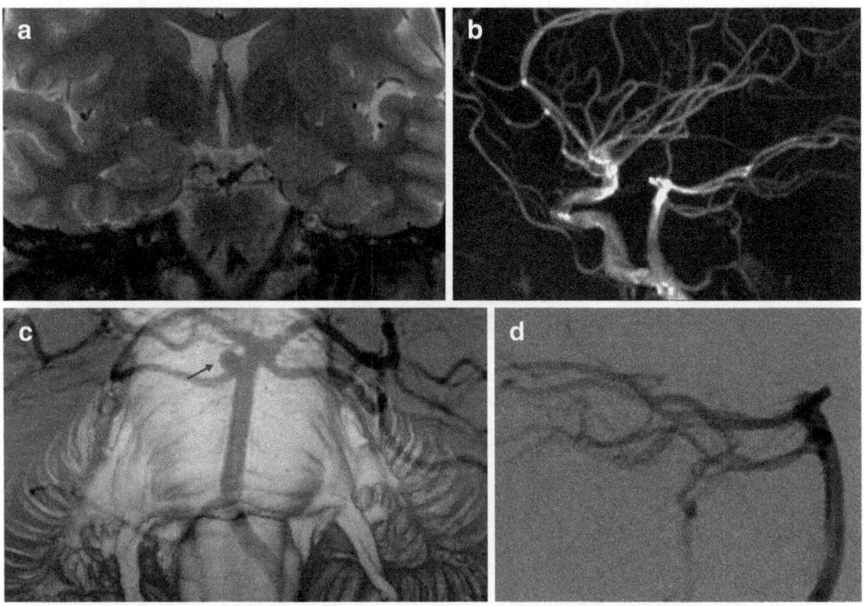

Fig. 10.13 Initial MRI and angiography. (**a**) A coronal T2-weighted image at the level of the anterior pons, showing a flow void in the origin of the right superior cerebellar artery. (**b**) Arterial 3D reconstruction showing a hyperintense image compatible with a superior cerebellar artery aneurysm. (**c**) Anteroposterior digital angiography of the vertebrobasilar system with an anatomical specimen to show the relationship of the aneurysm to the base of the pons. (**d**) Lateral digital subtraction angiography, which confirmed the presence of a 3-mm aneurysm

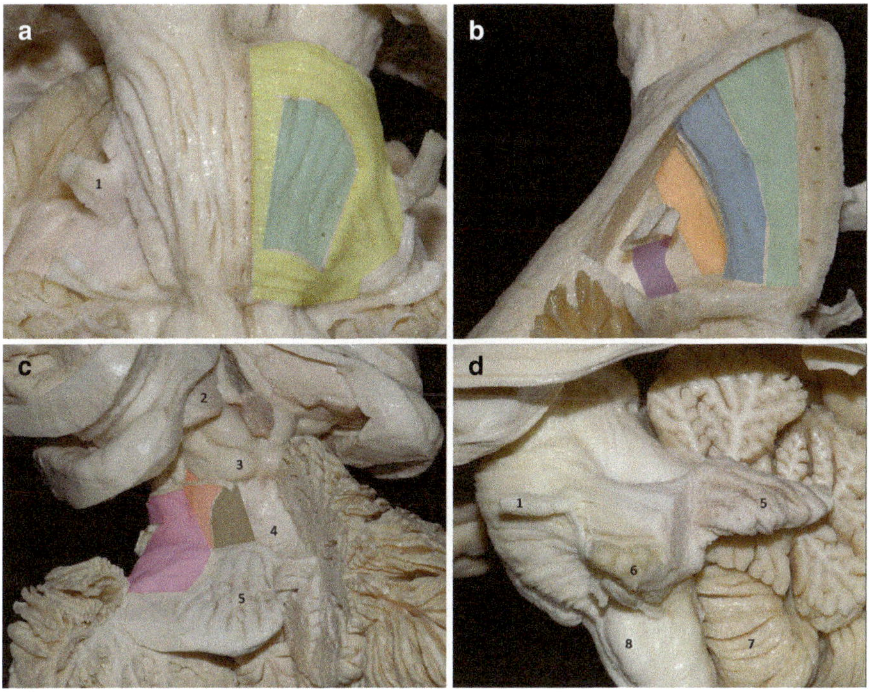

Fig. 10.14 Intrinsic anatomy of the pons. (**a**) On the anterior surface, the transverse fibers (yellow) of the pons are located anterior, between, and posterior to the corticospinal tract (green). (**b**) On the lateral surface, the base of the pons is located anterior to the medial lemniscus (orange). The corticopontine and corticonuclear fibers (blue) are located posteriorly and mixed with the corticospinal tract. (**c**) The posterolateral view after the left cerebellar hemisphere has been removed. The fibers of the dentate nucleus form the superior cerebellar peduncle and decussate at the level of the inferior colliculi of the mesencephalon, before entering the red nucleus. The trochlear nerve is one of the most important landmarks of the posterior fossa: it divides the lateral mesencephalic sulcus from the interpeduncular sulcus and the quadrigeminal plate from the superior cerebellar peduncle and superior medullary vellum. Medial to the interpeduncular sulcus is the superior cerebellar peduncle (brown), and above its anterior part, the lateral lemniscus (red) can be seen. Lateral to the interpeduncular sulcus is the middle cerebellar peduncle, from which the trigeminal nerve exits the pons. (**d**) The lateral surface, showing the relationship between the dentate nucleus and amygdala, cerebellar peduncles, and trigeminal nerve and the flocculus serving as an important landmark to find the vestibulocochlear nerve and the internal acoustic meatus. (1) Trigeminal nerve, (2) pulvinar, (3) inferior colliculi, (4) superior medullary vellum, (5) dentate nucleus, (6) flocculus, (7) amygdala, (8) inferior olive

Now, the next question arises. *How is it going to be possible to access the lesion?*

In this case, a pretemporal craniotomy is most suitable due to the anterosuperior location of the aneurysm immediately below the third nerve. With a pretemporal craniotomy, the third nerve is placed in the center of the visual field; therefore, the aneurysm can be fully exposed (Figs. 10.15 and 10.16).

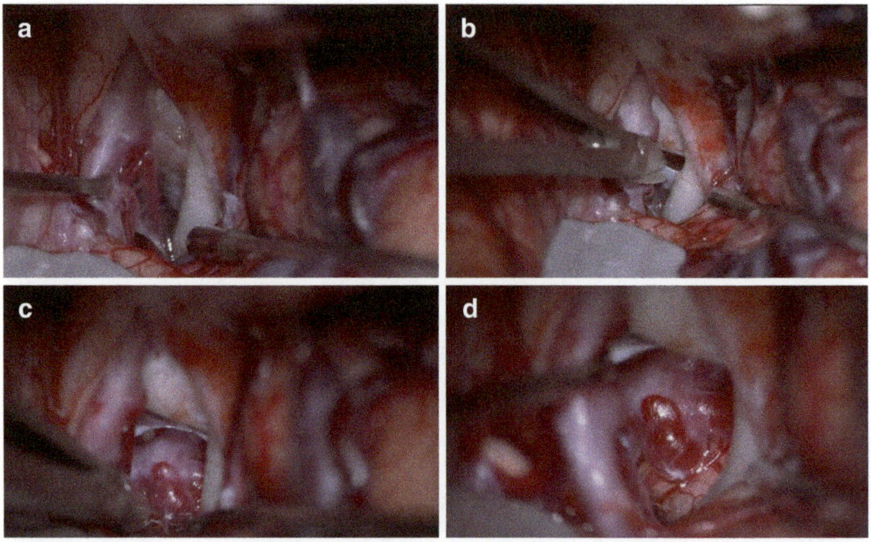

Fig. 10.15 Intraoperative photographs. (**a**) The posterior communicating artery and the anterior choroidal artery have been exposed. A widely opened carotid oculomotor window can be appreciated. (**b**) The Liliequist membrane was cut to access the prepontine cistern. (**c**, **d**) The aneurysm is widely dissected and exposed

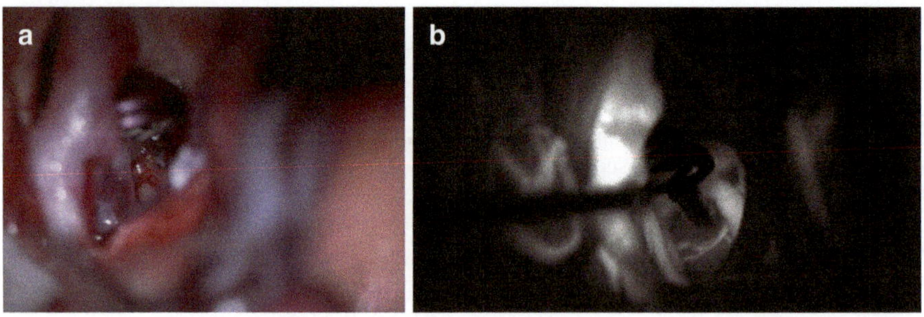

Fig. 10.16 Intraoperative photographs. (**a**) The final view after the aneurysm has been clipped. (**b**) Indocyanine green video angiography was performed to show patency of the superior cerebellar artery and the posterior cerebellar artery

10.4 Illustrative Case 3

A 25-year-old female patient presented with a clinical history of headaches, right central paralysis, and a sudden loss of consciousness. The patient underwent an MRI scan (Fig. 10.17) for investigation of the underlying pathology.

After having discussed the anatomy and surgical nuances in relation to the anterior pons, we now focus on the posterior aspect of the pons, the tegmentum.

What do we have to know regarding the extrinsic surgical anatomy of the pons tegmentum, before operating on this case?

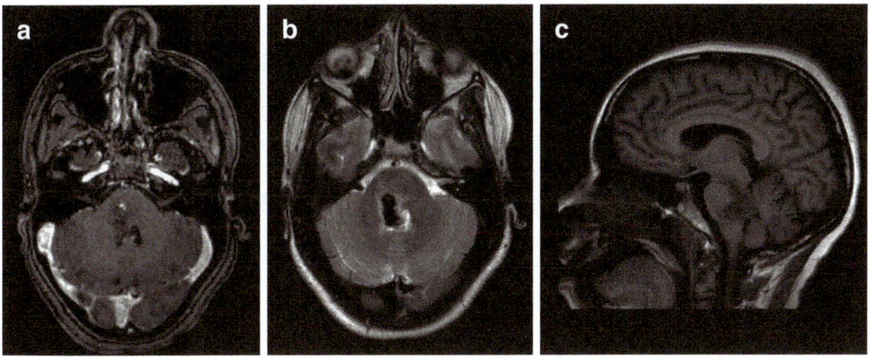

Fig. 10.17 Illustrative case 1. Initial brain magnetic resonance imaging. (**a**) An axial T1-weighted post-contrast image and (**b**) T2-weighted image, showing a mixed lesion with hypo- and hyperintense components, located in the tegmentum of the pons. (**c**) A sagittal T1-weighted image showing edema in the tegmentum of the pons resulting in partial occlusion of the aqueduct. These findings were highly suggestive of a pontine cavernoma

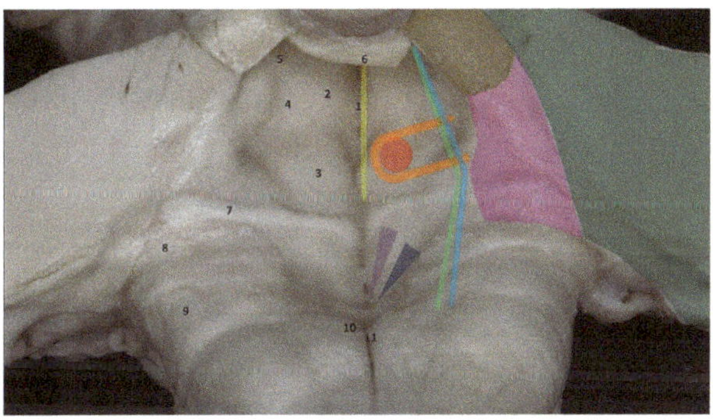

Fig. 10.18 Surgical perspective of the ventricular surface of the pons tegmentum. On the left side, the surgical landmarks have been exposed, and on the right side, the main fiber tracts and nuclei are shown, as follows: medial longitudinal fascicle (yellow); central tegmental tract (green); trigeminal mesencephalic and spinal tract (light blue); superior, inferior, and middle cerebellar peduncles (brown, pink, and dark green, respectively); facial nerve (orange); abducens nucleus (red); hypoglossal nucleus (purple); and vagal nucleus (dark blue). (1) median sulcus, (2) medial eminence, (3) facial colliculus, (4) sulcus limitans, (5) locus coeruleus, (6) superior medullary velum, (7) medullary striae, (8) vestibular area, (9) inferior cerebellar peduncle, (10) area postrema, (11) obex

The tegmentum of the pons is closely related to the fourth ventricle and forms the upper part of it. Its upper limit is the superior medullary velum that is at the same level as the fourth nerve and the inferior border of the inferior colliculi. The inferior limit is at the level of the lateral recess above the medullary striae (these medullary striae can be composed of 1–4 striae). It is important to note that the medullary stria is an important surgical landmark because the facial colliculus is sited in the tegmentum of the pons. Therefore, we can localize the suprafacial and infrafacial entry zones to safely access the floor of the fourth ventricle (Fig. 10.18).

The ventricular surface of the tegmentum of the pons is composed of, from medial to lateral, the median sulcus, medial eminence, sulcus limitans, and vestibular area.

However, this knowledge alone is insufficient to safely perform surgery. *It is critical to know the intrinsic anatomy of the pons tegmentum*; this means being able to identify all the fiber tracts surrounding the lesion (Figs. 10.18 and 10.19).

The pontine segment of the fourth ventricle floor has a triangular shape, bounded inferiorly by the medullary striae. In the superficial plane, the medial longitudinal fasciculus is next to the middle and behind the tectospinal tract, the abducens nucleus surrounded by the inferior and superior aspects of the intrapontine segment of the facial nerve located inside the facial colliculus, and laterally the central tegmental tract and the trigeminal mesencephalic tract can be found deep to the sulcus limitans. In the deep plane, the medial lemniscus is found anteriorly to the previously described fascicles. Lateral to the medial lemniscus, from medial to lateral, the spinothalamic tract and the lateral lemniscus are found. It is important to remember that only the superior and inferior cerebellar peduncles form the lateral wall of the fourth ventricle (Fig. 10.19).

It is especially important not only to be aware of the existing tracts but also to know the exact spatial relationship between them. Therefore, when one has to manage a lesion in this region, it is important to know exactly which tracts are surrounding the lesion to be treated.

At this point, we know the intrinsic and extrinsic anatomies of the pons. Therefore, we can choose less eloquent areas in order to access the pons surgically.

So, *what are the safe entry points to the pons?*

Seven safe entry zones have been described to safely access the pons. Surgeons must be aware of the surrounding structures and potential deficits if the deep tracts and nuclei are injured (Table 10.1; Fig. 10.20).

Therefore, considering the relationship of the cavernoma with the deep tracts (Fig. 10.19), a suprafacial approach was performed through a telovelar route (Fig. 10.21).

The postoperative imaging revealed complete resection of the cavernoma (Fig. 10.22).

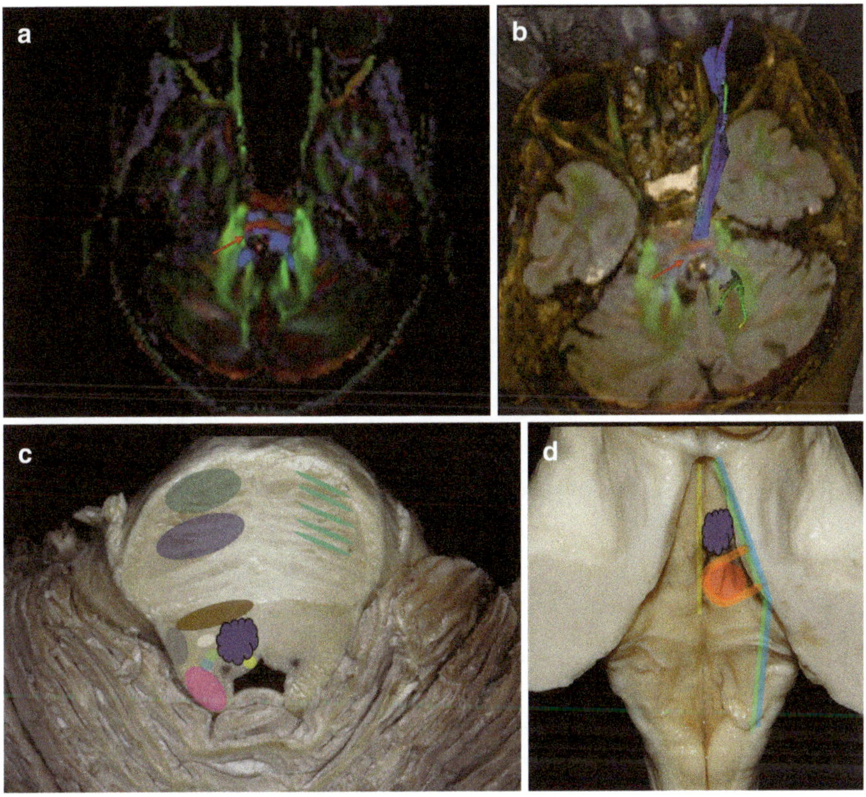

Fig. 10.19 MRI fiber tractography and anatomical specimen correlation of the lesion with the surrounding tracts. (**a**, **b**) The axial tractography at the level of the upper pons shows the lesion displacing the right medial lemniscus (red arrow). The corticospinal tract (blue descending fibers) is located anterior to the posterior transverse fibers (red) and is preserved. (**c**, **d**) The correlation of the lesion (dark purple cloud) with the surrounding tracts. The cavernoma is located in the pons tegmentum between the medial lemniscus (brown) and the medial longitudinal (yellow) fascicle, above the nucleus of the abducens (red) and the facial nerve (orange), and medial to the central tegmental tract (light green) and trigeminal mesencephalic tract (light blue). Transverse pontine fibers (turquoise), corticospinal tract (dark green), corticonuclear tract (dark blue), spinothalamic tract (white), lateral lemniscus (gray), and superior cerebellar peduncle (pink)

Table 10.1 Safe entry zones to the pons

Safe entry zone	Location	Suitable approaches	Pontine scope area	Main structures at risk	Potential deficits
Peritrigeminal	Between the origin of CN V and VII	Anterior transpetrous, Presigmoid, Retrosigmoid	Caudal ventrolateral	CST, trigeminal nuclei and spinal tract, CN VI and VII	Hemiparesis, trigeminal neuropathy, paresis of CN VI and VII
Lateral transpeduncular	MCP, anterior to the cerebellar peduncle-hemisphere junction	Retrosigmoid	Rostral ventrolateral	Intrapontine segment of CN V	Trigeminal neuropathy
Supratrigeminal	4 mm below the pontomesencephalic sulcus, at the exit level of CN III	Orbitozygomatic, Pretemporal transcavernous	Rostroventral	CST	Hemiparesis
Epitrigeminal	Above the apparent origin of CN V	Subtemporal transtentorial	Posterolateral	Intrapontine segment of CN V	Trigeminal neuropathy
Interpeduncular sulcus	Between MCP and SCP, extending 8.2 mm posterior to the LMS	Supracerebellar infratentorial, Subtemporal transtentorial	Posterolateral	SCP, LL	Extrapyramidal syndrome, hearing impairment
Suprafacial	Above facial colliculus	Suboccipital telovelar	Dorsorostral tegmentum	MLF, TST, CTT, CN VII, Abducens nuclei	Facial palsy, oculomotor disturbances
Infrafacial	Below facial colliculus	Suboccipital telovelar	Dorsocaudal tegmentum	TST, CTT, CN VII, Abducens nuclei	Facial palsy, oculomotor disturbances

Abbreviations: *CN* cranial nerve, *CST* corticospinal tract, *MCP* middle cerebellar peduncle, *SCP* superior cerebellar peduncle, *LMS* lateral mesencephalic sulcus, *MLF* medial longitudinal fasciculus, *TST* tectospinal tract, *CTT* central tegmental tract

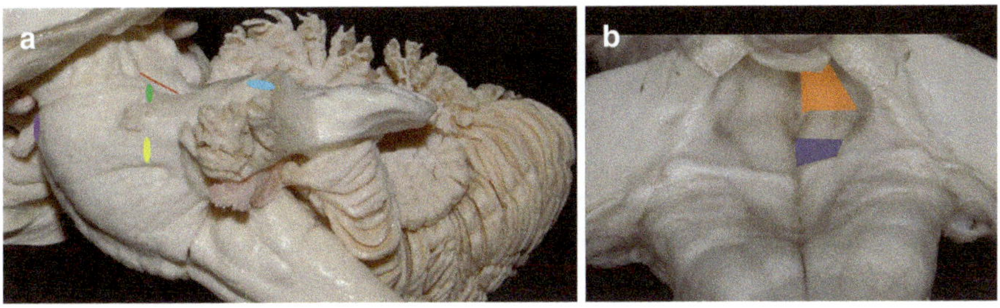

Fig. 10.20 Safe entry points to the pons. (**a**) Lateral view of the brainstem, supratrigeminal (purple), epitrigeminal (green), peritrigeminal (yellow), lateral transpeduncular (blue), interpeduncular sulcus (red). (**b**) View of the fourth ventricle floor: suprafacial (orange), infrafacial (dark blue)

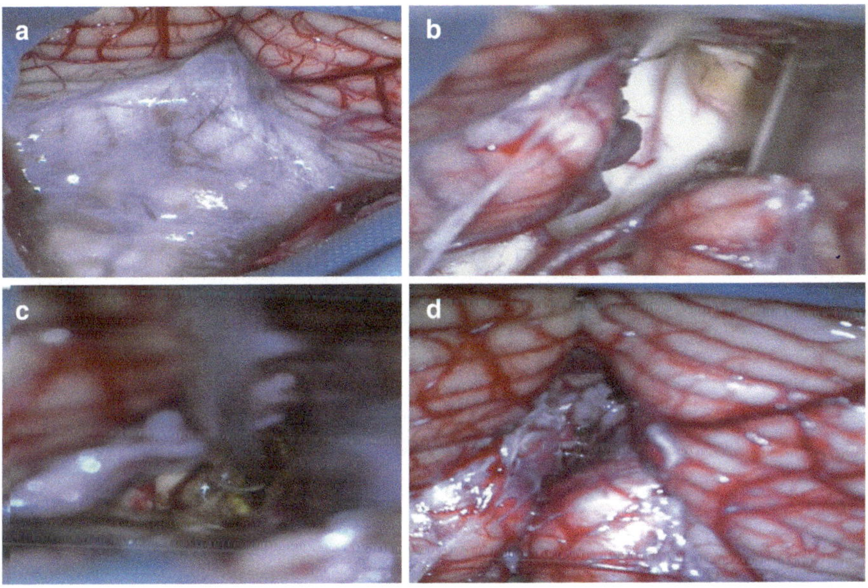

Fig. 10.21 Intraoperative photographs. (**a**) The surgical exposure of the cisterna magna and cerebellum. (**b**) The floor of the fourth ventricle has been exposed, and the hemosiderin can be seen on the suprafacial entry zone. (**c**) The cavernoma is resected. (**d**) The surgical view after resection of the cavernoma

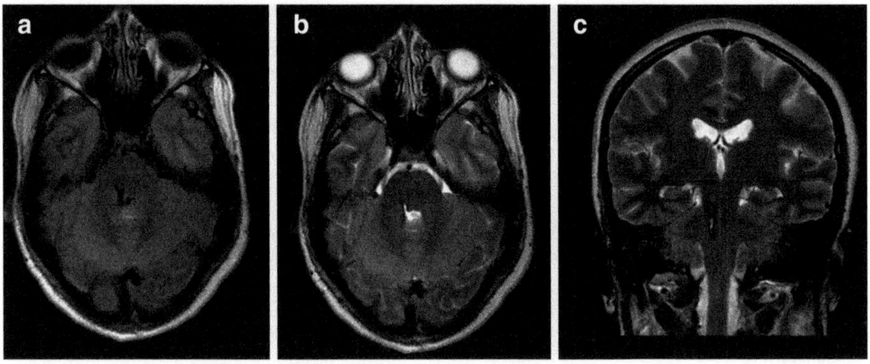

Fig. 10.22 Postoperative brain magnetic resonance imaging. (**a**) Axial T1- and (**b**) T2-weighted images showing the surgical trajectory. (**c**) A coronal T2-weighted image showing complete resection of the cavernoma

10.5 Illustrative Case 4

A 29-year-old woman presented with left-sided facial paresis that started in the preceding 10 days. The patient underwent an MRI scan, which demonstrated a lesion compatible with a cavernous malformation (Fig. 10.23a, b, d, e).

The cavernous malformation was situated in the inferior aspect of the pons tegmentum. Due to the age of the patient, promising prognosis, and a previous bleeding episode, it was decided to proceed to surgery. All surgical routes were studied, including the related fiber tracts (Fig. 10.23c, f). The best surgical route was determined to be the infrafacial corridor, due to the shortest distance of the cavernoma to the surface, and the absence of eloquent tracts in the long axis of the cavernoma. The safe entry zone was reached through a suboccipital craniotomy and telovelar approach (Fig. 10.24).

When operating in the region of the floor of the fourth ventricle, it is important to use electrophysiological monitoring of the related cranial nerves, particularly the facial, abducens, vestibulocochlear, and lower cranial nerves. Neuronavigation is also helpful and can be used to confirm the entry point as well as the extent of resection. Nevertheless, use of this technology does not substitute for good anatomical knowledge.

The cavernoma was completely resected (Fig. 10.25), and the patient was discharged without any additional deficit.

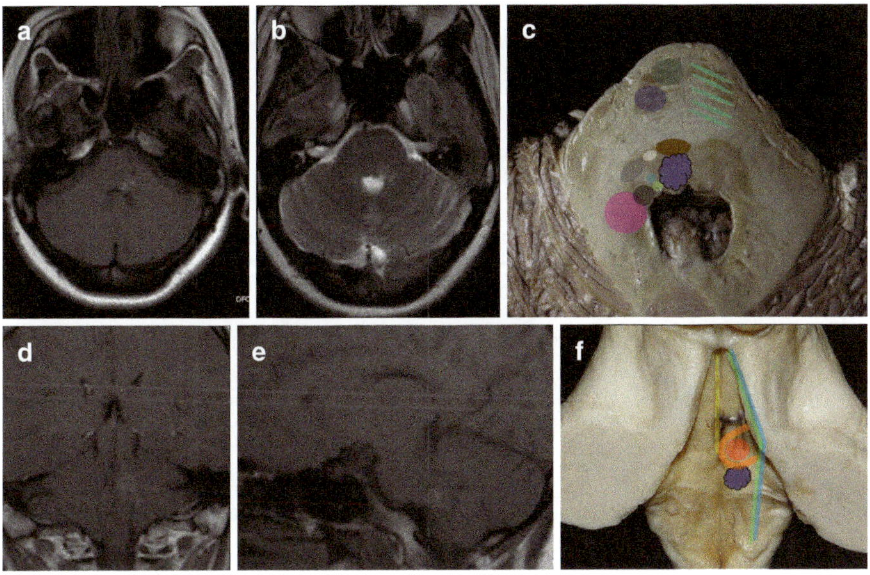

Fig. 10.23 Preoperative MRI and anatomical correlation. (**a**) An axial T1- and (**b**) T2-weighted image, showing a hypo- and hyperintense lesion in the lower pons tegmentum. (**c**) Anatomical correlation of the lesion (dark purple cloud) with the surrounding tracts. (**d**) A coronal T1-weighted post-contrast image and (**e**) sagittal T1-weighted image showing the lesion to be located in the floor of the fourth ventricle immediately below the facial colliculus. (**f**) Anatomical correlation of the lesion (dark purple cloud) and the related tracts of the fourth ventricle. Medial lemniscus (brown), medial longitudinal fascicle (yellow), nucleus of the abducens (red), facial nerve (orange), central tegmental tract (light green), and trigeminal mesencephalic tract (light blue). Transverse pontine fibers (turquoise), corticospinal tract (dark green), corticonuclear tract (dark blue), spinothalamic tract (white), lateral lemniscus (gray), vestibular nuclei (black), and the inferior cerebellar peduncle (pink)

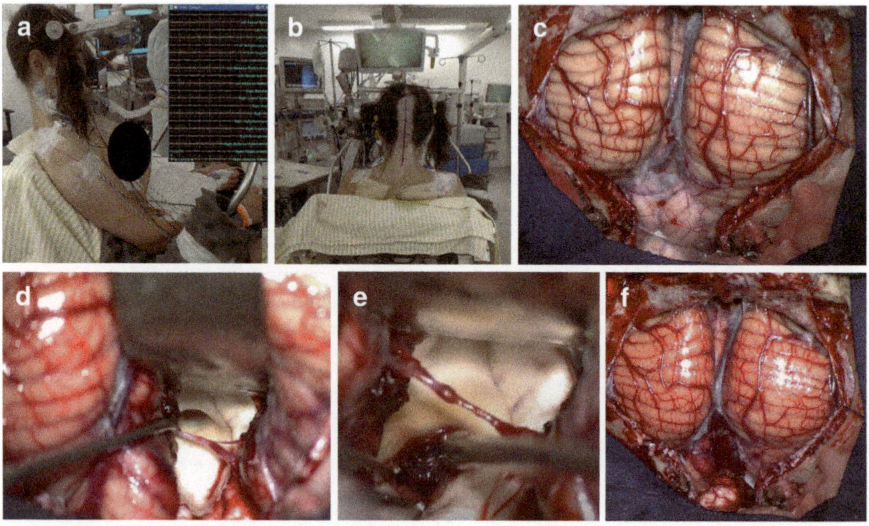

Fig. 10.24 Intraoperative photographs. (**a**, **b**) Patient positioning and planned incision for a telovelar approach through a median suboccipital craniotomy. Intraoperative monopolar stimulation (inset) of the floor of the fourth ventricle was used to confirm the absence of facial stimulation. (**c**) The suboccipital surface of the cerebellum was exposed widely, and the foramen magnum was opened. (**d**) The tela choroidea and inferior medullary velum were dissected. The right amygdala was retracted upwards to expose the floor of the fourth ventricle up to the level of the aqueduct. (**e**) The infrafacial entry zone was accessed to remove the cavernoma. (**f**) The postoperative surgical view

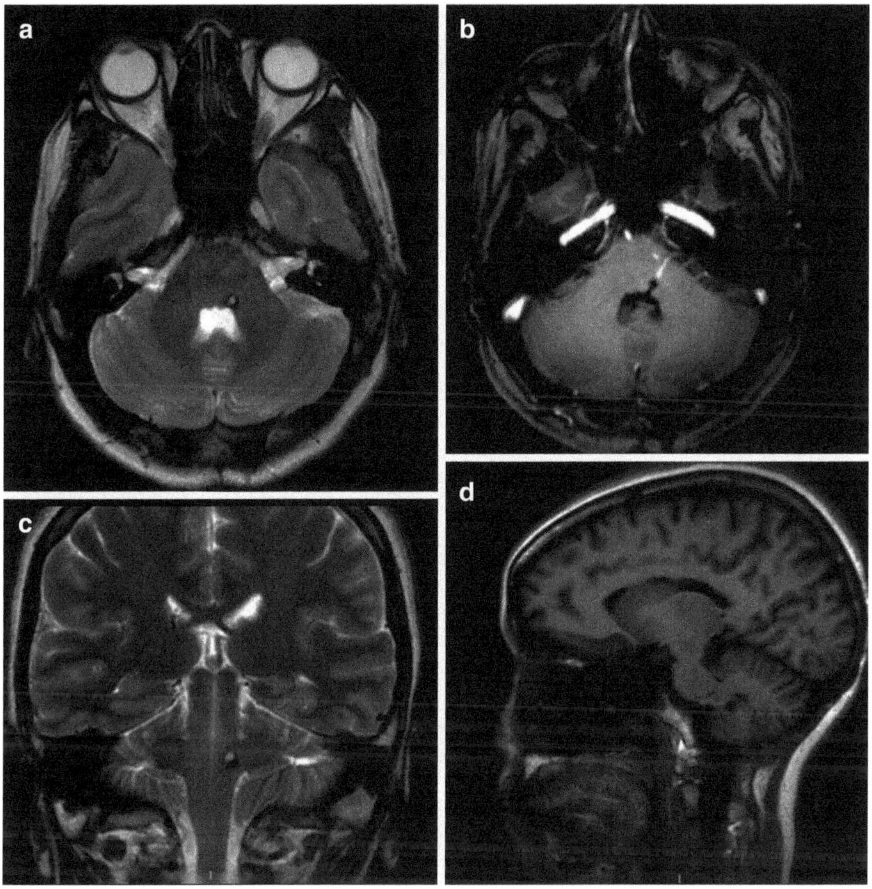

Fig. 10.25 Postoperative brain magnetic resonance imaging. (**a**) An axial T2-weighted image showing the surgical cavity. (**b**) An axial T1 weighted post-contrast image showing the preserved developmental venous anomaly. (**c**) A coronal T2-weighted image and (**d**) a sagittal T1-weighted image showing complete resection of the cavernous malformation

Suggested Bibliography

Akiyama O, Matsushima K, Nunez M, Matsuo S, Kondo A, Arai H, et al. Microsurgical anatomy and approaches around the lateral recess with special reference to entry into the pons. J Neurosurg. 2018;129(3):740.

Cavalheiro S, Serrato-Avila JL, Párraga RG, da Costa MDS, Nicácio JM, Rocha PR, et al. Interpeduncular Sulcus Approach to the Posterolateral Pons. World Neurosurg [Internet]. 2020 Jun 1 [cited 2020 Sep 6];138:e795–805. Available from: https://pubmed.ncbi.nlm.nih.gov/32217179.

Cavalheiro S, Yagmurlu K, da Costa MDS, Nicácio JM, Rodrigues TP, Chaddad-Neto F, et al. Surgical approaches for brainstem tumors in pediatric patients. Childs Nerv Syst. 2015;31(10):1815–40.

da Costa MDS, Braga VL, Yagmurlu K, Centeno RS, Cavalheiro S, Chaddad-Neto F. A technical guide for fiber tract dissection of the internal capsule. Turk Neurosurg. 2018;28(6):934–9.

Granados OS, da Costa MDS, Costa BL, González-Echeverría K, Paganelli SL, Caramanti RL, et al. Microsurgery for upper basilar tip aneurysm with intraoperative rupture: 3-dimensional operative video. Oper Neurosurg (Hagerstown, Md). 2019;16(2):43.

Párraga RG, Possatti LL, Alves RV, Ribas GC, Türe U, de Oliveira E. Microsurgical anatomy and internal architecture of the brainstem in 3D images: surgical considerations. J Neurosurg. 2015;124(5):1377–95.

Serrato-Avila JL, Paz Archila JA, Costa MDSD, Rocha PR, Marques SR, Moraes LOC, Cavalheiro S, Yağmurlu K, Lawton MT, Chaddad-Neto F. Microsurgical approaches to the cerebellar interpeduncular region: qualitative and quantitative analysis. J Neurosurg. 2021;22:1–14.

Yagmurlu K, Rhoton AL, Tanriover N, Bennett JA. Three-dimensional microsurgical anatomy and the safe entry zones of the brainstem. Neurosurgery. 2014;10 Suppl 4(12):602–19; discussion 619-20.

Chapter 11
Surgical Anatomy of the Medulla Oblongata

11.1 Introduction

The brainstem is routinely divided into three parts: the midbrain, pons, and medulla oblongata (MO). The MO, also called the *medulla*, is situated at the most inferior portion of the brainstem and connects the encephalon with the spinal cord. The MO has a conical shape with the base cranial, and its measurements are 27–30 mm long, 12–15 mm in anteroposterior diameter, 20–25 mm in transverse diameter at its upper extremity, and 10–12 mm in diameter at the lower extremity.

The medulla lies upon the basal portion of the occipital bone (clivus) and is ventral to the fourth ventricle and cerebellum. The upper portion is separated from the pons by the pontomedullary sulcus, and inferiorly, it is continuous with the spinal cord, from which it is separated by an imaginary line drawn just below the decussation of the pyramids and above the first cervical spinal nerve.

The anatomy of the MO is divided into external and internal configurations. The external configuration involves three surfaces: dorsal, ventral, and lateral. The internal configuration is divided into gray and white matter structures.

11.2 External Configuration

11.2.1 Ventral Surface

The anterior medial fissure is located at the midline of this surface; this fissure ends superiorly toward the pontomedullary sulcus in a triangularly shaped fossa called the *foramen cecum*. This area is known to have a very high concentration of

Supplementary Information The online version of this chapter (https://doi.org/10.1007/978-3-030-82747-2_11) contains supplementary material, which is available to authorized users.

F. Chaddad-Neto, M. D. Silva da Costa, *Microneuroanatomy and Surgery*,
https://doi.org/10.1007/978-3-030-82747-2_11

brainstem-perforating vessels. On each side of the fissure, there are two prominent columns, the pyramids, which carry the corticospinal tracts that descend from the cerebral cortex and end inferiorly with the decussation of the pyramids corresponding to the partial crossing of the corticospinal and cortico-bulbar tracts. This decussation has an extension between 6 and 8 mm (Fig. 11.1). The abducens cranial nerve (CN) has an apparent origin in the pontomedullary sulcus anterior to the pyramids. Lateral to the pyramids are two eminences known as the olives, overlying the lower olivary nuclei. The olives are delimited by two sulci, one medial or anterolateral sulcus (preolivary), from where the hypoglossal (XII) CN emerges and another lateral or posterolateral sulcus (postolivary) from where the glossopharyngeal (IX), vagus (X), and accessory (XI) CNs emerge (Fig. 11.1).

11.2.2 Lateral Surface

The lateral surface has two fossae in the upper portion. The *supraolivary fossa* is anterior and separates the medulla from the pons; the facial and intermediate (or Wrisberg) nerves emerge in this fossa. The *medullary fossa* is a narrow depression located behind the supraolivary fossa, where the vestibulocochlear CN emerges. The olives are located at the anterior portion of the lateral surface (Fig. 11.1).

11.2.3 Dorsal Surface

The dorsal surface of the medulla is divided into an upper portion, called the "open" portion, which represents the lower part of the floor of the fourth ventricle, and a lower portion, called the "closed" portion, which contains the rostral continuation of the central canal of the spinal cord. These two portions are delimited by the obex, i.e., the point where the fourth ventricle opens.

In the closed portion of the medulla, the midline is marked by a posterior median sulcus, similar to the spinal cord. In each half of the caudal part of the medulla, the dorsal columns and fasciculi gracilis and cuneatus continue rostrally from the spinal cord to their termination in the nuclei gracilis and cuneatus, the locations of which are marked by two small elevations, the gracile and cuneate tubercles, being the gracile medial and cuneate lateral separated by the posterior intermediate sulcus (Fig. 11.2).

The floor of the fourth ventricle, also called the *rhomboid fossa*, is formed by a depression of the dorsal surface of the pons and medulla. The lower triangle of the rhomboid fossa forms the open portion of the medulla. This triangle is a downwardly directed apex, named the calamus scriptorius (as its shape resembles that of a writing quill nib). The widest portion of this triangle is toward the pontomedullary

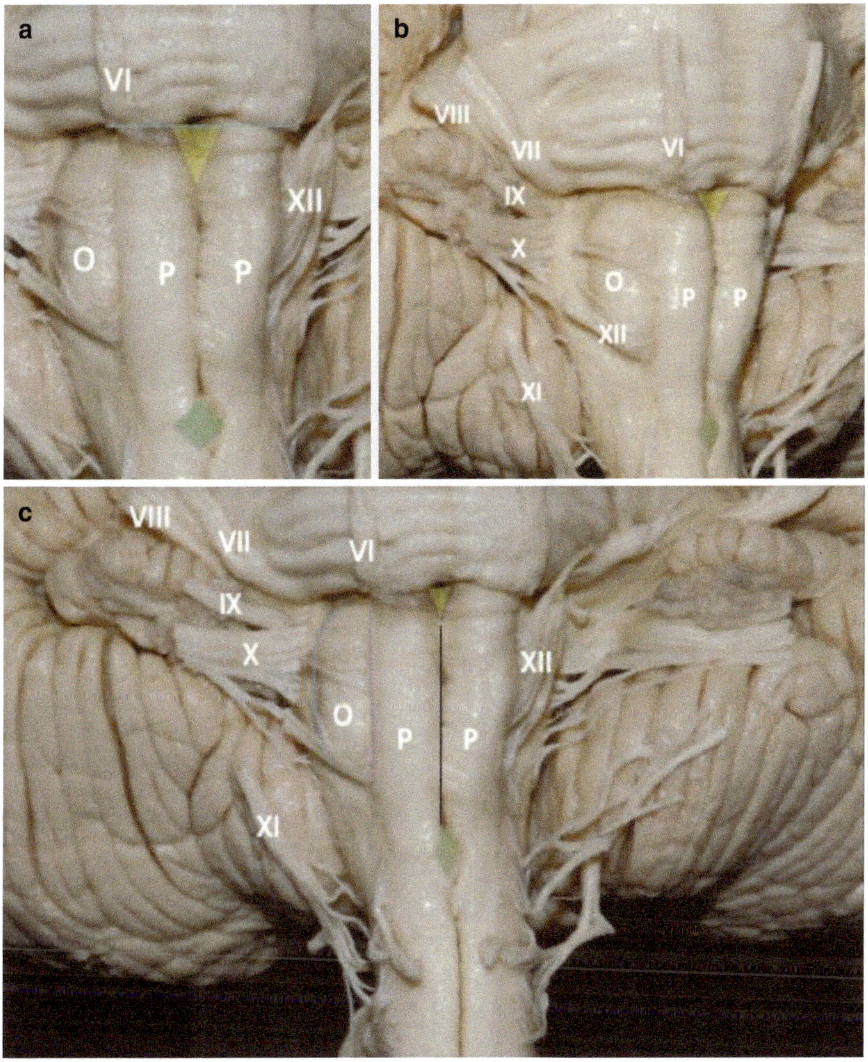

Fig. 11.1 (**a–c**) Anterior and lateral surfaces of the medulla oblongata. On the anterior surface of the medulla is the anterior medial fissure (black line), which is a continuation of the anterior medial fissure of the spinal cord. In the upper termination of the anterior medial fissure is the foramen caecum (yellow triangle), which is separated from the pons by the pontomedullary sulcus (sky blue line). The pyramids (P) contain the pyramidal tract (corticospinal tract) and end at the decussation of the pyramid (green diamond). Olives (O) elevated by underlying inferior olivary nucleus. The anterolateral sulcus or preolivary sulcus (purple line) contains the rootlets of the hypoglossal nerve (XII). The posterolateral sulcus or retroolivary sulcus (blue line) contains the rootlets of the glossopharyngeal (IX), vagus (X), and accessory (XI) nerves (images courtesy of Richard Gonzalo Párraga)

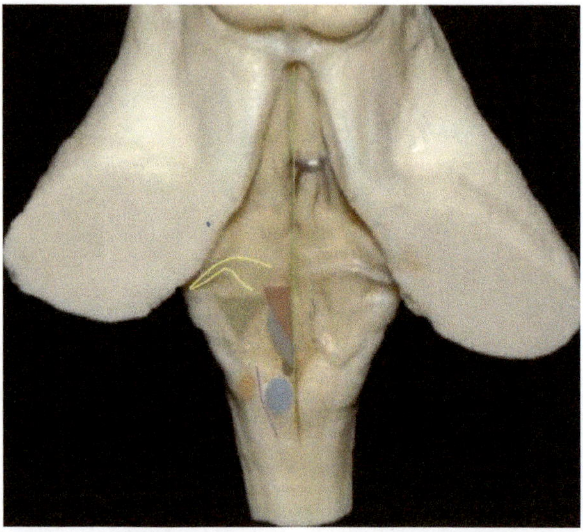

Fig. 11.2 Gray matter of the medulla oblongata. On the dorsum of the medulla (floor of the fourth ventricle), the median sulcus divides the floor longitudinally in the midline (green line). In the open portion, the hypoglossal and vagal nuclei and the area postrema are stacked one above the other in the lower part of the floor to give the configuration of a pen nib, and thus, the area is referred to as the *calamus scriptorius*. Lateral to these nuclei are the terminal nuclei of the vestibular nerve in the vestibular area. In the closed portion of the medulla, the gracilis (blue medial circle) and cuneatus (orange lateral circle) nuclei are separated by the posterior intermedial sulcus (purple line). The stria medullaris crosses the floor at the level of the lateral recess (yellow lines). (Images courtesy of Richard Gonzalo Párraga)

junction, where the lateral recess is located on each side, and the lateral recess ends in a small opening called Luschka's foramen. The cerebrospinal fluid passes from the fourth ventricle into the subarachnoid space that surrounds the brain through this foramen.

This lower triangle is divided into two by the posterior medial sulcus. Each half of the triangle has three areas: The *hypoglossal trigone* is superior and medial and corresponds to the internal triangular white eminence with a superior base that covers the hypoglossal nucleus (Fig. 11.2). The *vestibular area* is superior and lateral and corresponds to the external triangular white eminence with the superior base formed by the acoustic tubercle (Fig. 11.2). Beneath these eminences, between the hypoglossal trigone and the lower part of the area acustica, is a triangular dark field with an inferior base, the *vagal trigone*, which corresponds to the sensory nuclei of the vagus and glossopharyngeal nerves. The area postrema is a smaller tongue-shaped eminence found below and medial to the trigone of the vagus, between the funiculus separans, a narrow translucent ridge, and the gracile nucleus. The striae medullares are white strands that cross the acoustic area and the medial eminence, which form a portion of the cochlear division of the acoustic nerve and disappear into the median sulcus (Fig. 11.2).

11.3 Internal Configuration

The MO represents a transition between the spinal cord and the encephalon, and therefore, its internal configuration resembles the internal configuration of the spinal cord, but it has some typical changes characteristic of this region of the brainstem in both gray and white matters.

11.3.1 Gray Matter

During embryological development, the caudal portion of the neural tube gives rise to the spinal cord where the alar and basal plates are formed. The alar plate gives rise to *sensory neurons* and the basal plate to *motor neurons*. In the open portion of the MO (because of the presence of the fourth ventricle), the alar plate moves laterally, leaving the basal plate medially, being separated by the limiting sulcus, and then the sensory nuclei are found lateral to the limiting sulcus, and the motor nuclei are located medial to the limiting sulcus. Among the *motor nuclei* is the *inferior olivary nucleus*, which is the largest structure of the olivary complex. Its structure is laminar with medial concavity and has connections with the red nucleus and cerebellum through cerebellar-olivary fibers that pass through the inferior cerebellar peduncle. This nucleus influences the coordination of previously learned movements. The medial and dorsal accessory olivary nuclei form the inferior olivary complex with the inferior olivary nucleus and share the same connections and functions with the inferior olivary nucleus.

The *hypoglossal nucleus* inside the hypoglossal trigone gives rise to the hypoglossal nerve, which is directed ventrally and emerges through the preolivary sulcus. Its function is to innervate the extrinsic and intrinsic muscles of the tongue (general somatic efferent) (Fig. 11.2).

The *dorsal nucleus of the vagus* is lateral to the nucleus of the hypoglossus nerve. It gives rise to some of the fibers of the vagus nerve. These are preganglionic parasympathetic fibers that are directed and emerge through the retroolivary sulcus. It functions to provide parasympathetic innervation to all organs of the thorax and part of the abdominal organs up to the left colic flexure (general visceral efferent) (Fig. 11.2).

Dorsal to the inferior olivary nucleus is the *ambiguous nucleus*, which innervates the muscles derived from the third and fourth branchial arches, which are the palate, pharynx, larynx, and upper third of the esophagus (voluntary portion) through the glossopharyngeal (IX), vagus (X), and accessory (XI) CNs. This nucleus (special visceral efferent) gives rise to the cranial portion of the accessory nerve.

One of the *sensory nuclei* of the alar plate is the *solitary nucleus*, which is located lateral to the dorsal nucleus of the vagus in relation to the floor of the fourth ventricle. Its characteristic shape comprises a column of neurons that are divided into a cephalic and a caudal portion. The cephalic portion receives gustatory information through the fibers of the ganglionic neurons of the three nerves; the geniculate ganglion of the facial nerve, which collects gustatory signals from the anterior 2/3 of the tongue; the glossopharyngeal ganglion, which receives gustatory signals from the posterior 1/3 of the tongue; and the vagus nerve, which collects gustatory signals from the epiglottis. The caudal portion receives information on visceral sensitivity (distention or pressure) from the organs of the thorax and abdomen through the vagus nerve and from the two carotid receptors, i.e., the carotid sinus, a baroreceptor perceiving changes in pressure, and the carotid body, a chemoreceptor perceiving changes in pH; both are innervated by fibers of the glossopharyngeal nerve.

The *vestibular nuclei* (inferior, medial, and lateral) are found at the junction of the MO and the pons forming the vestibular area in the lateral portion of the rhomboid fossa (Fig. 11.2). These nuclei receive fibers from the vestibular part of the VIII CN through the vestibular ganglion (scarp ganglion), which provides information on acceleration and deceleration in an angular and linear manner (horizontal and vertical). Lateral to the vestibular nuclei and in relation to the inferior cerebellar peduncle are the cochlear nuclei, one ventral and the other dorsal to the inferior cerebellar peduncle. They receive auditory information from the cochlear ganglion.

In the lateral portion of the medulla, the *trigeminal spinal nucleus* extends to the first cervical segments of the spinal cord, in the segment two of the lamina in the gelatinous substance. This nucleus has three portions, oral, interpolar, and caudal, which receive fibers carrying signals of sensation of pain and temperature from the head through the trigeminal (V), facial (VII), glossopharyngeal (IX), and vagus (X) CNs.

In the most superficial part of the pyramids, the *arcuate nucleus* forms the medullary or bulbar striae that cross in the lower triangle of the rhomboid fossa. These nuclei are the extension of the pontine nuclei; they receive fibers from the corticospinal tract and send their axons through the anterior external arcuate fibers and medullary striae to the cerebellum via the inferior cerebellar peduncle. These nuclei have chemosensitivity function and help control the breathing rate.

In the closed portion of the medulla, dorsal to the ependymal duct, are the *nuclei of the gracilis* (medial) and *cuneiform* (lateral) fasciculi (Fig. 11.2). These receive sensory information of vibration, fine touch, conscious proprioception, weight, and discrimination of two points. The gracilis nucleus carries information from the lower portion of the body (lower extremities, pelvis, abdomen, and lower part of the thorax), and the cuneate nucleus carries information from the upper part (upper thorax, upper extremities, and neck) of the body. Lateral to the cuneiform nucleus, there is a small nucleus called the *accessory cuneate nucleus*, which receives unconscious proprioceptive information from the upper limb.

11.3.2 White Matter

At the level of the pyramids, the pyramidal pathway carries corticospinal and cortico-bulbar fibers. *Corticospinal fibers* carry voluntary motor impulses from the primary motor area in the cerebral cortex, descending through the anterior portion of the pyramids, where they form an incomplete decussation (90%) and continue into the spinal cord as the anterior (ipsilateral) corticospinal tract, which continues along the anterior column of the spinal cord, and the lateral (contralateral) corticospinal tract, which continues along the lateral column of the spinal cord. The *cortico-bulbar fibers* originate in the cerebral cortex, descend posteromedially to the corticospinal fibers, and reach the nuclei of the motor CNs of the MO for their respective motor functions.

The fibers of the gracile and cuneiform nuclei decuse to form the *medial lemniscus pathway* (Fig. 11.3). This is located toward the midline, behind the corticospinal tract on each side of the MO. Its fibers are second-order and ascend at the ventral posterolateral nucleus of the thalamus to terminate in the primary somatosensory region of the brain cortex.

The *anterior spinocerebellar tract* (Gower's tract) and the *posterior spinocerebellar tract* (Flechsig's tract) are found in the superficial lateral portion of the MO. Gower's tract is located between the olivary nucleus and the posterior spinocerebellar tract. It carries contralateral information of vibration, fine touch, conscious proprioception, sensation of weight, and discrimination of two points of the

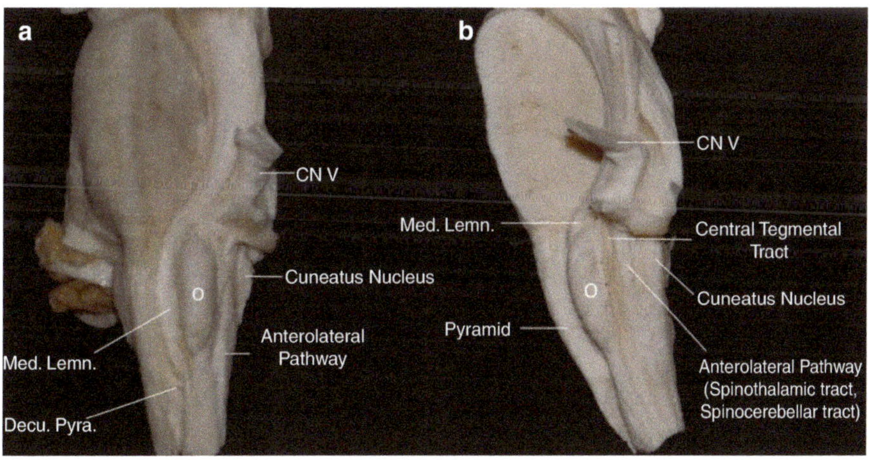

Fig. 11.3 (**a**, **b**) White matter of the medulla oblongata. The left medullary pyramid has been removed to identify the medial lemniscus (Med. Lemn.). At the end of the pyramidal pathway (corticospinal and cortico-bulbar fibers), there is the decussation of the pyramids (Decu. Pyra.). The spinothalamic and spinocerebellar tracts run in the lateral surface of the medulla, posterior to the olive (O). (Images courtesy of Richard Párraga)

lower extremities through the superior cerebellar peduncle and then crosses again and ends in the ipsilateral cerebellar hemisphere. Flechsig's tract, located between the anterior spinocerebellar tract and the inferior cerebellar peduncle, passes through the inferior cerebellar peduncle carrying unconscious proprioception information from the trunk and lower extremities to the ipsilateral cerebellar hemisphere (Fig. 11.3).

In the lateral portion of the MO medial to the anterior spinocerebellar tract is located the *spinothalamic tract*, which consists of two adjacent pathways: the anterior and lateral pathways (Fig. 11.3). The above spinothalamic tract carries information regarding coarse touch and firm pressure. The lateral spinothalamic tract conveys information on pain and temperature. Both tracts carry information to the ventral posterolateral nucleus in the thalamus, which is relayed upward to the somatosensory cortex of the postcentral gyrus.

The *tectospinal*, *rubrospinal*, *vestibulospinal*, and *reticulum spinal tracts* correspond to tracts of the *extrapyramidal system*. These tracts originate in different areas of the brain and descend directly to the spinal cord.

The *solitary tract* is a compact fiber bundle that extends longitudinally through the posterolateral region of the medulla. The solitary tract is formed by afferent fibers of the VII, IX, and X CNs and descends to the upper cervical segments of the spinal cord.

11.4 Vascular Anatomy of the Medulla

The blood supply to the MO is distinct from that of other areas of the brainstem, which may be associated with the distinct pattern of arterial supply to the MO.

The medulla receives blood supply from the vertebral artery and branches of the anterior spinal artery. The *vertebral artery* may be divided into four parts: preforaminal (V1), foraminal (V2), atlantic or extradural (V3), and intradural (V4). The fourth (intradural or intracranial) part crosses the dura mater and inclines medialward to the front of the MO; it is located between the hypoglossal nerve and the anterior root of the first cervical nerve and beneath the first digitation of the ligamentum denticulatum. At the lower border of the pons, it unites with the vessel on the opposite side to form the basilar artery. The *anterior spinal artery* is a midline vessel lying in the anterior median fissure and is formed by the union of a branch from each vertebral artery. The *anterolateral perforating arteries*, branches of the vertebral artery and the anterior spinal artery, perfuse the pyramidal tract and inferior olivary nuclei (Fig. 11.4).

The *posteroinferior cerebellar artery*, by definition, arises from the vertebral artery near the inferior olive and passes posteriorly around the medulla. At the anterolateral margin of the medulla, it passes rostral or caudal to or between the rootlets of the hypoglossal nerve, and at the posterolateral margin of the medulla, it courses rostral to or between the fila of the glossopharyngeal, vagus, and accessory nerves. After passing the latter nerves, it courses around the cerebellar tonsil and

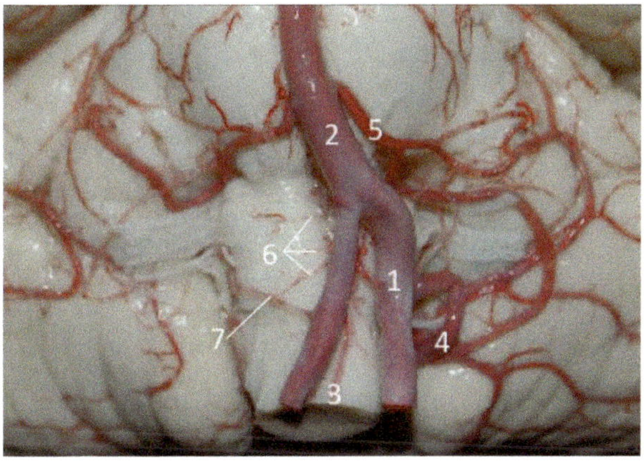

Fig. 11.4 Arterial supply of the medulla oblongata. (1) Vertebral artery, (2) basilar artery, (3) anterior spinal artery, (4) posteroinferior cerebellar artery, (5) anterior inferior cerebellar artery, (6) anterolateral perforating arteries, (7) lateral arteries (images courtesy of Richard Párraga)

enters the cerebellomedullary fissure and passes posterior to the lower half of the roof of the fourth ventricle. The *lateral arteries* are branches of the anterior inferior cerebellar artery and vertebral and basilar arteries. They perfuse the inferior cerebellar peduncle, spinothalamic tract, spinocerebellar tract, spinal trigeminal nucleus, central reticular formation, dorsal motor nucleus of the vagus, nucleus and tractus solitarius, and the hypoglossal, vestibular, cochlear, cuneate, and ambiguous nuclei (Fig. 11.4). The gracile and cuneate nuclei, area postrema, and vagal, solitary, and medial vestibular nuclei are supplied by these branches.

The veins in the medulla (similar to those in the brainstem) are named on the basis of the surface of the drainage and the direction of their course (longitudinal or transverse). The longitudinal veins are the *median anterior medullary vein*, which courses in the midline; the *medullary veins*, which course in the anterolateral surface of the brainstem; and the *lateral medullary veins*, which course in the lateral surface of the brainstem. The transverse veins running in the sulci of the pons and medulla are the *pontomedullary sulcus veins* and the *transverse veins* across the anterior and lateral surfaces of the medulla (Fig. 11.5).

11.5 Surgical Considerations

The medulla is perhaps the most difficult structure of the brainstem to approach because of the high density of nuclei located therein, CN pairs IX to XII. Lesions affecting the medulla include neoplasms, vascular malformations, demyelinating/inflammatory lesions, and infection. The criteria for operating on brainstem lesions, especially at the medulla, mainly depend on the nature of the lesion, on its topography, and on the surgeon's experience. Preoperative planning of the surgical strategy requires careful evaluation of the functions of the CNs, nerve tracts, and cerebellar function. For this objective, intraoperative monitoring includes auditory evoked potentials (testing the lateral lemniscus), somatosensory evoked potentials (testing

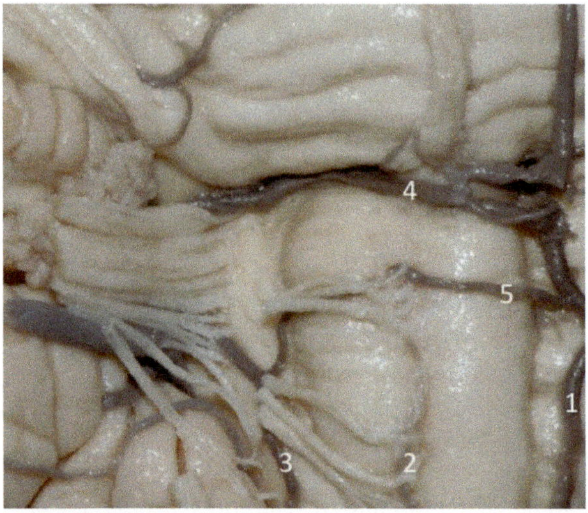

Fig. 11.5 Veins of the medulla oblongata. (1) Median anterior medullary vein, (2) medullary vein, (3) lateral medullary vein, (4) pontomedullary sulcus vein, (5) transverse vein (images courtesy of Richard Párraga)

the medial lemniscus), motor evoked potentials for the evaluation of the corticospinal pathway, and mapping of the VII, IX, X, and XII CNs to identify the safest entry zones. It is important to keep in mind that medulla mapping does not detect lesions of the corticospinal and corticonuclear pathways, which can be detected by monitoring motor evoked potentials. There is no technique for monitoring cerebellar function.

The approaches to the MO depend on where the lesions are located. For this purpose, we can divide the medulla in two different areas to be approached: the anterolateral surface and posterior surface.

11.5.1 Anterolateral Medulla

Lesions located in the anterolateral portion of the medulla are accessed via a far-lateral approach, which requires a lateral suboccipital craniectomy with resection of the posterior arch of the atlas (C-1), and there was several approach variations according to the part of the condyle to be removed: transcondylar, supracondylar, and paracondylar exposures.

In adults, we can perform a far-lateral approach with partial drilling of the posterior third of the occipital condyle. In children, it is possible to access the anterior portion of the medulla without removing the condyles by performing the section of the dentate ligament next to the entry of the vertebral artery that facilitates mobility of the medulla, facilitating lateral access, as opening of the condyle is avoided.

Access to the medulla may be anterior to the olive, posterior to the olive, or sometimes through the olivary body. To enter the medulla via the anterolateral sulcus, the entry zone is along the preolivary sulcus, between the caudal hypoglossal and the rostral C1 rootlets, very near the pyramidal tract, next to its decussation, and should be used only for exophytic lesions. The retroolivary sulcus is a safe entry area located between the olive and the inferior cerebellar peduncles ventral to the glossopharyngeal and vagus rootlets and is safe because the olivary body offers a surgical space of approximately 13.5 mm in the craniocaudal axis, 7 mm in the transverse axis, and 2.5 mm in the anterodorsal axis (Fig. 11.6).

11.5.2 Posterior Medulla

Surgical approaches to the posterior surface of the MO can be divided into superior or intraventricular and inferior.

Intraventricular lesions in the posterior part of the medulla are difficult to approach because of the large number of nuclei in that region. The approach through the ventricle poses a greater risk of neurological deficits, including dysphagia due to damage to the hypoglossal nucleus and cardiovascular dysfunction due to injury to the calamus scriptorius. It is important to evaluate the invasion of the ventricular floor by the tumor and preoperatively plan a subtotal radical resection so that the surgeon can avoid any temptation during surgery to manipulate the tumor that invades the floor. Any alterations in vital signs while working near the floor of the fourth ventricle should be considered a serious warning sign to stop manipulation, and a sheet of adherent tumor should be left on the floor.

Medullary lesions inferior to the obex may be accessed via the midline through the posterior median sulcus, as are intramedullary lesions, also through the posterior

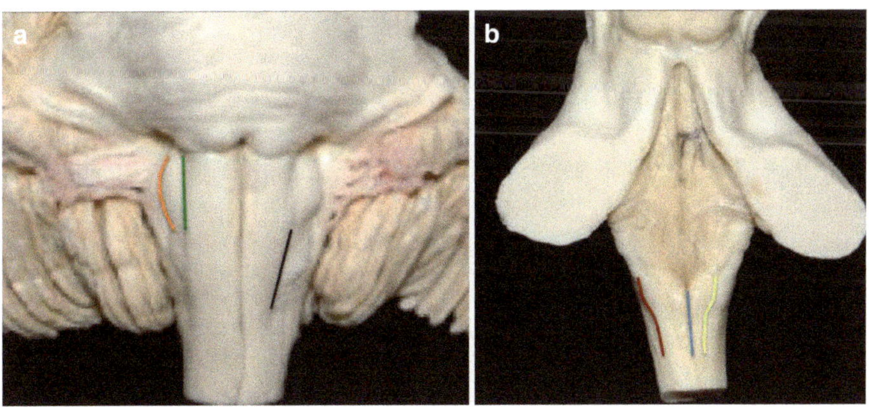

Fig. 11.6 Approaches to the medulla oblongata. In (**a**), we observe the anterolateral surface with the entry zones. Preolivary sulcus (green line), postolivary sulcus (orange line), anterolateral sulcus (black line). In (**b**), we show the entry zones of the posterior surface. Posterior medial sulcus (blue line), posterior intermediate sulcus (yellow line), posterolateral sulcus (red line). (Images courtesy of Richard Gonzalo Párraga)

intermediate sulcus, and the posterolateral sulcus; all these entry points can be accessed with a median suboccipital craniotomy (Fig. 11.6).

During the intraoperative period, severe vegetative alterations may occur, such as hypertension and tachycardia in the case of medullary lesions on the right side and bradycardia for medullary lesions on the left side.

11.6 Far Lateral Approach

11.6.1 Positioning

We prefer to place the patient in a semisitting position. Initially, when the patient is in the supine position on the operative table, the legs should be dressed with antithrombotic socks, and the articular body edges should be protected.

The head should be supported by a (three or four pins) skull fixation device (Mayfield or Sugita). The pins should be placed on both sides of the superior temporal line, always avoiding placing them through the temporal muscles. The shoulders must be at the same level parallel to each other and the head inflected until the chin reaches a one-finger distance, approximately 3 cm, from the chest.

11.7 Trichotomy

After general anesthesia has been induced, trichotomy should be performed up to 2 cm from the region of the surgical incision.

11.7.1 Marking, Antisepsis, and Scalp Incision

Once the area has been shaved, it is treated with ether-soaked gauze to remove the fat of the scalp and facilitate the fixation of the fields and the marking of the incision area with methylene blue.

The inion, asterion, C2 spinous process, and mastoid apex should be marked; they should be linked in an arched "horse shoe" skin incision with its edges approximately 5 cm below the mastoid process, laterally, and at the C2 spinous process, medially.

The incision begins in the midline, approximately 5 cm below the inion, and moves straight upward until 3 cm above the external occipital protuberance. Then, it turns laterally to the asterion, and finally, it turns downward and laterally over the sternocleidomastoid muscle posterior edge, approximately 5 cm below the mastoid apex. The use of bipolar coagulation helps avoid bleeding of the scalp arteries. The placement of wet gauze while applying traction of the scalp flap can spare the use of hemostatic clips and specific staples for this purpose.

Abundant irrigation of the operative field with physiological solution is imperative for all long procedures to avoid gaseous embolization.

11.8 Craniotomy

A retromastoid craniotomy is performed with complete exposure of the transverse and sigmoid sinuses as the superior and lateral limits, respectively. Inferiorly, the craniotomy should be extended until the edges of the foramen magnum and the posterior arch of C1 can be resected.

The greater the size of the lesion to be approached and the need for cerebellar medial retraction, the more medial the craniotomy. Sometimes, we should even cross over the midline to avoid great retraction against the bone edge and impairment of the cerebellar parenchymal tissue.

The far lateral approach starts by performing trepanations in the lower portion of occipital bone, one medial and another lateral close to the mastoid. Afterward, we should perform two trepanations near the inion (external occipital protuberance) medially and the other two trepanations at the asterion, both at the superior and inferior edges of the transverse sinus. These points should be connected to conclude a lateral suboccipital craniotomy.

The second step involves drilling the inferior portion of the occipital bone (jugular process) as the posterior margin of the jugular foramen, lateral to the condyle. Bone resection should be performed anteriorly until the jugular foramen. Then, it should be performed laterally following the entire sigmoid and transverse sinuses, performing a mastoidectomy. Care should be taken to avoid damage to the bridge vein between the sinus and the mastoid. Medially, the bone may be resected to open the magnum foramen.

Resection of the condyle should be performed depending on the target of the approach, varying from the paracondylar, supracondylar, and transcondylar. The condylar resection is performed by drilling its inner portion first until the cortical bone of the hypoglossal canal is discovered. Theoretically, the posterior two-thirds of the condyle should be drilled. Nevertheless, in practical surgery, the condyle should be drilled until the limit of the hypoglossal canal.

Finally, after exposing the hypoglossal canal above the occipital condyle, the bone of the jugular tubercle, situated above the hypoglossal canal, may be removed extradurally to gain additional exposure.

The jugular tubercle is a rounded prominence located at the junction of the basilar and condylar parts of the occipital bone. The glossopharyngeal, vagus, and accessory nerves cross the posterior portion of the jugular tubercle extending from the brainstem to the jugular foramen. The jugular tubercle blocks access to basal cisterns and inferior clivus anterior to the lower nerves.

11.9 Illustrative Case

A 47-year-old man with a history of worsening headaches and episodes of nausea and vomiting. The neurologic examination did not show pathological findings.

The MRI showed a tumor dorsal medulla oblongata without radiological evidence of hydrocephalus at presentation (Fig. 11.7).

Based on the previous magnetic resonance images, the following questions arise: Where is the tumor located? Which structures should be taken in mind at the surgical resection?

The tumor is situated in the midline and insinuated between the inferior aspects of the floor of the fourth ventricle. One of the most important features to consider is the location of the tumor because we must know that the MO at the posterior surface can be divided into superior or intraventricular and inferior areas. If we see that most of the lesion is located in the lower half of the fourth ventricle, the *midline suboccipital telovelar approach* is most commonly used for tumor resection and, in combination with surgical adjuncts (neuronavigation, neurophysiologic monitoring, and intraoperative MRI), can reduce the likelihood of morbidity. In this patient, we consider the *preservation of the floor of the fourth ventricle* because this region contains numerous nuclei and is essential yet challenging in patients with no clear boundary between the tumor tissue and the brain.

The semisitting position was selected for the surgery because it offers a very clear operative field, as blood and cerebrospinal fluid drain out of the operative site (Fig. 11.8). Previously, a central venous catheter was placed, transesophageal echocardiography was performed, and sequential compression stockings were applied.

The midline suboccipital telovelar approach was adopted, observing a tumor protruding the midline of the posterior surface of the medulla compromising the inferior part of the floor of the fourth ventricle (Fig. 11.9).

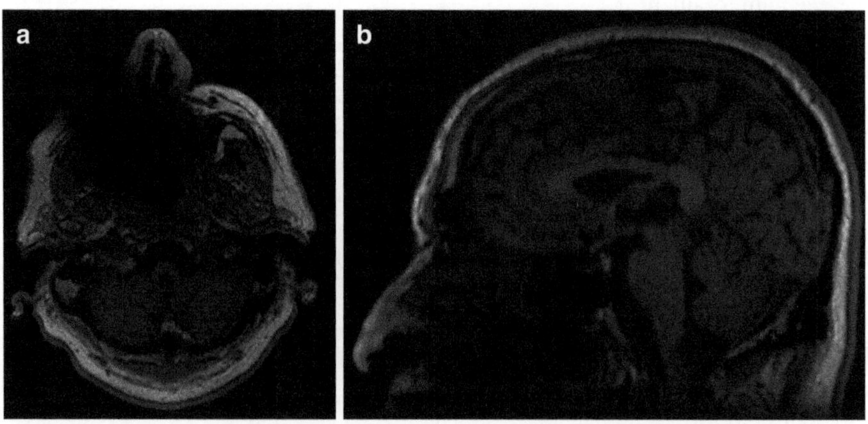

Fig. 11.7 Preoperative magnetic resonance imaging. (**a**) Axial T1-weighted image. (**b**) Sagittal T1-weighted image revealing a solid-cystic lesion in the inferior and dorsal medulla oblongata

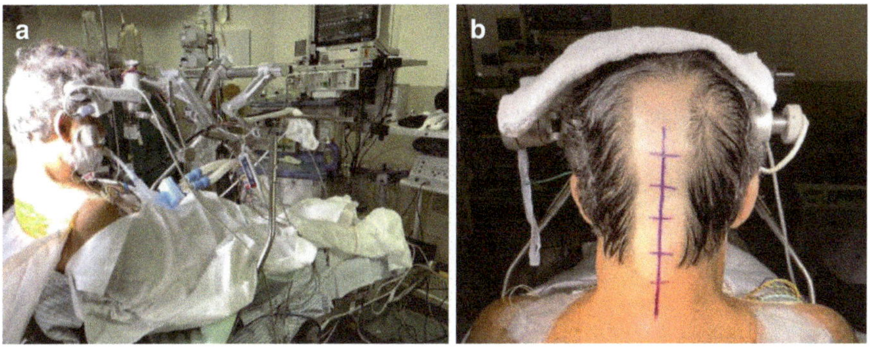

Fig. 11.8 The patient's head position and skin incision mark. (**a**) Semisitting position with the head supported by a (three or four pins) skull fixation device (Mayfield) and with both of the patient's legs flexed at the knees to prevent postoperative sciatica. (**b**) The incision begins in the midline, 6 cm above the inion, and extends to the C2 spinous process

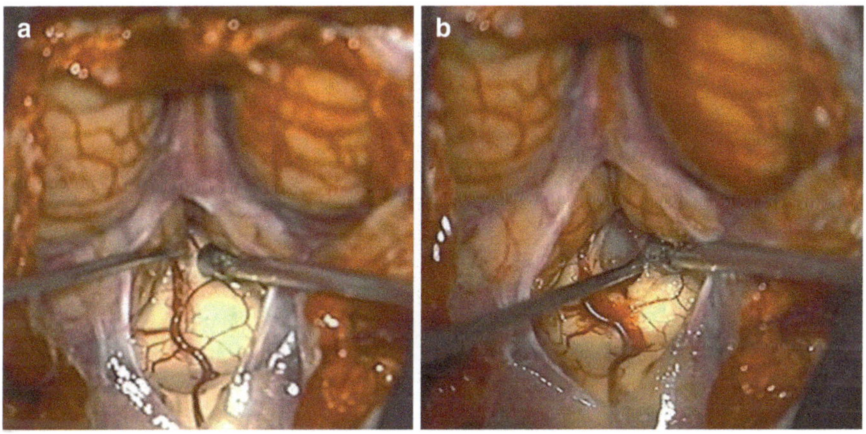

Fig. 11.9 Intraoperative images. (**a**) The surgical field after a midline suboccipital approach, observing the cerebellar tonsils and the magna cistern with the tumor at the most anterior region between them and the protrusion of the posterior surface of the medulla. (**b**) Image showing the blunt dissection without using the bipolar coagulation; gentle dissection and aspiration can separate the tumor from its surroundings

Total resection was successfully performed with careful dissection using microsurgical scissors avoiding the use of electrocoagulation to the extent possible (Fig. 11.10).

The postoperative MRI showed total resection (Fig. 11.11). Histological evaluation of the lesion confirmed a diagnosis of glioma. The patient was subsequently discharged after an uneventful postoperative course.

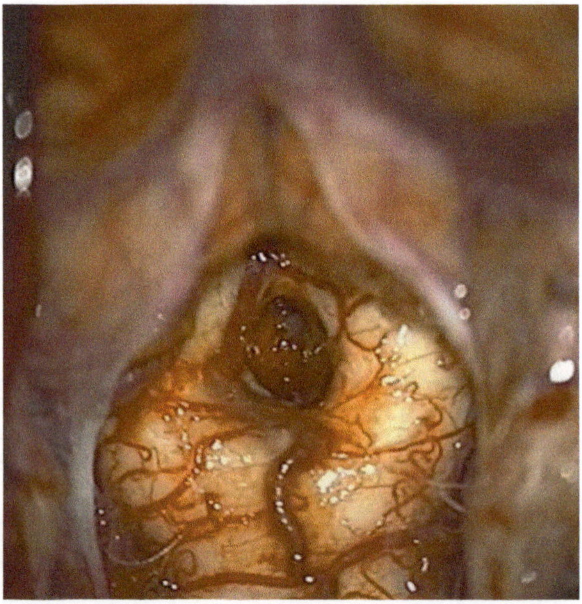

Fig. 11.10 Intraoperative image. Final image of the medulla and cerebellar anatomy before the closure, evidencing the anatomy preservation of the medulla oblongata and the surgical cavity without tumor

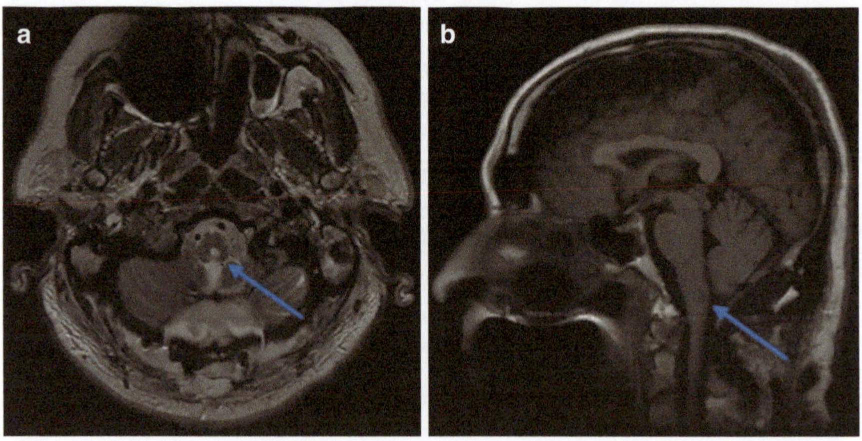

Fig. 11.11 Postoperative magnetic resonance imaging. (**a**) Axial T2-weighted image showing the postoperative changes in the cerebellar hemispheric tissue around the surgical area at the midline with the absence of tumor. (**b**) Sagittal T1-weighted image showing the complete resection and postoperative changes after the resection through a suboccipital craniotomy

Suggested Bibliography

Caramanti RL, da Costa MDS, Hernández YAU, Lancellotti CLP, Granados OS, Echeverria KG, Doria Netto HL, de Campos Filho JM, Chaddad-Neto F. Microsurgery for Cervicomedullary Tumor: 2-Dimensional Operative Video. Oper Neurosurg (Hagerstown). 2018 Dec 1;15(6):725.

Cavalheiro S, Yagmurlu K, da Costa MD, Nicácio JM, Rodrigues TP, Chaddad-Neto F, et al. Surgical approaches for brainstem tumors in pediatric patients. Childs Nerv Syst. 2015;31:1815–40.

Chaddad Neto F, Doria-Netto HL, de Campos Filho JM, Neto MR, Rothon AL Jr, de Oliveira E. The far-lateral craniotomy: tips and tricks: a craniotomia extremo-lateral: dicas e truques. Arquivos de Neuro-Psiquiatria. 2014;72:699–705.

Matsushima T, Rhoton AL Jr, Lenkey C. Microsurgery of the fourth ventricle: part 1. Microsurg Anat Neurosurg. 1982;11:631–67.

Párraga R, Possatti L, Alves R, Ribas G, Türe U, de Oliveira E. Microsurgical anatomy and internal architecture of the brainstem in 3D images: surgical considerations. J Neurosurg. 2016;5:1377–95.

Chapter 12
Surgical Anatomy of the Anterior Basal Cisterns

12.1 Introduction

Microsurgical navigation into the subarachnoid space consists of the sequential incision of the leptomeningeal trabeculae. This critical procedure provides access to structures and lesions located not only into the subarachnoid space but also near its pial boundaries. The following processes occur simultaneously: (1) the cerebrospinal fluid (CSF) is drained from the subarachnoid space, reducing the volume and increasing the compliance of the brain; (2) the ventricular system may be punctured (e.g., lamina terminalis, corpus callosum), allowing additional CSF drainage; (3) the frontal and temporal lobes are untethered and mobilized with minimal retraction pressure; and (4) critical traction forces transmitted from the cerebral surface to the major arteries can be avoided (e.g., early intraoperative rupture of a posterior communicating aneurysm after temporal lobe retraction).

12.2 The Cisterns

12.2.1 Hemispheric Cistern

The hemispheric cistern is a thin compartment of the subarachnoid space that surrounds the superior-lateral, inferior, and medial surfaces of the brain, extending into its fissures and sulci. Unlike other cisterns, it contains cortical branches of the major arteries and lacks major or perforating arteries (Fig. 12.1).

Supplementary Information The online version of this chapter (https://doi.org/10.1007/978-3-030-82747-2_12) contains supplementary material, which is available to authorized users.

F. Chaddad-Neto, M. D. Silva da Costa, *Microneuroanatomy and Surgery*,
https://doi.org/10.1007/978-3-030-82747-2_12

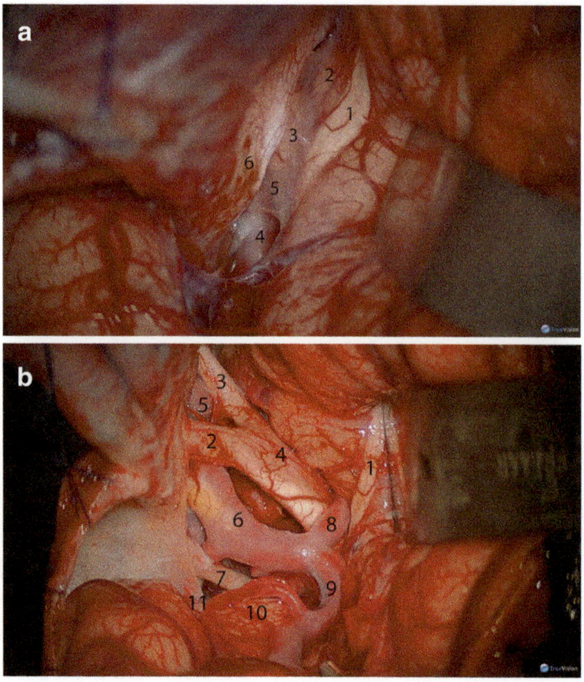

Fig. 12.1 Operative photograph of the anterior supratentorial cisterns (left pterional approach). (**a**) The dissection of the subfrontal subdural plane reveals part of the hemispheric cistern. The arachnoid mater and pia mater are very close, except at the level of the fissures and sulci. The volume of cerebrospinal fluid (CSF) available for drainage is low. In the midline and major fissures, the arachnoid mater and pia mater are distant and define compact cisterns containing major arteries. The incision of the arachnoid mater opened the Sylvian and carotid cisterns. (1) Left olfactory tract. (2) Gyrus rectus. (3) Chiasmatic cistern and optic nerve. (4) Left supraclinoid internal carotid artery (ICA). (**b**) The sequential incision of leptomeningeal trabeculae provides a panoramic view of the six compact anterior supratentorial cisterns: carotid, chiasmatic, Sylvian, olfactory, lamina terminalis, and pericallosal. The limen insulae, rhinal incisure, and lateral borders of the optic chiasm are fundamental microsurgical landmarks. Except for the ophthalmic artery, the branches of the supraclinoid ICA are exposed: superior hypophyseal, posterior communicating, anterior choroidal, middle cerebral, and anterior cerebral arteries. The arteries originate from the ICA into the carotid cistern and migrate to the neighboring cisterns. The superior aspect of the ICA and the anterior aspect of the A1-anterior cerebral artery (ACA) and M1-middle cerebral artery (MCA) lack critical branches and provide a safe plane for dissection into the carotid, lamina terminalis, and Sylvian cisterns. (1) Left olfactory tract. (2) Left optic nerve. (3) Right optic nerve. (4) Optic chiasma. (5) Right supraclinoid segment of the ICA. (6) Left supraclinoid segment of the ICA. (7) Left oculomotor nerve. (8) Left A1. (9) Left M1. (10) Left uncus. (11) Rhinal incisure

12.2.2 Carotid Cistern

The carotid cistern is a paired anterior supratentorial cistern containing the supraclinoid internal carotid artery (ICA) as its central element (Figs. 12.1 and 12.2).

The branches of the supraclinoid ICA have their origin in the carotid cistern and migrate to the neighboring cisterns: ophthalmic, superior hypophyseal, posterior communicating, anterior choroidal, middle cerebral, and anterior cerebral arteries. Similarly, several veins converge into the upper part of the carotid cistern to form the striate segment of the basal vein of Rosenthal: anterior cerebral, inferior striate, olfactory, fronto-orbital, and deep middle cerebral veins.

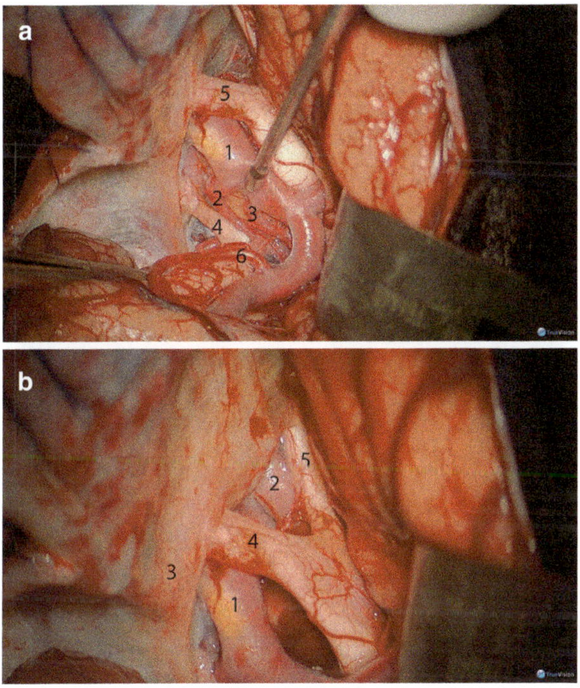

Fig. 12.2 Operative photographs of the anterior compartment of the carotid cistern. (**a**) Left pterional approach. A condensation of leptomeningeal trabeculae joining the inferior aspect of the optic nerve and the basal arachnoid mater, closely related to the hypophyseal artery, forms the posterior part of the medial wall of the anterior compartment of the carotid cistern. (1) Left supraclinoid internal carotid artery (ICA). (2) Left posterior communicating artery. (3) Left anterior choroidal artery. (4) Left oculomotor nerve. (5) Left optic nerve. (6) Early temporal branch of left M1. (**b**) Left pterional approach. When the interoptic space is wide (e.g., post-fixed optic chiasm), the anterior compartment of the contralateral carotid cistern is accessible through a narrow working channel. (1) Left supraclinoid ICA. (2) Right supraclinoid ICA and right carotid cistern. (3) Left anterior clinoid process. (4) Left optic nerve. (5) Right optic nerve. (**c**) Left pretemporal approach. The microsurgical accessibility to the anterior compartment of the ipsilateral carotid cistern is hindered by the anterior clinoid process, laterally, and the optic nerve, superiorly. (1) Left anterior clinoid process. (2) Left optic nerve. (3) Carotid cistern. (4) Left supraclinoid ICA. (**d**) Left pretemporal approach. Anterior clinoidectomy, complemented by division of the arachnoid cuff and the distal dural ring, supporting the ICA, provides a wide exposure of the anterior compartment of the carotid cistern. Gentle mobilization of the optic nerve and chiasm, after section of the falciform dural fold, further expands the working channel. (1) Left optic nerve. (2) Clinoidal segment of left ICA. (3) Carotid cistern

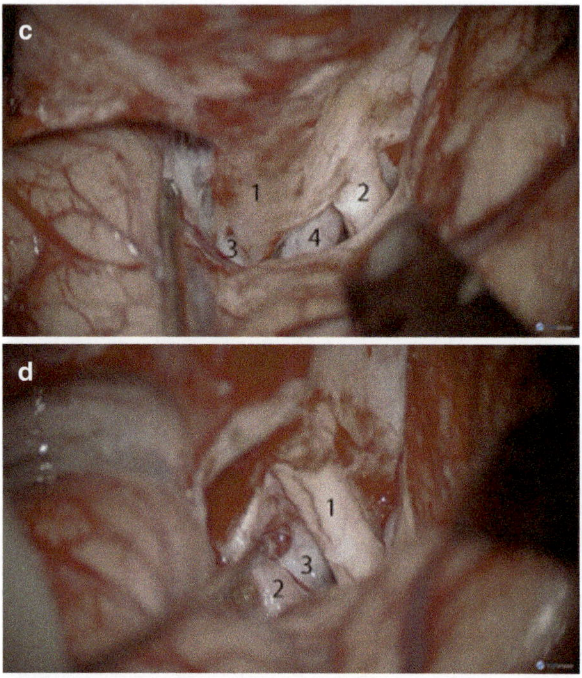

Fig. 12.2 continued

The carotid cistern may be subdivided into a small anterior and a wide posterior compartment.

12.2.2.1 Anterior Compartment

The anterior compartment is a tight space encased by osseous structures and the optic nerve. It contains the ophthalmic segment of the ICA as its central element (Figs. 12.1 and 12.2). The ophthalmic and superior hypophyseal arteries migrate from the carotid cistern to the chiasmatic cistern. The lateral wall is the arachnoid mater apposed to the medial border of the anterior clinoid process. The arachnoid mater apposed to the optic strut is the anterior border.

The medial wall has two parts: (1) anteriorly, the arachnoid mater apposes to a small area of the sphenoid body and sinus; (2) posteriorly, a condensation of leptomeningeal trabeculae joining the inferior aspect of the optic nerve and the basal arachnoid mater closely attaches to the superior hypophyseal artery. An arachnoid cuff surrounds the ICA at its entrance into the subarachnoid space. The optic nerve contained into the optic-chiasmatic cistern is the roof of the carotid cistern. Posteriorly, the anterior and posterior compartments of the carotid cistern communicate. The boundaries of the anterior compartment of the carotid cistern limit the microsurgical accessibility.

12.2.2.2 Posterior Compartment

The posterior compartment of the carotid cistern presents several leptomeningeal walls and contains the communicating segment of the supraclinoid ICA supported by dense leptomeningeal trabeculae as its central element (Figs. 12.1 and 12.3).

The artery gives off several branches: (1) the posterior communicating artery migrates to the transition between the interpeduncular and ambient cisterns; (2) the anterior choroidal artery migrates to the crural cistern; (3) the anterior cerebral artery migrates to the lamina terminalis cistern; and (4) the middle cerebral artery (MCA) migrates to the Sylvian cistern.

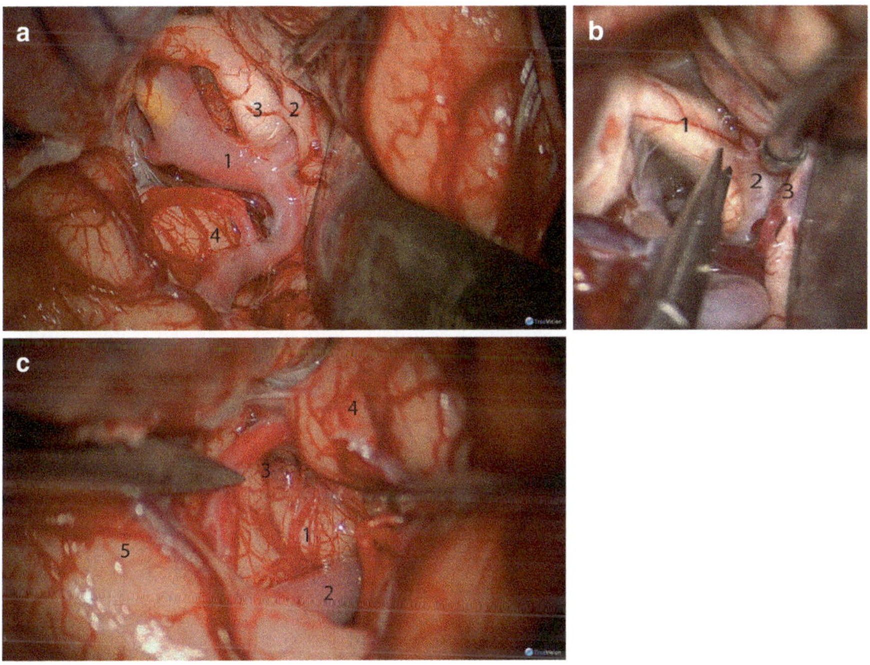

Fig. 12.3 Operative photographs of the posterior compartment of the carotid cistern (left pterional approach). (**a**) On the anterior wall, bright and thin condensations of leptomeningeal trabeculae bridge between the olfactory trigone, the lateral border of the optic chiasm, and the anterior part of the hippocampal uncus. (1) Supraclinoid left internal carotid artery (ICA). (2) Left olfactory tract and trigone. (3) Optic chiasma. (4) Anterior uncus. (**b**) On the medial wall, the ipsilateral border of the optic chiasm is the fundamental microsurgical landmark. Anteriorly and superiorly, condensations of leptomeningeal trabeculae bridge between the olfactory trigone and the optic chiasm, investing the A1-anterior cerebral artery (ACA) and separating the carotid and lamina terminalis cisterns. Inferiorly and posteriorly, the ipsilateral border of the optic chiasm is attached to the basal arachnoid mater condensations of leptomeningeal trabeculae separating the carotid and chiasmatic cisterns. (1) Left optic nerve. (2) Left A1. (3) Left olfactory tract. (**c**) The rhinal incisure is the fundamental microsurgical landmark of the lateral boundary. The arachnoid mater and leptomeningeal trabeculae join the frontal-orbital gyri and hippocampal uncus, investing the M1-middle cerebral artery (MCA) and separating the carotid and Sylvian cisterns. (1) Anterior uncus. (2) M1 segment of left MCA. (3) Rhinal incisure. (4) Frontal operculum. (5) Temporal operculum

The rhinal incisure and the ipsilateral border of the optic chiasm are the most relevant structures forming the lateral and medial borders of the posterior compartment of the carotid cistern, representing the fundamental microsurgical landmarks.

The central part of the anterior perforated substance is the roof of the cistern. On the anterior perforated substance, the lenticulostriate arteries, branches of the ICA, anterior cerebral artery (ACA), and MCA, enter the brain, and the inferior striate veins leave the brain to join the basal vein of Rosenthal.

The arachnoid mater apposes the oculomotor triangle, crossed by CN 3 and 4 in the way from the subarachnoid space to the cavernous sinus, and is the floor of the cistern. The oculomotor triangle is part of the roof of the cavernous sinus and is formed by three dural folds related to the tentorial attachments: (1) laterally, the anterior petroclinoid fold connects the anterior clinoid process and the petrous apex; (2) posteriorly, the posterior petroclinoid fold connects the posterior clinoid process and the petrous apex; and (3) medially, the interclinoid fold connects the anterior and posterior clinoid processes.

The anterior wall is formed by condensations of leptomeningeal trabeculae that bridge the olfactory trigone (superiorly), the lateral border of the optic chiasm (inferiorly and medially), and the anterior part of the hippocampal uncus (inferiorly and laterally), attaching to the superior aspect of the communicating segment and bifurcation of the ICA. The superior part of the anterior wall links the carotid and olfactory cisterns over the ipsilateral optic nerve. The inferior part of the anterior wall connects with the anterior and posterior compartments of the carotid cistern, under the ipsilateral optic nerve.

On the lateral wall, the anterior segment of the hippocampal uncus is attached to the frontal-orbital gyri by the arachnoid mater superficially and more so by condensations of the arachnoid trabeculae. The latter are particularly developed in the region of the rhinal incisure, supporting the M1-MCA and separating the carotid from the Sylvian cistern.

The medial wall is divided into two parts by the ipsilateral border of the optic chiasm. Anteriorly and superiorly, the ipsilateral border of the optic chiasm is attached to the olfactory trigone by condensations of leptomeningeal trabeculae, which surround the A1-ACA and separate the carotid and lamina terminalis cisterns. Inferiorly and posteriorly, the ipsilateral border of the optic chiasm is attached to the basal arachnoid mater by condensations of leptomeningeal trabeculae separating the carotid and chiasmatic cisterns.

Posteriorly, the cistern is confluent with interpeduncular, crural, and ambient cisterns. Condensations of leptomeningeal trabeculae separate into different sleeves, namely, the posterior communicating and anterior choroidal arteries, as they leave the carotid cistern to enter the interpeduncular and crural cisterns, respectively.

12.2.3 Chiasmatic Cistern

The chiasmatic cistern is an unpaired anterior supratentorial cistern containing the optic nerves and chiasm, the infundibulum, and the pituitary stalk as the central elements (Figs. 12.1 and 12.4).

The optic apparatus divides the cistern into a small anterior and a wide posterior compartments.

Unlike the other cisterns, the chiasmatic cistern lacks major arteries and veins. Instead, it contains the ophthalmic artery and several small but critical perforating branches from the internal carotid, posterior communicating, and anterior cerebral arteries.

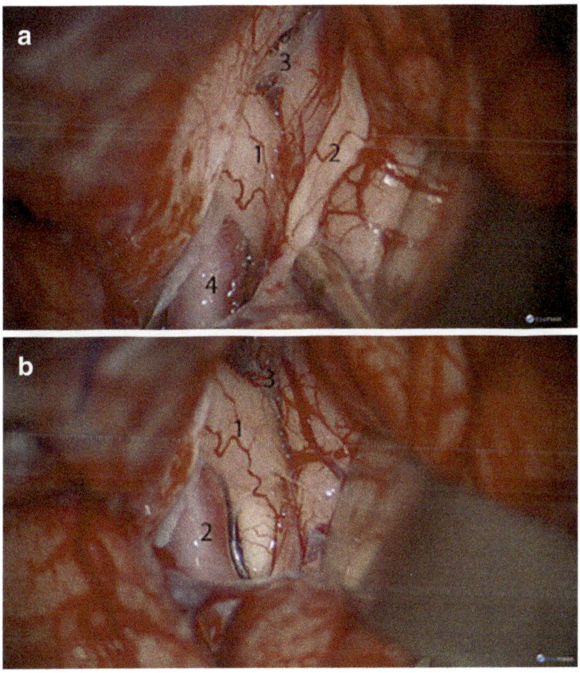

Fig. 12.4 Operative photographs of the optic-chiasmatic cistern. (**a**) Left pterional approach. The incision of the arachnoid mater over the left optic nerve reveals condensations of leptomeningeal trabeculae organized into a bright, thin, and almost continuous membrane closely apposed to the superior aspect of the optic apparatus, defining the superior boundary of the optic-chiasmatic cistern. (1) Left optic nerve. (2) Left olfactory tract. (3) Optic-chiasmatic cistern. (4) Left supraclinoid segment of the internal carotid artery (ICA). (**b**) Left pterional approach. This condensation of leptomeningeal trabeculae was incised through the left optic-carotid space, near the border of the optic chiasm, opening the optic-chiasmatic cistern. (1) Left optic nerve. (2) Left supraclinoid segment of ICA. (3) Optic-chiasmatic cistern. (**c**) Right pterional approach. The wide right optic-carotid triangle reveals the arachnoid mater apposed to the diaphragma sellae, dorsum sellae, and posterior clinoid process, forming the floor of the optic-chiasmatic cistern. Posteriorly, the Liliequist membrane, a thick and continuous condensation of leptomeningeal trabeculae, separates the optic-chiasmatic and interpeduncular cisterns. The anterior-inferior angle of the Liliequist membrane attaches to the CN3, hippocampal uncus, and free tentorial edge. (1) Right optic nerve. (2) Right supraclinoid ICA. (3) Liliequist membrane. (4) Right oculomotor nerve. (5) Posterior clinoid process. (**d**) Left pterional approach. The internal carotid artery (ICA) is medially mobilized. Inferior and laterally to the internal carotid and posterior communicating arteries, the anterior and inferior angle of the Liliequist membrane firmly attaches to the CN3, hippocampal uncus, and free tentorial edge. (1) Left oculomotor nerve. (2) Left supraclinoid ICA. (3) Liliequist membrane. (4) Left anterior uncus. (5) Left M1 segment of ICA

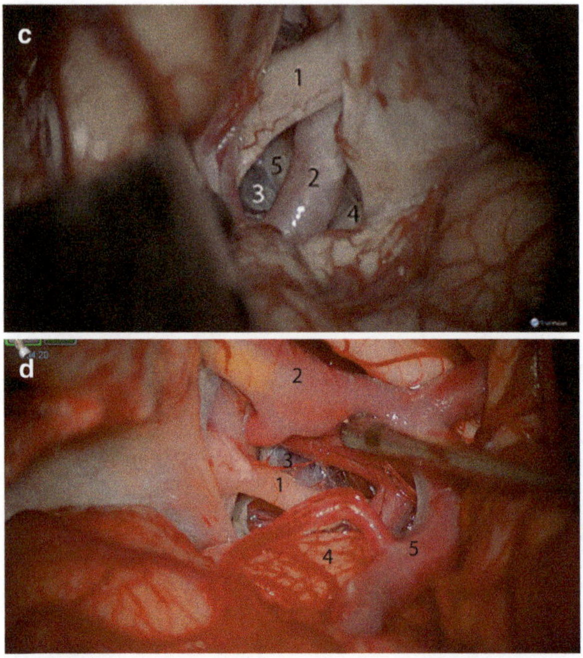

Fig. 12.4 continued

The ipsilateral and contralateral borders of the optic apparatus are the most relevant structures forming the lateral borders of the optic-chiasmatic cistern, representing the fundamental microsurgical landmarks.

The anterior border of the cistern are (1) in the midline, the arachnoid mater apposed to the optic sulcus and, (2) laterally, the arachnoid mater forming a sleeve that surrounds the optic nerves into the optic canals and orbits.

Superiorly, the leptomeningeal trabeculae form an almost continuous membrane closely apposed to the optic apparatus that separates the optic-chiasmatic and the lamina terminalis cisterns.

Inferiorly, the cistern is bordered by the arachnoid mater apposed to the diaphragma sellae, dorsum sellae, and posterior clinoid processes. The arachnoid mater and the cistern may extend through the diaphragma sellae into the pituitary fossa (sellar arachnoidocele).

Posteriorly, the Liliequist membrane (sellar part and diencephalic leaflet) separates the chiasmatic from the interpeduncular cistern. These thick condensations of the leptomeningeal trabeculae bridge from the arachnoid mater apposed to the anterior dorsum sellae and inferior posterior clinoid processes to the posterior mammillary bodies and adjacent superior hypothalamus. The anterior-inferior angle of the Liliequist membrane firmly attaches to the CN3, the hippocampal uncus, and the free tentorial border. Dense leptomeningeal trabeculations join the inferior surface of the optic chiasm and the superior surface of the Liliequist membrane surrounding the infundibulum and pituitary stalk.

Laterally, irregular leptomeningeal trabeculae join the optic nerves and the chiasm to the basal arachnoid mater, closely connected to the medial aspect of the supraclinoid ICA, the posterior communicating artery, and perforating branches, separating the chiasmatic and carotid cisterns.

12.2.4 Sylvian Cistern

The Sylvian cistern is a paired supratentorial cistern containing the MCA supported by dense leptomeningeal trabeculations, which is its central element (Figs. 12.1 and 12.5).

The MCA originates from the bifurcation of the ICA into the carotid cistern. The M1 segment, contained in the carotid and sphenoidal compartments of the Sylvian cistern, is directed laterally and slightly upward. Except the uncal, temporal-polar, and anterior temporal arteries, the anterior aspect of the artery provides a safe plane for dissection, both into the carotid and Sylvian cisterns. Several lenticulostriate arteries emerge from the posterior aspect. The M2 segment is directed posteriorly, superiorly, and laterally, and the M3 segment is directed laterally, toward the hemispheric cistern. Both segments are contained in the opercular-insular compartment.

The superficial Sylvian vein complex, in its pathway to the sphenoparietal, sphenobasal, cavernous, or superior petrosal sinuses, is close to the arachnoid mater.

The insular and inferior striate veins are major tributaries of the basal vein of Rosenthal, crossing the deep aspect of the Sylvian and carotid cisterns.

The Sylvian fissure is the superficial aspect of the Sylvian cistern, representing the fundamental microsurgical landmark on the basal and lateral surfaces of the brain. On the basal surface, the stem is directed laterally and upward, paralleling the sphenoid ridge, and separates the frontal-orbital part, related to the anterior cranial base, from the temporal-occipital part, related to the middle cranial base and tentorium cerebelli. The anterior Sylvian point, related to the craniometric point pterion, is characterized by the division of the stem into anterior horizontal, anterior ascending, and posterior rami. The anterior Sylvian point is coronally related to the insular apex and limen. The small anterior horizontal and anterior ascending rami separate the orbital, triangular, and opercular parts of the inferior frontal gyrus. The long posterior ramus, directed posteriorly and upward, grossly paralleling the anterior part of the squamous suture, represents the posterior continuation of the stem and separates the frontal-parietal and temporal opercula. At its posterior end, the posterior ramus divides into an ascending terminal ramus, which merges into the supramarginal gyrus, and the descending terminal ramus, which merges into the superior temporal gyrus.

The Sylvian cistern is subdivided into small sphenoidal and wide opercular-insular compartments.

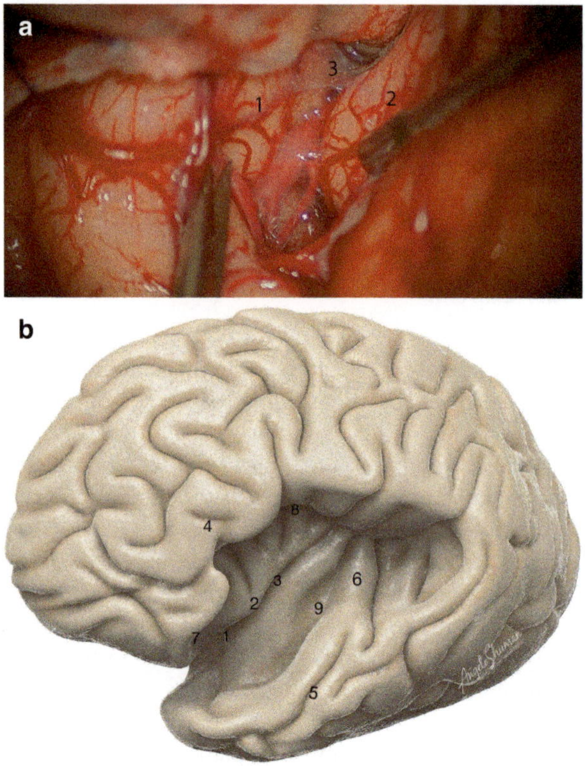

Fig. 12.5 The Sylvian cistern. (**a**) Operative photograph, pterional approach. The sphenoidal compartment contains the lateral half of the M1-middle cerebral artery (MCA). The rhinal incisure and the limen insulae are the fundamental microsurgical landmarks regarding the medial and lateral boundaries. The limen insulae is related to the 90 degrees turn that defines the M1–M2 transition. Frequently, the MCA bifurcation occurs proximally to the limen insulae. (1) Temporal operculum. (2) Frontal operculum. (3) Sylvian cistern sphenoidal compartment. (**b**) Anatomical photograph. The opercular-insular compartment contains the M2 and M3 segments of the MCA. The frontal-parietal and temporal opercula facing each other and the insular lobe border two perpendicular clefts. From anterior to the posterior, the area of the opercular cleft increases and the area of the insular cleft decreases, impairing the microsurgical accessibility to the opercular-insular compartment of the Sylvian cistern. (1) Limen insulae. (2) Apex of insulae. (3) Central sulcus of insulae. (4) Frontal operculum. (5) Temporal operculum. (6) Heschl's gyrus. (7) Anterior limiting sulcus. (8) Superior limiting sulcus. (9) Inferior limiting sulcus

12.2.4.1 Sphenoidal Compartment

The sphenoidal compartment of the Sylvian cistern contains the lateral half of the M1 segment of the MCA supported by dense leptomeningeal trabeculae, as the central element.

The rhinal incisure and limen insulae are the fundamental microsurgical landmarks of the medial and lateral boundaries of the sphenoidal compartment of the Sylvian cistern.

Dense leptomeningeal trabeculae joining the posterior frontal-orbital gyri and the rhinal incisure region form the medial limit and communicate with the carotid cistern.

The limen insulae, the prominence overlying the uncinate and inferior fronto-occipital fasciculi connecting the frontal and temporal lobes, including the temporal stem, form the lateral limit and communicate with the opercular-insular compartment. The limen insulae is related to the 90° turn that characterizes the M1–M2 transition.

The arachnoid mater apposed to the sphenoid ridge forms the anterior wall.

The posterior part of the frontal-orbital gyri and the anterior part of the planum temporale overlying the temporal pole are closely apposed and form the roof and floor, respectively.

On the posterior wall, the lateral part of the anterior perforated substance and the posterior part of the planum temporale overlying the temporal pole converge and form the Sylvian vallecula, a tunnel through which the M1-MCA passes.

Condensations of leptomeningeal trabeculae bridging between the frontal and temporal lobes form an almost continuous membrane anterior to the M1-MCA segment. Additional leptomeningeal trabeculae connect the MCA-M1 and the lenticulostriate arteries to the cisternal walls (*a case example is also demonstrated in the* Video 12.1).

12.2.4.2 Opercular-Insular Compartment

The opercular-insular compartment of the Sylvian cistern contains the M2 and M3 segments of the MCA, supported by leptomeningeal trabeculae, as the central elements.

The frontal-parietal and temporal opercula, which face each other, and the insular lobe border two perpendicular clefts.

The frontal operculum includes the orbital, triangular, and opercular parts of the inferior frontal gyrus, the inferior-lateral part of the pre- and postcentral gyri, and the supramarginal gyrus.

The temporal operculum corresponds to the superior temporal gyrus.

The inferior surface of the frontal-parietal operculum faces the superior surface of the temporal operculum, bordering a superficial opercular cleft that is grossly perpendicular to the cerebral surface. The superficial border of the opercular cleft communicates with the hemispheric cistern, underneath the arachnoid mater lining the Sylvian fissure. At the deep border, the opercular and insular clefts communicate. The medial surface of the frontoparietal and temporal opercula faces the insular lobe, bordering a deep cleft that is grossly parallel to the cerebral surface.

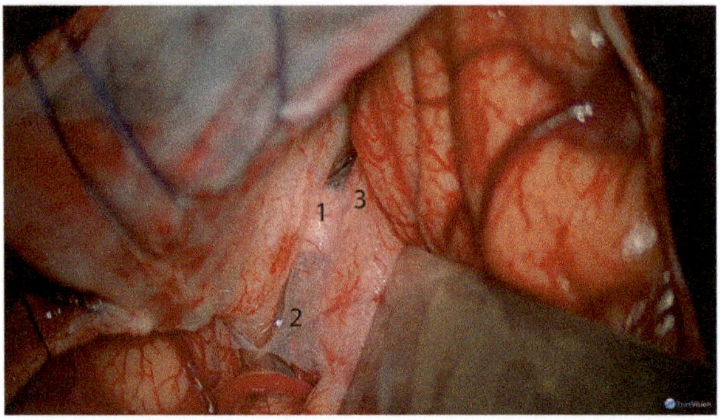

Fig. 12.6 Operative photograph of the olfactory cistern (pterional approach). The olfactory cisterns lacks significant CSF content. The olfactory bulb and tract are the fundamental microsurgical landmarks of the frontal-orbital cerebral surface, separating the gyrus rectus from the orbital gyri. The frontal retraction is limited by the firm attachment of the olfactory bulb to the ethmoidal cribriform plate and of the posterior part of the olfactory tract to the olfactory sulcus. (1) Optic nerve. (2) Carotid cistern. (3) Left olfactory tract

12.2.5 Olfactory Cistern

The olfactory cistern is a paired anterior supratentorial cistern containing the olfactory bulb and tract, the olfactory artery and vein, and a sulcal loop of the fronto-orbital artery (Figs. 12.1 and 12.6).

The leptomeningeal trabeculae form an almost continuous sleeve overlapping the olfactory sulcus that separates the olfactory and hemispheric cisterns. Posteriorly, the olfactory cistern joins the carotid cistern.

The olfactory bulb and tract are the fundamental microsurgical landmarks of the frontal-orbital cerebral surface, separating the gyrus rectus from the orbital gyri.

12.2.6 Lamina Terminalis Cistern

The lamina terminalis cistern is an unpaired anterior supratentorial cistern that contains the complex of the anterior communicating artery, supported by dense leptomeningeal trabeculae, as its central element (Figs. 12.1 and 12.7).

The complex of the anterior communicating artery includes 12 arteries: ipsi- and contralateral A1-ACA segments; ipsi- and contralateral A2-ACA segments; the ACoA; ipsi- and contralateral medial frontal-orbital arteries; ipsi- and contralateral frontal-polar arteries; ipsi- and contralateral recurrent arteries of Heubner; and the perforators originating from the A1, anterior communicating artery, and proximal A2.

The ipsilateral and contralateral borders of the optic chiasm are the fundamental microsurgical landmarks of the medial and lateral boundaries of the lamina terminalis cistern.

The cistern is bordered by the arachnoid mater apposed to the limbus sphenoidale.

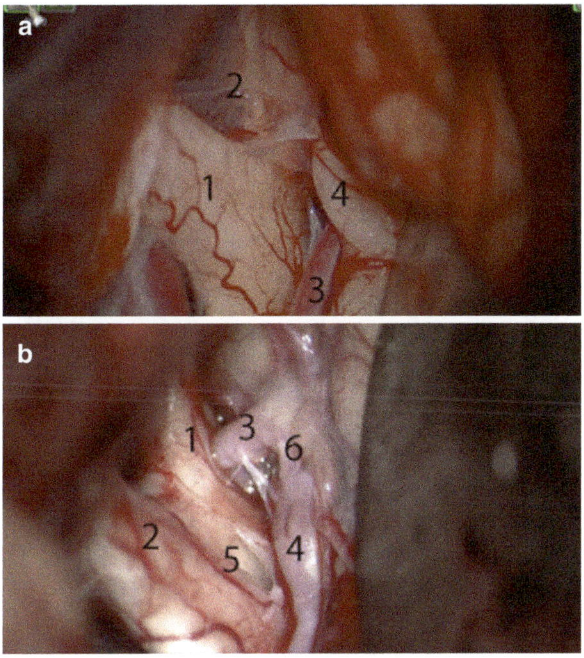

Fig. 12.7 Operative photographs of the lamina terminalis cistern (left pterional approach). (**a**) The ipsilateral and contralateral borders of the optic chiasm are the fundamental landmarks. Starting from the ipsilateral carotid cistern, the anterior aspect of the A1-anterior cerebral artery (ACA) provides a safe plane for dissection into the lamina terminalis cistern. (1) Left optic nerve. (2) Optic-chiasmatic cistern. (3) Left A1. (4) Gyrus rectus. (**b**) The complex of the ACoA is exposed and the lamina terminalis fenestrated. (1) Right optic nerve. (2) Left optic nerve. (3) Right A1. (4) Left A1. (5) Lamina terminalis fenestrated. (6) AcoA

Superiorly, the gyri recti are joined by dense leptomeningeal trabeculations that separate the lamina terminalis and pericallosal cisterns, investing the A2-ACA at the transition between cisterns.

Inferiorly, the leptomeningeal trabeculae organize into an almost continuous membrane that separates the lamina terminalis and optic-chiasmatic cisterns.

Posteriorly, the cistern is bordered by the lamina terminalis, a potential point for ventricular access.

Laterally, the ipsilateral border of the optic chiasm is attached to the olfactory trigone by condensations of leptomeningeal trabeculae that enclose the A1-ACA and separate the carotid and lamina terminalis cisterns.

12.2.7 Pericallosal Cistern

The pericallosal cistern is a supratentorial cistern located at the deep part of the longitudinal fissure, containing the paired anterior and posterior pericallosal arteries as the central elements (Fig. 12.8).

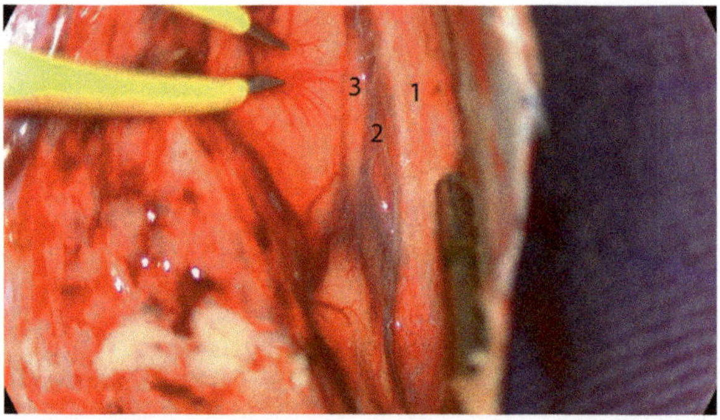

Fig. 12.8 Operative photograph of the pericallosal cistern (left frontoparietal interhemispheric approach). The lateral mobilization of the left frontal lobe reveals the bright arachnoid mater, apposed to the free edge of the falx cerebri, bordering superficially the pericallosal cistern. (1) Falx cerebri. (2) Pericallosal cistern. (3) Cingulate gyrus

The cistern is superficially bordered by the arachnoid mater apposed to the planum sphenoidale and the free border of the falx cerebri and the rostrum, body, and splenium of the corpus callosum.

Variable extensions of the medial surface of the cerebral hemispheres form lateral walls. The cistern is wide anteriorly and tight posteriorly.

Anteriorly, the pericallosal and lamina terminalis cisterns share a wall formed by dense leptomeningeal trabeculae that join the paired rectus gyri and surround the A2-ACA at the transition between the cisterns.

Posteriorly, the pericallosal cistern continues around the splenium of the corpus callosum without demarcation with the quadrigeminal and velum interpositum cisterns.

12.3 Conclusion

The subarachnoid space provides oriented landmarks and minimally traumatic pathways to a wide range of possible extrinsic and intrinsic pathologies involving deep intracranial areas. Microsurgical navigation through the cisterns provides a unique opportunity to appreciate the fine fundamental aspects of the anatomy in vivo, in real time, and in physiological conditions.

Suggested Bibliography

Almeida JP, Reghin Neto M, Chaddad Neto F, De Oliveira E. Anatomical considerations in the treatment of intracranial aneurysms. J Neurosurg Sci. 2016;60(1):27–43.

Chaddad-Neto F, Campos Filho JM, Dória-Netto HL, Faria MH, Ribas GC, Oliveira E. The pterional craniotomy: tips and tricks. Arq Neuropsiquiatr. 2012;70(9):727–32.

Chaddad-Neto F, Doria-Netto HL, Campos-Filho JM, Ribas ES, Ribas GC, de Oliveira E. Head positioning for anterior circulation aneurysms microsurgery. Arq Neuropsiquiatr. 2014;72(11):832–40.

Osborn AG. Osborn's brain: imaging, pathology, and anatomy. Salt Lake City: Amirsys Publishing, Inc.; 2012.

Rhoton AJ. General and micro-operative techniques. In: Youmans JR, editor. Neurological surgery, vol. 2. 4th ed. Philadelphia: WB Saunders; 1996. p. 724–66.

Rhoton AL Jr. Tentorial incisura. Neurosurgery. 2000;47(Suppl 1):S131–53.

Rhoton AL Jr. The supratentorial arteries. Neurosurgery. 2002;51(4 Suppl):S53–S120.

Rhoton AL Jr. Rhoton's cranial anatomy and surgical approaches. Philadelphia: Lippincott Williams & Wilkins; 2003.

Riechelmann GS, da Costa MDS, Caramanti RL, Goiri MAA, Costa BL, González-Echeverría K, Chaddad-Neto F. Microsurgical Clip Placement for a Giant Anterior Communicating Artery Aneurysm With Intraluminal Thrombus: 2-Dimensional Operative Video. Oper Neurosurg (Hagerstown). 2019;16(3):E92–E93.

Yasargil M. Microneurosurgery. In: Microsurgical anatomy of the basal cisterns and vessels of the brain, diagnostic studies, general operative techniques, and pathological considerations of intracranial aneurysms, vol. 1. New York: George Thieme Verlag; 1984.

Chapter 13
Surgical Anatomy of the Posterior Basal Cisterns

13.1 Interpeduncular Cistern

The interpeduncular cistern is located between the cerebral peduncles and dorsum sellae, between the temporal lobes. It is considered a basal transition cistern, since it is a posterior basal cistern that relates to the anterior basal and posterior fossa cisterns. The Liliequist membrane has an important role in this cisternal transition (Fig. 13.1).

The Liliequist membrane is an arachnoid membrane that lies in the posterior clinoid process and the dorsum sellae, extending into the space between the oculomotor nerves, which are the pillars onto which their leaflets are attached. The Liliequist membrane has two arachnoid sheets: a diencephalic and a mesencephalic one. The diencephalic sheet attaches to the posterior edge of the mammillary bodies, separating the optic-chiasmatic and interpeduncular cisterns. The mesencephalic sheet attaches to the midbrain-pontine transition, separating the interpeduncular and pre-pontine cisterns. Moreover, the interpeduncular cistern is laterally related to the cistern of the lateral fissure of the brain (Sylvian cistern) (Fig. 13.2).

The interpeduncular cistern is related to the anterior portion of the tentorial incisure, at the confluence of the supra- and infratentorial parts of the subarachnoid space. Its superior limit is the posterior edge of the mammillary bodies (transition with the optic-chiasmatic cistern), and its inferior limit is at the midbrain-pontine transition (correlation to the pre-pontine cistern). As previously stated, these limits are defined by the Liliequist membrane. The oculomotor nerves run along the lateral wall of the interpeduncular cistern, while the posterior wall is formed by the posterior perforated substance.

Supplementary Information The online version of this chapter (https://doi.org/10.1007/978-3-030-82747-2_13) contains supplementary material, which is available to authorized users.

F. Chaddad-Neto, M. D. Silva da Costa, *Microneuroanatomy and Surgery*,
https://doi.org/10.1007/978-3-030-82747-2_13

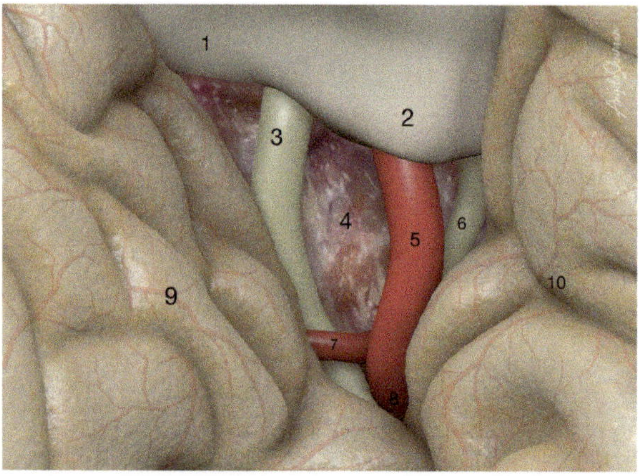

Fig. 13.1 Right transsylvian surgical approach. (1) Planum sphenoidale, (2) anterior clinoid process, (3) optic nerve, (4) Liliequist membrane, (5) ICA displaced laterally (opening of the optic-carotid space), (6) oculomotor nerve, (7) ACA, (8) MCA, (9) frontal lobe, (10) temporal lobe

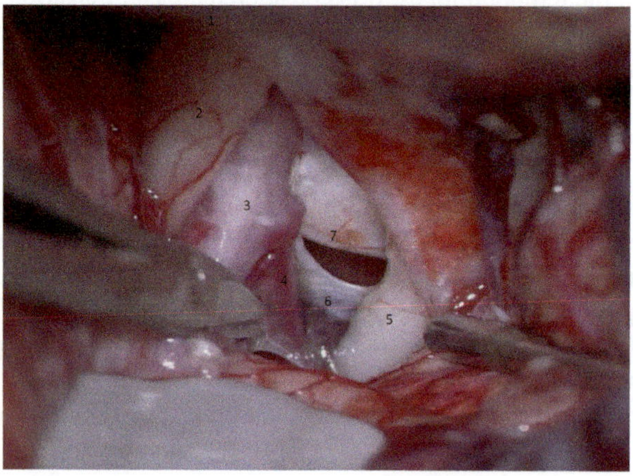

Fig. 13.2 Right transsylvian surgical approach. The falciform ligament (1) is a dural fold above the optic nerve (2) proximal to the entrance of the optic canal. The ICA (3) is inferolateral to it and runs over the carotid cistern, and the AChA (4) emerges laterally in this cerebrospinal fluid space. Located deep down, and between the oculomotor nerve (5) and the ICA, the Liliequist membrane (6) is an arachnoid membrane that lies over the dorsum sellae (7)

The interpeduncular cistern contains several vascular structures essential to the brain: the bifurcation of the basilar artery, peduncular segments of the posterior cerebral arteries, peduncular segments of the superior cerebellar arteries, perforating branches of the posterior cerebral arteries, posterior communicating arteries, basal vein of Rosenthal, interpeduncular vein, and posterior communicating vein (Fig. 13.3).

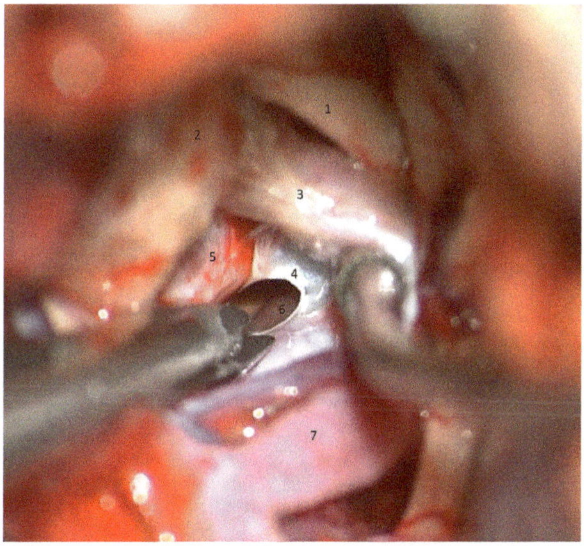

Fig. 13.3 Left transsylvian surgical approach. Opening the Liliequist membrane and displacing the ICA medially (carotid-oculomotor space), it is possible to access the structures in the interpeduncular cistern as the bifurcation of the basilar artery and P1 segment of PCA. (1) Optic nerve, (2) anterior clinoid process, (3) ICA, (4) Liliequist membrane, (5) dorsum sellae, (6) basilar artery, (7) MCA

13.1.1 Clinical Case 1

A 45-year-old woman presented to the emergency room with headache and a brief loss of consciousness 24 hours prior to consulting. A clinical examination revealed a Glasgow coma scale of 15 with nuchal rigidity, with no other neurological deficits. Brain tomography revealed no signs of edema or subarachnoid hemorrhage. A lumbar puncture was performed to obtain CSF, pointing to subarachnoid hemorrhage (Figs. 13.4, 13.5, and 13.6).

13.2 Crural Cistern

The crural cistern is located between the cerebral peduncle (medial wall) and the temporal lobe uncus (lateral wall) contained in the middle incisural space (Fig. 13.7). The cerebral crus (*crus cerebri*) usually refers to the most anterior portion of the cerebral peduncle, which contains white fibers of the motor tract. The crural cistern consists of a posterolateral extension of the interpeduncular cistern and communicates posteriorly with the ambient cistern. The roof of the crural cistern is formed by the optic tract.

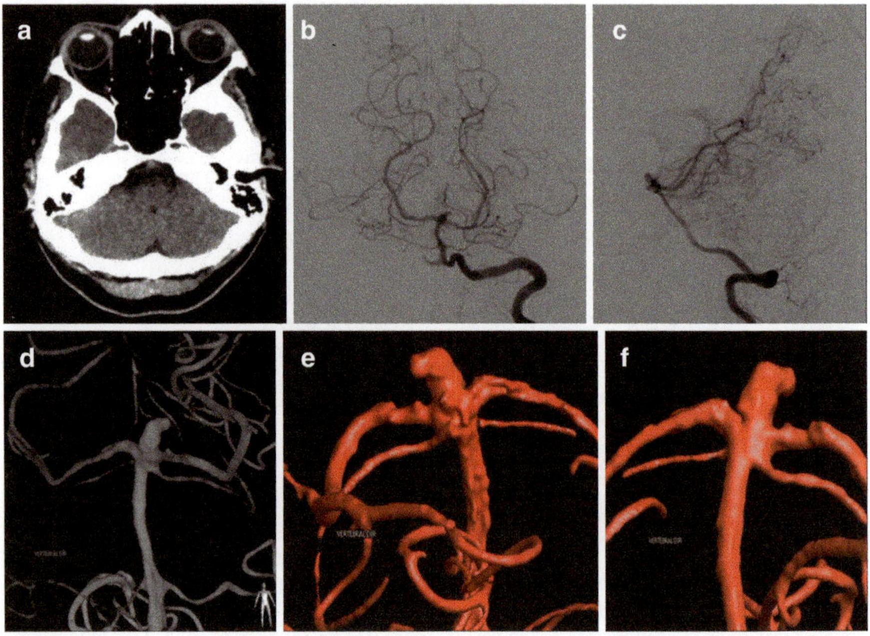

Fig. 13.4 Preoperative images. (**a**) Brain tomography shows an hyperdense image at the lower level of the interpeduncular cistern. (**b**) Posterior circulation angiography in anteroposterior (**b**) and lateral views (**c**) show an aneurismatic lesion located at the basilar tip. (**d–f**) Digital subtraction and tridimensional representation of the lesion show an irregular conformation with a lobule at the tip, likely corresponding to the rupture point of the aneurysm

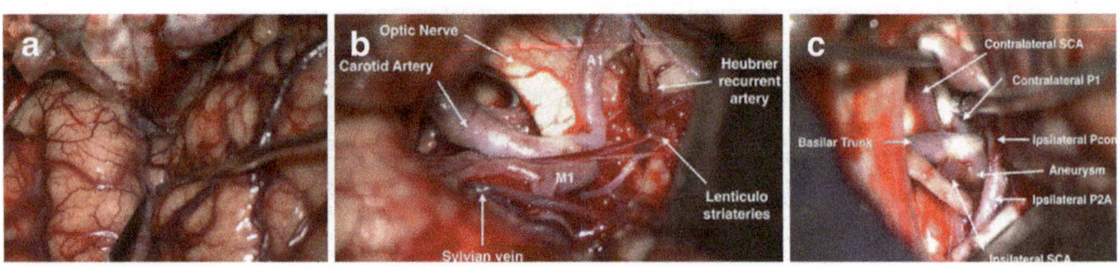

Fig. 13.5 Intraoperative images. An orbitozygomatic approach was used to reach the interpeduncular cistern, where the basilar bifurcation (high) and aneurysm were located. (**a**) Sylvian fissure opening. (**b**) Opening of the carotid, chiasmatic, and lamina terminalis cisterns, showing the contents. (**c**) Approach to the interpeduncular cistern, where the aneurysm was located, shows all anatomical features related to the approach

The P2 segment of the posterior cerebral artery is an important structure in the crural cistern (see below). Other vascular structures of this cistern include the anterior choroidal artery, the medial posterior choroidal artery, and the basal vein of Rosenthal.

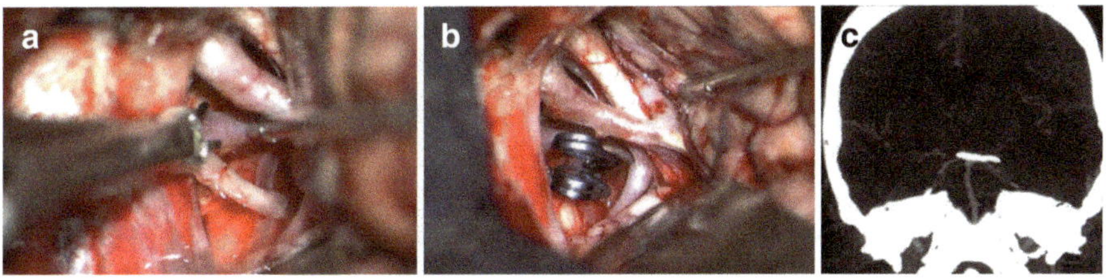

Fig. 13.6 Transoperative and control images. (**a** and **b**) Aneurysm clipping. (**c**) Brain angiotomography showing the surgical clips and the complete exclusion of the aneurysm from normal brain circulation

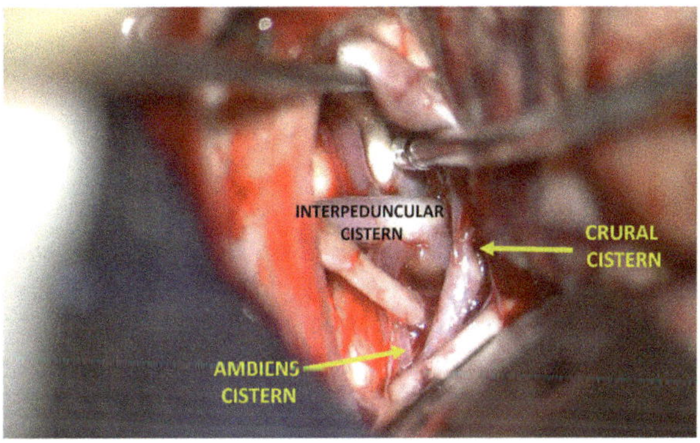

Fig. 13.7 Left transsylvian surgical approach and visualization of the posterior cisternal spaces

13.3 Ambient Cistern

The ambient cistern is limited anteromedially by the midbrain, superiorly by the pulvinar, and laterally by the parahippocampal and dentate gyri (Fig. 13.7). It communicates anteriorly with the crural cistern and posteriorly with the quadrigeminal cistern. The roof of this cistern is formed by the inferolateral portion of the pulvinar of the thalamus and by the geniculate bodies. The ambient cistern extends below the free margin of the tentorium to the cerebellomesencephalic fissure, above the origin of the trigeminal nerve.

The supratentorial compartment of the ambient cistern contains the basal vein and posterior cerebral artery. The infratentorial compartment contains the superior cerebellar artery and fourth nerve.

13.3.1 Clinical Case 2

A 48-year-old man presented with a history of intermittent emetic episodes lasting 3 months, preceded by a progressive loss of muscular strength in the legs and arms 6 months prior to admission. A clinical examination revealed that the patient was alert, attentive, and oriented and showed signs of tetraparesis (Medical Research Council grade 3/5 on both arms and 2/5 on both legs) (Figs. 13.8, 13.9, 13.10, and 13.11). (*This case is also demonstrated in the* Video 13.1)

13.4 Quadrigeminal Cistern

The quadrigeminal cistern (also known as collicular cistern or cistern of the great cerebral vein) is located posteriorly to the tectal plate of the midbrain, between the splenium of the corpus callosum and the superior surface of the cerebellum (Fig. 13.12). The tectum is the dorsal portion of the midbrain, located posteriorly to the mesencephalic aqueduct. The quadrigeminal plate, also known as the tectal plate or tectum, is constituted by prominent eminences, including the superior and inferior colliculi. Inferiorly to the tectal plate and immediately above the middle cerebellar peduncles, the trochlear nerves emerge.

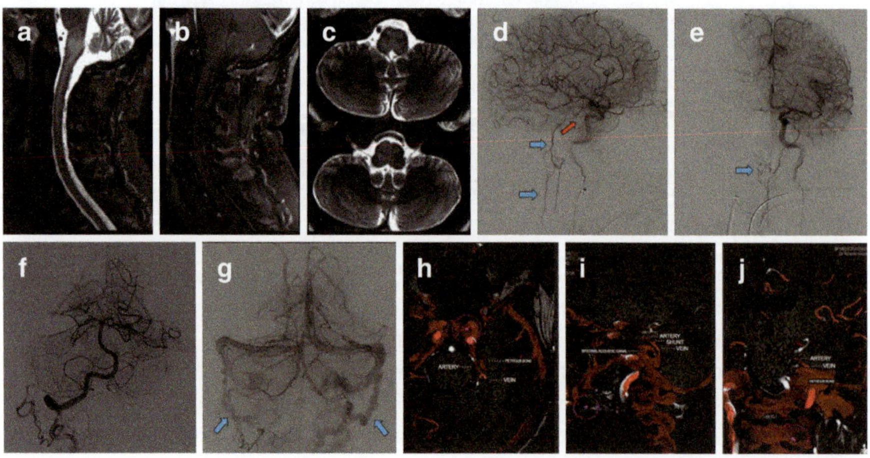

Fig. 13.8 Preoperative images. (**a–c**) T2-, flair-, and STIR-weighted images showing hyperintensity of the medulla oblongata and high cervical spinal cord, correlated with edema due to venous congestion. In addition, vascular hypertrophy was noticed on the anterior and posterior cervical pial surfaces. (**d–g**) Cerebral angiography showed a dural arteriovenous fistula on the left free margin of the tentorium with arterial irrigation by a meningeal branch from the meningohypophyseal trunk (red arrow). Lower venous drainage was identified through the lateral mesencephalic and perimedullary veins, resulting in venous congestion (blue arrows). (**h–j**) Fusion of cerebral angiography and 3D TOF magnetic resonance angiography showing the vascular and osseous relationships

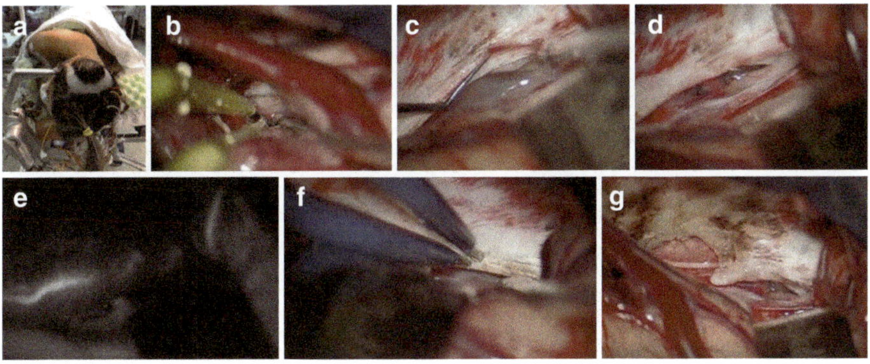

Fig. 13.9 Intraoperative images. (**a**) Lateral decubitus position for subtemporal approach. (**b**) Subtemporal approach with inferior temporal gyrus resection for mayor exposure. (**c** and **d**) Dissection and aperture of the ambient cistern exposing the fourth cranial nerve. (**e**) Indocyanine green intraoperative angiography showing the fistula. (**f** and **g**) Tentorial dural arteriovenous fistula resection and tentorial coagulation

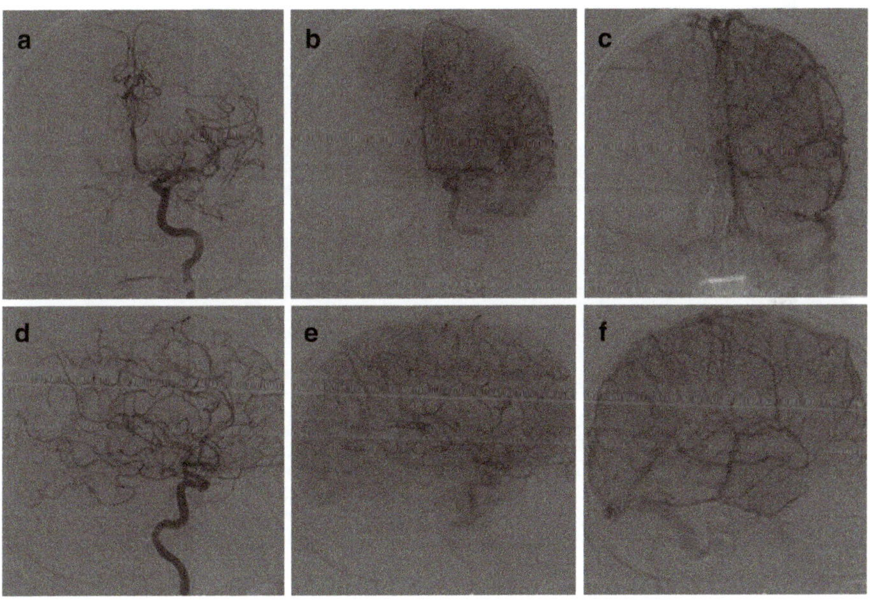

Fig. 13.10 Postoperative angiography. (**a–f**) Cerebral angiography images showing complete resolution of the dural arteriovenous fistula

The quadrigeminal cistern contains the vein of Galen, the posterior pericallosal arteries, the superior cerebellar arteries, perforating branches of the posterior cerebral and superior cerebellar arteries, and the P3 segment of the posterior cerebral arteries.

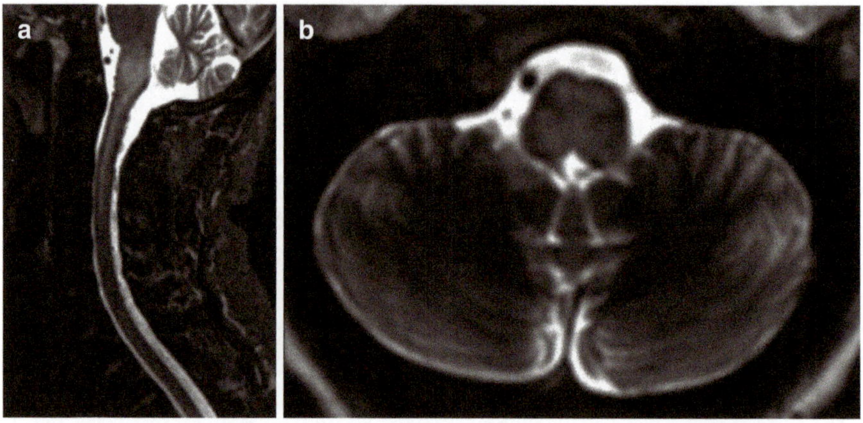

Fig. 13.11 One-month postoperative MRI. (**a** and **b**) T2-weighted axial and sagittal images showing improvement of the edema located at the medulla oblongata and high cervical spinal cord

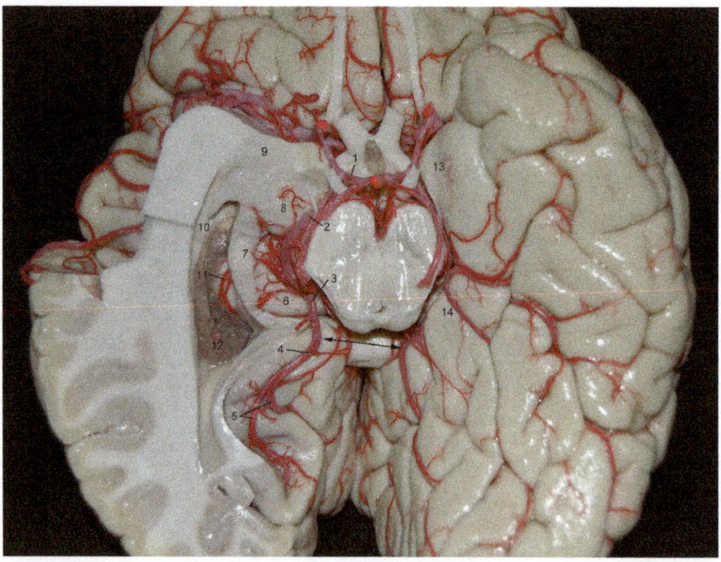

Fig. 13.12 Cadaveric dissection illustrating the posterior basal cisterns, which are cerebrospinal fluid spaces around the midbrain. The P1 segment (1) of the PCA occupies the interpeduncular cistern and is limited laterally in the posterior communicant artery (PCOM). Related to the crus cerebri, the P2A segment (2) occupies the crural cistern, from which the MPChA emerges. After the lateral limit of the midbrain, the PCA curves later in the P2P segment (3), occupying the ambient cistern and finally reaching the quadrigeminal cistern in the P3 segment (4). 1. Posterior cerebral artery, P1 Segment; 2. Posterior cerebral artery, P2A Segment; 3. Posterior cerebral artery, P2P Segment; 4. Posterior cerebral artery, P3 Segment; 5. Posterior cerebral artery, P4 Segment; 6. Pulvinar of the thalamus; 7. Dentate gyrus; 8. Uncus; 9. Amygdala; 10. Temporal horn of the lateral ventricle; 11. Posterolateral choroidal arteries; 12. Atrium of the lateral ventricle; 13. Uncus; 14. Parahippocampal gyrus

13.5 Cistern of the Velum Interpositum

The velum interpositum is located on the roof of the third ventricle, below the fornix, and between the superomedial surfaces of the thalamus. It is arranged between two layers of the tela choroidea and is the area to which many ventricular veins converge prior to reaching the internal brain veins. It may consist of a closed space or an opening (cistern of the velum interpositum) filled with cerebrospinal fluid between the splenium and the pineal gland region, communicating with the quadrigeminal cistern.

13.6 Tentorial Incisura Relations

In addition to the midbrain, an important anatomical reference for cisternal localization is the tentorial incisura (tentorial notch), which establishes the only communication channel between the supratentorial and infratentorial spaces. The area between the superior part of the brainstem and the tentorium is divided into the anterior, middle, and posterior incisural spaces (Fig. 13.13).

The anterior incisural space is located anterior to the brainstem and contains the interpeduncular cistern, which is located between the brain peduncles and the dorsum sellae. The anterior incisural space contains all the arteries that form the circle of Willis. The middle incisural space is located laterally to the brainstem and is closely related to the medial portion of the temporal lobe. The supratentorial portion

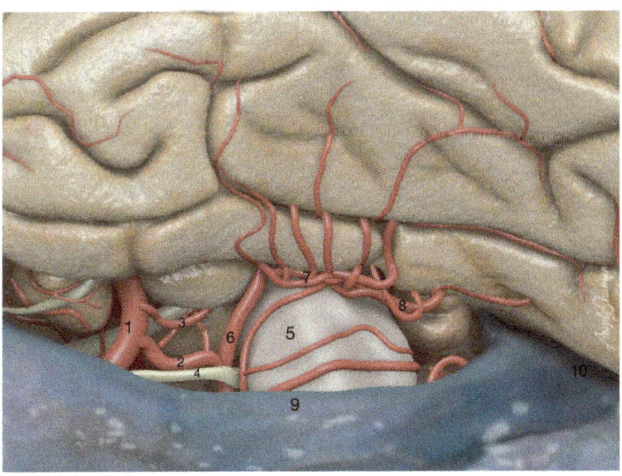

Fig. 13.13 Illustration of the left cerebral hemisphere. Retraction of the temporal lobe. The tentorial incisura is a large oval opening of the cerebellar tentorium (9–10) through which pass the cerebral peduncles (5), connecting the supratentorial and infratentorial spaces. (1) ICA, (2) Pcomm, (3) AChA, (4) oculomotor nerve, (5) midbrain, (6) P2A, (7) segments of P2P, (8) P3, (9) tentorial edge, (10) cerebellar tentorium

of the middle incisural space contains the crural and ambient cisterns. The posterior incisural space is located posterior to the midbrain, corresponding to the region in which the pineal gland is found. This space contains the convergence of the internal cerebral veins and the basal veins, in addition to many tributaries of the great cerebral vein (vein of Galen). The quadrigeminal cistern, located posteriorly to the tectum, is the largest cistern in the posterior incisural space.

13.7 Arterial Relations

The most important arteries in the brain run through cistern spaces. The arteries that cross the posterior basal cisterns are highlighted below.

The anterior choroidal artery (AChA) arises on the lateral face of the C4 segment of the internal carotid artery (ICA) and runs posteromedially, passing below and medially to the optic tract until it reaches the lateral margin of the cerebral peduncle (Fig. 13.14). At the anterior margin of the lateral geniculate body, it runs along a posterolateral route, through the crural cistern, towards the choroid plexus of the temporal horn of the lateral ventricle. The AChA thereafter has two segments: (1) a cisternal segment, divided into the anterior (from the origin to the lateral geniculate body) and posterior parts (from the lateral geniculate body to the origin of the choroidal fissure) and (2) a plexal segment, composed of the choroidal fissure and the choroid plexus of the temporal horn). The branches of

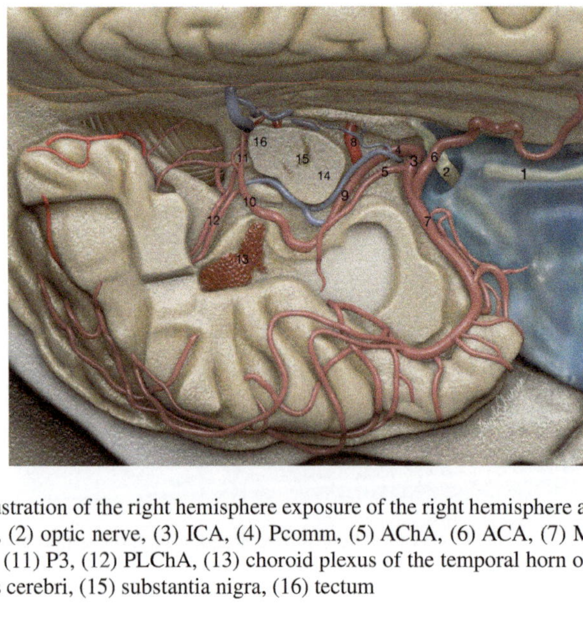

Fig. 13.14 Illustration of the right hemisphere exposure of the right hemisphere and midbrain. (1) Olfactory tract, (2) optic nerve, (3) ICA, (4) Pcomm, (5) AChA, (6) ACA, (7) MCA, (8) P1, (9) P2A, (10) P2P, (11) P3, (12) PLChA, (13) choroid plexus of the temporal horn of the lateral ventricle, (14) crus cerebri, (15) substantia nigra, (16) tectum

the cisternal segment irrigate the optic tract, the lateral geniculate body, the posterior arm of the internal capsule, the globus pallidus, and the medial part of the cerebral peduncle.

The posterior cerebral artery (PCA) originates from the bifurcation of the basilar artery and joins the posterior communicating artery on the lateral margin of the interpeduncular cistern (P1 segment, pre-communicating). It then lies posteriorly, involving the midbrain through the crural cisterns (P2A segment) and ambiens (P2P segment), after which it runs through the quadrigeminal cistern (P3 segment) to the posterior and cortical portion of the hemisphere (P4 segment) (Fig. 13.14). Thus, it is possible to understand the segmentation of the PCA in relation to the posterior basal cisterns: P1, interpeduncular; P2A, crural; P2P, ambiens; P3, quadrigeminal; and P4, cortical.

The posterior thalamo-perforating arteries are branches of P1. They enter the brain through the posterior perforated substance and the medial portion of the cerebral peduncle. Therefore, they are vessels that occupy the interpeduncular cistern. They irrigate the anterior portion and part of the posterior portion of the thalamus and hypothalamus as well as the subthalamus, substantia nigra, red nucleus, nucleus, proximal part of the oculomotor nerve, and the reticular formation of the midbrain.

The posterior choroidal arteries are important branches of the P2 segment. The posteromedial choroidal artery (MPChA) is a P2A branch surrounding the brainstem medially to the P2 and P3 segments, occupying the crural, ambient, and quadrigeminal cisterns. It then projects laterally to the pineal gland, reaching the roof of the third ventricle. The MPChA emits branches to the cerebral peduncle, medial geniculate body, colliculi, pulvinar, and medial thalamus. The posterolateral choroidal artery (PLChA) is usually a P2P branch, although it may be a P2A or P3 branch, and it runs through the choroidal fissure between the fimbria and thalamus, towards the choroid plexus of the temporal horn and the atrium.

The perforating peduncular branches also originate from P2, in the crural cistern, and run directly to the cerebral peduncle. The circumflex branches are another group of arteries that originate from P2. The short circumflex arteries supply the geniculate bodies and the tegmentum of the midbrain, while the long circumflex arteries reach the quadrigeminal cistern and irrigate the colliculi. In P2, the thalamogeniculate arteries emerge, originating below the lateral thalamus and penetrating the roof of the cistern to irrigate the lateral thalamus, the posterior arm of the internal capsule, and the optic tract.

The PCA has cortical branches. The hippocampal artery is the first cortical branch of P2, originating in the crural cistern or in the ambient cistern. This artery irrigates the uncus, the parahippocampal gyrus, the hippocampal formation, and the dentate gyrus. The anterior, middle, and posterior temporal arteries originate in the crural cistern or ambient cistern and irrigate the inferior surface of the temporal lobe. The parietooccipital and calcarine arteries are also cortical branches of the PCA, originating in the quadrigeminal cistern in most cases.

13.8 Venous Relations

The internal cerebral veins originate immediately behind the interventricular foramen and project posteriorly into the vellum interpositum, leaving it immediately above the pineal gland, entering the quadrigeminal cistern to thereafter empty into the great cerebral vein.

The deep veins are structures that also occupy the posterior basal cisterns, with drainage through the interpeduncular, crural, ambient, and quadrigeminal cisterns. The veins that drain the structures anterior to the quadrigeminal cistern drain into the basal Rosenthal vein; those in this cistern drain mainly into the internal cerebral and great cerebral veins. The area drained by the cisternal venous group can also be divided into three regions, in relation to the brainstem, namely, the anterior, medium, and posterior regions. It is worth noting that the quadrigeminal cistern has a particular relation to the venous structures, while the anterior cistern is predominated by arterial vessels, and the quadrigeminal cistern is predominated by venous vessels. The major basal veins drain to the vein of Galen, which produces a huge venous complex that is crucial to successful surgical manipulations in this area.

Suggested Bibliography

Chaddad Neto F, Doria Netto HL, Campos Filho JM, Reghin Neto M, Silva-Costa MD, Oliveira E. Orbitozygomatic craniotomy in three pieces: tips and tricks. Arq Neuropsiquiatr. 2016;74(3):228–34.

Gazi Yaşargil M, Kasdaglis K, Jain KK, Weber H-P. Anatomical observations of the subarachnoid cisterns of the brain during surgery. J Neurosurg. 1976;44(3):298–302.

Granados OS, da Costa MDS, Costa BL, González-Echeverría K, Paganelli SL, Caramanti RL, Palmiero H, Chaddad-Neto F. Microsurgery for Upper Basilar Tip Aneurysm With Intraoperative Rupture: 3-Dimensional Operative Video. Oper Neurosurg (Hagerstown). 2019 Feb 1;16(2):43.

Rhoton AL. The posterior fossa cisterns. Neurosurgery. 2000;47(suppl_3):S287–97.

Wainberg RC, da Costa MDS, Marchiori M, Soder RB, de Campos Filho JM, Netto HLD, Neto EP, Chaddad-Neto F. Microsurgical Clipping of Low-Riding Basilar Bifurcation Aneurysm. World Neurosurg. 2018 Dec 31:S1878-8750(18)32916–4.

Yağmurlu K, Kalani MYS, Chaddad-Neto F, Cevik OM, Bozkurt B, Belykh E, Doria-Netto HL, Grande AW, Preul MC, Spetzler RF. Anterior temporal artery to posterior cerebral artery bypass for revascularization of the posterior circulation: An anatomical study. J Clin Neurosci. 2018;47:337–40

Chapter 14
Surgical Anatomy of the Posterior Fossa Cisterns

14.1 Posterior Fossa Cisterns

1. *Unpaired*
   1.1 Cisterna Magna
   1.2 Interpeduncular Cistern
   1.3 Prepontine Cistern
   1.4 Premedullary Cistern
   1.5 Quadrigeminal Cistern
2. *Paired*
   2.1 Cerebelopontine Cistern
   2.2 Cerebelomedullary Cistern

14.1.1 Cisterna Magna

It is located dorsal to the medulla and cerebellar vermis (Fig. 14.1). Its posterior wall is formed by the arachnoid membrane that covers the inner surface of the occipital bone above the foramen magnum. It contains a dense arachnoid mesh that extends from the cerebellar tonsils to the medulla and to the margin of the foramen of Magendie.

The lower portion of the cisterna magna is located behind the medulla. Superiorly, it protrudes anteriorly and posteriorly into the cerebellar vermis. Anteriorly, it opens into the cerebellomedullary fissure and posteriorly into the posterior cerebellar

---

Supplementary Information The online version of this chapter (https://doi.org/10.1007/978-3-030-82747-2_14) contains supplementary material, which is available to authorized users.

F. Chaddad-Neto, M. D. Silva da Costa, *Microneuroanatomy and Surgery*,
https://doi.org/10.1007/978-3-030-82747-2_14

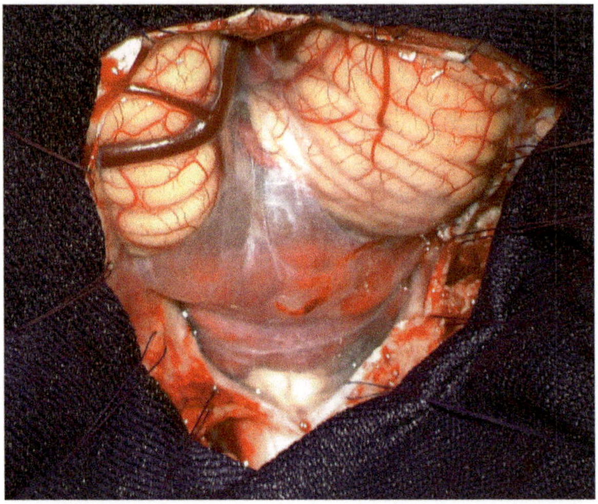

Fig. 14.1 Surgical view of a suboccipital craniotomy showing the cisterna magna at the craniocervical junction

incisura. The arachnoid membrane covering the incisura is reflected around the falx cerebelli. The upper limit of the extension behind the vermis is formed by the tentorium. The upper part of the cistern may be wider if the falx cerebelli is small or absent. A median arachnoid leaflet can extend from the dorsal surface of the medulla to the outer arachnoid membrane to divide the cistern into two sagittal halves. Inferiorly, the cisterna magna communicates, without obstruction, with the posterior spinal cistern.

The posterior cerebellar arteries pass posteriorly around the medulla. They enter the cisterna magna at a point where they commonly divide into a lateral trunk, which irrigates the hemisphere and tonsil, and a medial trunk, which irrigates the vermis.

14.1.2 Interpeduncular Cistern

The interpeduncular cistern is located between the cerebral peduncles and the leaves of the Liliequist membrane, which form its upper and lower limits (Fig. 14.2). The posterior wall of the cistern is formed by the posterior perforated substance and its upper limit, by the posterior border of the mammillary bodies. The lower border lies between the midbrain and the pons.

The Liliequist membrane arises from the arachnoid membrane of the posterior clinoid process and dorsum of the sella turcica and has two leaflets: the diencephalic membrane, between the diencephalon and posterior border of the mammillary bodies, which separates the chiasmatic and interpeduncular cisterns, and the mesencephalic membrane, between the junction of the midbrain and pons, which separates the interpeduncular and pre-pontine cisterns. The lateral edges are attached to the arachnoid membrane of the two oculomotor nerves. The diencephalic membrane is

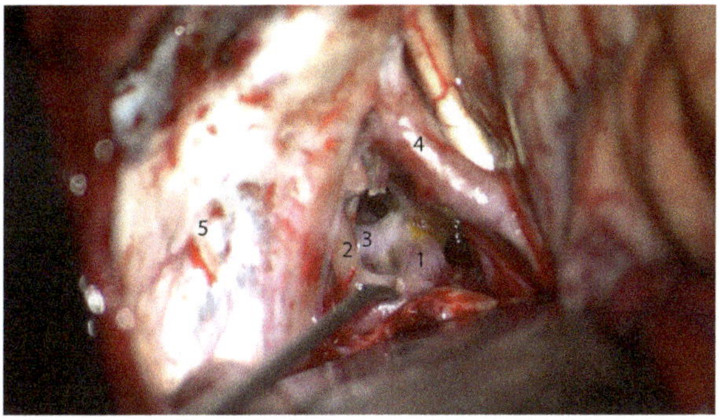

Fig. 14.2 Surgical view of a left pretemporal craniotomy using the temporopolar route for accessing a basilar tip aneurysm, after dissection of the interpeduncular cistern at the origin of the PCA and SCA, and the prepontine cistern, which contains the basilar artery. (1) Left posterior cerebral artery, segment P1; (2) left oculomotor nerve, (3) basilar artery; (4) Left supraclinoid segment of the ICA, (5) medial fossa of the skull base. PCA posterior cerebellar artery, SCA superior cerebellar artery, ICA internal carotid artery

thicker and well developed, while the mesencephalic membrane is thinner and extends to surround the basilar artery until it reaches the interpeduncular fossa. Several trabeculae of the diencephalic membrane are attached to the pituitary stalk, mammillary bodies, and posterior cerebral communicating arteries. The interpeduncular cistern communicates with the crural and ambient cisterns.

The oculomotor nerves form the lateral wall of the interpeduncular cistern and also separate the supra and infratentorial cisterns. The following membranes converge toward the oculomotor nerves: mesencephalic membrane (between the interpeduncular and prepontine cisterns), diencephalic (between the interpeduncular and chiasmatic cisterns), anterior pontine membrane (between the cerebellopontine and prepontine cisterns), lateral mesencephalic membrane (between the cisterna ambiens and cerebellopontine cistern), medial carotid membrane (between the chiasmatic and carotid cisterns), and lateral wall of the carotid cistern.

The interpeduncular cistern contains the thalamoperforating arteries, bifurcation of the basilar artery, origins of the posterior cerebral artery, superior cerebellar and posteromedial choroidal arteries; and peduncular, posterior communicating, anterior median pontomesencephalic veins and pontomesencephalic sulcus.

14.1.3 Prepontine Cistern

The prepontine cistern is located between the clivus anteriorly and the anterior surface of the pons posteriorly. Its upper limit is formed by the mesencephalic portion of the Liliequist membrane, and the lower limit is formed by the medial pontomedullary membrane in the pontomedullary sulcus, which surrounds the junction between the basilar and vertebral arteries. The lateral limits are formed by the anterior pontine membranes, which separate it from the cerebellopontine cisterns.

The anterior pontine membranes are closely related to the abducens nerves and become thinner as they extend caudally. The prepontine cistern does not contain cranial nerves but contains the basilar artery and origins of the anterior inferior cerebellar arteries.

14.1.4 Premedullary Cistern

The premedullary cistern is located between the anterior surface of the medulla and the arachnoid that covers the lower portion of the clivus. Its upper portion is located at the junction between the pons and medulla. It is separated from the prepontine cistern by the medial pontomedullary membrane. It is limited laterally by the cerebellomedullary cistern on the dorsal margin of the inferior olive in front of the glossopharyngeal, vagus, and accessory nerves, where the arachnoid becomes denser. Inferiorly, it is continuous with the anterior spinal cistern. The roots of the hypoglossal nerves arise within the posterior wall of the cistern, between the medullary pyramids and inferior olives.

The vertebral arteries enter the cistern as they ascend through the foramen magnum, crossing the cistern obliquely to the junction of the premedullary and prepontine cisterns. The anterior spinal arteries originate from the vertebral arteries and form a single trunk that runs along the midline in the anterior portion of the spinal cord.

14.1.5 Quadrigeminal Cistern

The quadrigeminal cistern encloses the space that corresponds to the pineal region (Figs. 14.3, 14.4, and 14.5). The quadrigeminal plate is located at the center of the anterior wall of the cistern. The anterior wall rostral to the colliculi in the midline is formed by the pineal gland. The pineal recess of the third ventricle protrudes into

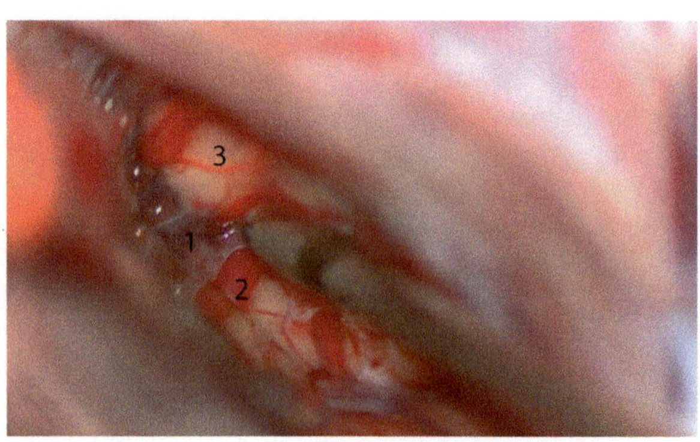

Fig. 14.3 Surgical view of a posterior interhemispheric approach, with the left medial parietal surface retracted, showing the quadrigeminal cistern and the contralateral hemisphere – right hemisphere. (1) Quadrigeminal cistern, (2) parieto-occipital artery, (3) cingulate gyrus of the right hemisphere

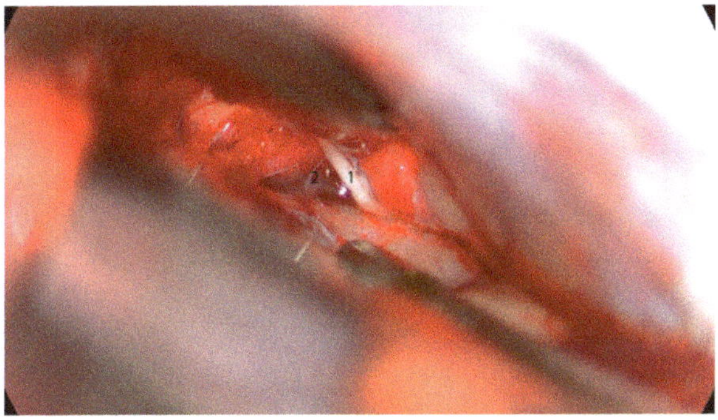

Fig. 14.4 Sequence of Fig. 14.3 after dissecting the quadrigeminal cistern. (1) Trochlear nerve, (2) opened quadrigeminal cistern

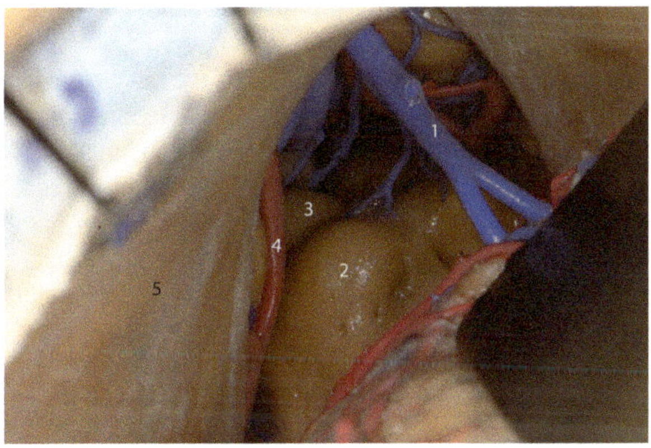

Fig. 14.5 Cadaveric dissection using the supracerebellar route to expose the region of the pineal gland and contents of the quadrigeminal cistern. (1) Precentral cerebellar vein, (2) inferior colliculus, (3) superior colliculus, (4) posterior cerebral artery, (5) tentorium

the cistern above the gland. Laterally, the anterior wall is formed by a part of the pulvinar, which is located medial to the point where the crus of the fornix surrounds the pulvinar. The medial half of the pulvinar forms the anterior wall of the cistern, and its lateral half forms the anterior wall of the atrium of the lateral ventricle.

Each lateral wall of the cistern has an anterior and posterior part. The anterior part is formed by the segment of the crus of the fornix surrounding the pulvinar. The posterior is formed by the part of the occipital cortex located below the splenium. The cistern extends to the cerebellomesenchephalic fissure below the colliculi.

The roof of the cistern is formed by the lower surface of the splenium and the broad membranous envelope that surrounds the great vein and its tributaries. This envelope is fixed to the inferior surface of the splenium and is continuous with the preceding choroidal mesh around the velum interpositum. The density of the venous structures is the highest within this envelope, in the superomedial part of the cistern. On the other hand, arteries are found in greater density in the inferolateral portion of the cistern.

The quadrigeminal cistern communicates with the posterior pericallosal cistern, which extends around the splenium. It opens inferolaterally below the pulvinars, into the ambient cistern, which is located between the midbrain and temporal lobes. It can communicate with the velum interpositum. The trochlear nerves originate in the quadrigeminal cistern below the inferior colliculi and traverse the midbrain and below the pulvinars to enter the cistern.

The trunks and branches of the posterior cerebral and superior cerebellar arteries enter the anterior inferior portion of the cistern and run lateral to and beneath the arachnoid envelope around the vein of Galen and its tributaries. The posterior cerebral arteries commonly bifurcate into the calcarine and parieto-occipital branches inside the cistern. The posteromedial choroidal arteries arise from the posterior cerebral arteries in front of the midbrain and surround the brainstem to enter the quadrigeminal cistern, where they turn back beside the pineal body to reach the velum interpositum. The upper cerebellar arteries run through the part of the cistern extending into the cerebellomesencephalic fissure. The perforating branches of the posterior cerebral arteries irrigate the cistern walls located above the shallow groove between the superior and inferior colliculi, and the superior cerebellar arteries irrigate the cistern walls below this groove.

The venous relationships in the cistern are the most complex in the skull because the cistern is the site of convergence of the internal cerebral vein, basal veins, and multiple tributaries of the vein of Galen. The internal cerebral veins exit the velum interpositum, and the basal veins exit the cistern to reach the quadrigeminal cistern, where they drain into the vein of Galen. The latter pass below the splenium to enter the sinus rectus at the tentorium apex. The veins that converge in the cistern to drain into the great, basal, or internal cerebral veins include the posterior pericallosal veins, which run around the splenium; the atrial veins, which drain the walls of the atrium; the internal occipital veins, which originate at or near the calcarine and parieto-occipital sulci; and the vein of cerebellomesencephalic fissure, which originates in the superior cerebellar peduncles and ends with the superior vermian vein, which in turn drains into the great vein of Galen.

14.2 Superior Cerebellar Cistern

The superior cerebellar cistern is located between the upper part of the vermis and the arachnoid that rests on the lower edge of the straight sinus. It first opens into the quadrigeminal cisterna and communicates posteriorly with the cisterna magna, below the torcula. Laterally, it communicates with the subarachnoid space above the cerebellar hemispheres. It contains the median and paramedian branches of the superior cerebellar arteries and superior vermian vein.

14.2.1 Cerebellopontine Cistern

The cerebellopontine cistern is located between the anterolateral surface of the pons and cerebellum and the arachnoid membrane attached to the petrous bone (Figs. 14.6 and 14.7). It is separated superiorly from the ambient cistern by the lateral

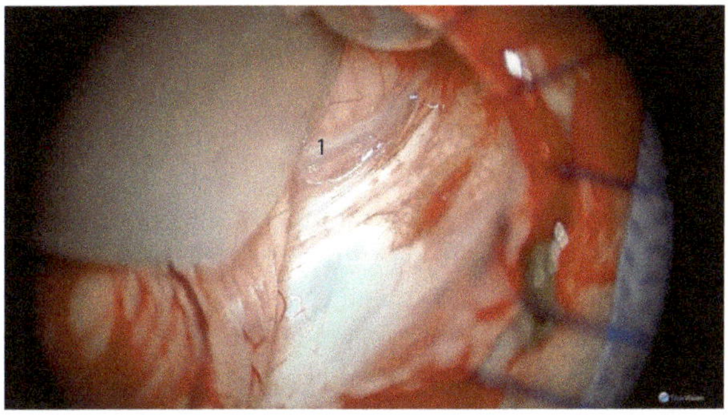

Fig. 14.6 Surgical view of the lateral suboccipital craniotomy–retrosigmoid approach showing the cerebellopontine cistern (1), before its opening

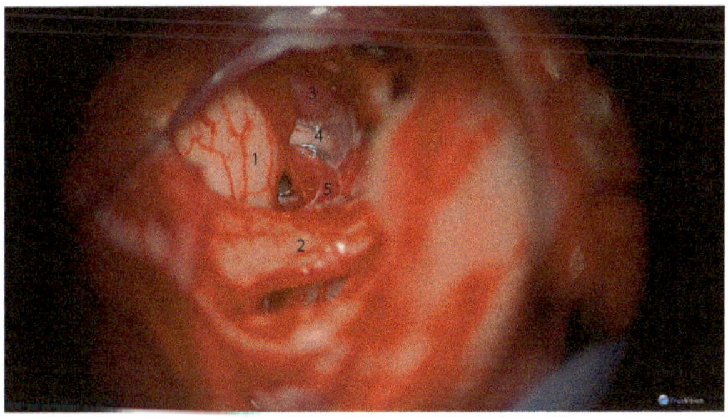

Fig. 14.7 Microsurgical view after resection of the right vestibulocochlear (eighth cranial) nerve schwannoma using the retrosigmoid approach. (1) Pons, (2) complex right seventh and eighth cranial nerves, (3) basilar artery, (4) prepontine cistern, (5) right anterior inferior cerebellar artery

pontomesencephalic membrane, which is fixed at the junction between the midbrain and pons and also at the tentorium. It intersects the oculomotor nerves and separates the posterior cerebral and superior cerebellar arteries. Inferiorly, it is separated from the cerebellomedullary cistern by the lateral pontomedullary membrane, which crosses the spaces between the vestibulocochlear and glossopharyngeal nerves. It is separated from the prepontine cistern by the anterior pontine membrane medially and extends laterally to the cerebellopontine fissure.

The trigeminal nerve arises at the middle portion of the pons and crosses the superolateral portion of the cistern. The abducens nerve originates at the level of the pontomedullary sulcus and ascends laterally to the anterior pontine membrane. The facial and vestibulocochlear nerves arise from the lower portion of the cistern, above the lateral medullary membrane. The external arachnoid lamina extends to the internal auditory canal and surrounds the intracanalicular segments of the facial and vestibulocochlear nerves. The flocculus protrudes into the cistern, behind the facial and vestibulocochlear nerves.

The cerebellopontine cistern is the central cistern in the lateral suboccipital craniotomy–retrosigmoid approach, which serves to guide the surgeon to the complex of the seventh and eighth cranial nerves. The flocculus can support the retractor to facilitate the exposure of the cerebellopontine cistern. A practical tip to find the proximal segment of the facial nerve (i.e., the seventh cranial nerve) in the retrosigmoid approach involves the identification of the ninth cranial nerve and tracing of an imaginary perpendicular line from its apparent origin in the brainstem to the intersection with the seventh cranial nerve.

The cerebellopontine cistern contains the upper cerebellar and anterior inferior cerebellar arteries. The superior cerebellar artery enters the cistern through the junction of the anterior pontine membrane and the oculomotor nerves and travels below the trochlear nerve and lateral mesencephalic membrane, and above the trigeminal nerve, as it passes through the cistern. The bifurcation of the superior cerebellar artery at its rostral and caudal trunk may be located in the prepontine or cerebellopontine cistern. The anterior inferior cerebellar artery enters the cistern through or below the anterior pontine membrane and commonly bifurcates into the rostral and caudal trunks within the cistern. The veins in this cistern converge around the trigeminal nerve and combine to form the upper petrous veins, which in turn drain into the superior petrous sinus.

14.2.2 Cerebellomedullary Cistern

The cerebellomedullary cistern is located inferior to the pontomedullary junction (Fig. 14.8). It is separated from the cerebellopontine cistern by the lateral pontomedullary membrane to form the premedullary cistern by the trabeculae resting against the glossopharyngeal, vagus, and accessory nerves. Its lower limit is formed by the foramen magnum. This cistern extends behind the dorsal margin of the inferior olive around the dorsolateral portion of the medulla to the biventeral lobe of the cerebellum.

The glossopharyngeal and vagus nerves and the medullary portion of the accessory nerve originate inside the cistern and traverse it to reach the jugular foramen. The spinal portion of the accessory nerve arises from the posterior spinal cistern

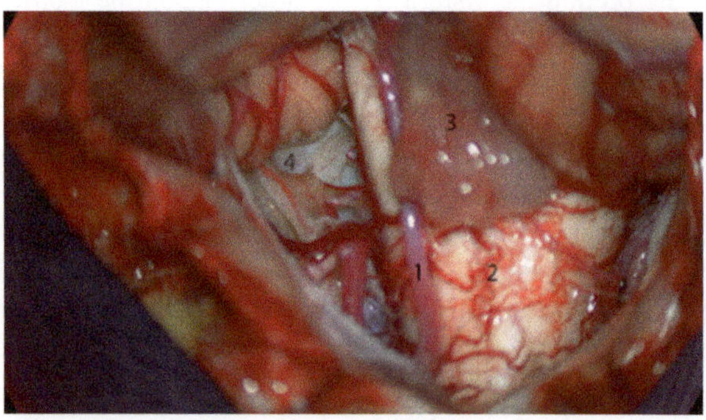

Fig. 14.8 Surgical view through a median suboccipital craniotomy using the telovelar approach, highlighting the relation of the left cerebellomedullary cistern to the lower cranial nerves and posterior inferior cerebellar artery. (1) Tonsillomedullary segment of PICA, (2) posterior surface of the medulla, (3) tumor of the fourth ventricle protruding through the foramen of Magendie, (4) lower cranial nerves after opening the cerebellomedullary cistern

until it reaches the cerebellomedullary cistern. The lateral recess of the fourth ventricle communicates with the cistern through the foramen of Luschka. The choroid plexus protruding from this foramen is located on the posterior surface of the glossopharyngeal and vagus nerves.

The vertebral artery pierces the dura mater at the lowest portion of the cistern and immediately exits it to enter the premedullary cistern. The posterior inferior cerebellar artery enters the cistern to reach the anterior surface of the roots of the glossopharyngeal, vagus, and accessory nerves, from where it travels dorsally between the roots of these nerves, around the medulla, to enter the cisterna magna.

14.3 Conclusion

Knowledge of the microneurosurgical anatomy of the posterior fossa cisterns and their close relations with the arterial, venous, and neural structures constitute a sine qua non condition for the proper management of lesions affecting this region. The presence of Lesions (tumors, aneurysms, hematomas, abscesses, cysts, vascular handles, etc.) may involve the structures associated with one, two, or more cisterns. These cisterns provide important information regarding the structures related to these lesions, choice of the best craniotomy and approach, as well as the relevant intraoperative strategy, all of which are aimed at satisfactory cerebrospinal fluid drainage, safe and complete access to the lesion, and minimization of the risk of injury to the adjacent structures

Suggested Bibliography

Chaddad Neto F, Doria-Netto HL, de Campos Filho JM, et al. The far-lateral craniotomy: tips and tricks. Arq Neuropsiquiatr. 2014;72(9):699–705. https://doi.org/10.1590/0004 282X20140130.

de Oliveira E, Tedeschi E, Rhoton A, Peace D. Microsurgical anatomy of the posterior circulation: vertebral and basilar arteries. In: Carter L, Spetzler R, editors. Neurovascular surgery. New York: McGraw Hill, Inc.; 1995. p. 25–34.

Osborn AG. Osborn's brain: imaging, pathology, and anatomy. Salt Lake City: Amirsys Publishing, Inc.; 2012.

Rhoton AL. The posterior cranial fossa: microsurgical anatomy and surgical approaches. Neurosurgery. 2000;47(suppl_3):S5–6. https://doi.org/10.1097/00006123-200009001-00005.

Yasargil M. Microneurosurgery, volume I: microsurgical anatomy of the basal cisterns and vessels of the brain, diagnostic studies, general operative techniques and pathological considerations of the intracranial aneurysms. New York: George Thieme Verlag; 1984.

Chapter 15
Surgical Anatomy of the Sellar Region

15.1 Osseous Relationships

The endocranial surface of the skull base faces the brain and is divided into anterior, middle, and posterior fossa (Fig. 15.1).

The anterior limit of the middle fossa is marked by the sphenoid ridge joined medially by the chiasmatic sulcus; the posterior limit includes the petrous ridge of the temporal bone joined medially by the posterior clinoid processes and dorsum sellae. The medial cranial base has medial and lateral parts: the medial part consists of the body of the sphenoid bone (sellar region), and the lateral part includes the temporal fossa (Fig. 15.2).

15.2 Sphenoid Bone

The lesser wings of the sphenoid bone are connected medially to the sphenoid body (Figs. 15.3 and 15.4); they form the roof of the optic canal and are continuous with the sphenoid planum. The optic strut (Fig. 15.5) (posterior root of the anterior clinoid process) separates the optic canals from the superior orbital fissure.

The chiasmatic sulcus and endocranial openings of the optic canals are located posterior and bilateral to the planum sphenoidale, respectively. Posteriorly, the chiasmatic sulcus is separated from the sellar cavity by the tuberculum sellae (Fig. 15.4). The posterior limit of the sellae is formed by the posterior clinoid processes and the dorsum sellae (continuous with the clivus), which is also the limit between the middle and posterior fossa.

Supplementary Information The online version of this chapter (https://doi.org/10.1007/978-3-030-82747-2_15) contains supplementary material, which is available to authorized users.

F. Chaddad-Neto, M. D. Silva da Costa, *Microneuroanatomy and Surgery*,
https://doi.org/10.1007/978-3-030-82747-2_15

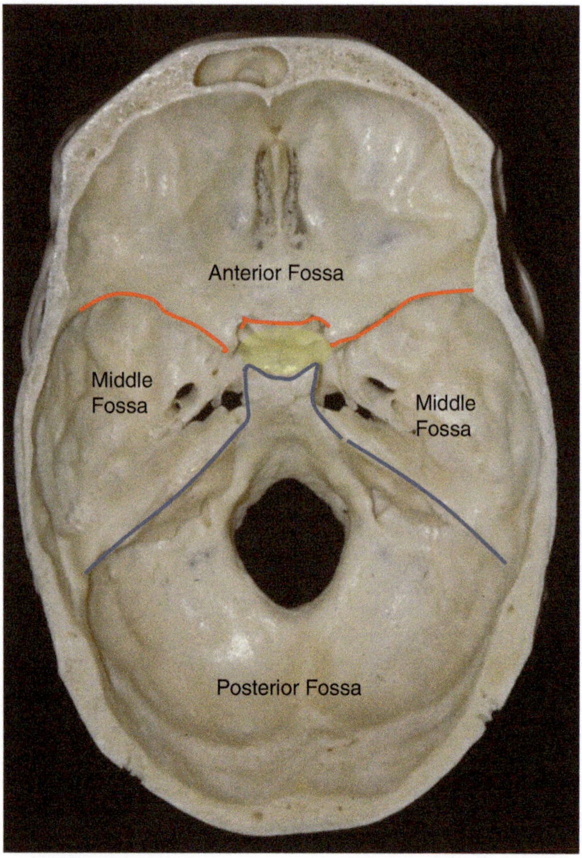

Fig. 15.1 Cranial base: endocranial surface. The border between the anterior and middle fossa is marked by the sphenoid ridge, joined medially by the chiasmatic sulcus (red line). The border between the middle and posterior fossa is marked by the petrous ridge joined by the dorsum sellae and posterior clinoid process (blue line). The sella is located on the endocranial surface of the middle cranial base over the body of the sphenoid bone (yellow shadow)

15.3 Sphenoid Sinus

The sphenoid sinus (Figs. 15.6 and 15.7), one of the four paranasal sinuses (cavity within the body of the sphenoid bone), is located in the upper back portion of the nasal cavity and separates the cavernous sinuses, cavernous segments of the carotid arteries, and optic and trigeminal nerves. It also separates the pituitary gland from the nasal cavity.

It has a wide variation in size and pneumatization (Fig. 15.8). There are three types of sphenoid sinus pneumatization in sphenoid sinus: conchal, sellar, and suprasellar. In the conchal type, the pneumatization is incomplete, and the region bellow the sellar is comprised of the solid bone, making it difficult to recognize the normal anatomy during surgery. The presellar type also presents incomplete pneumatization, and the air cavity does not penetrate beyond the anterior sellar wall. The

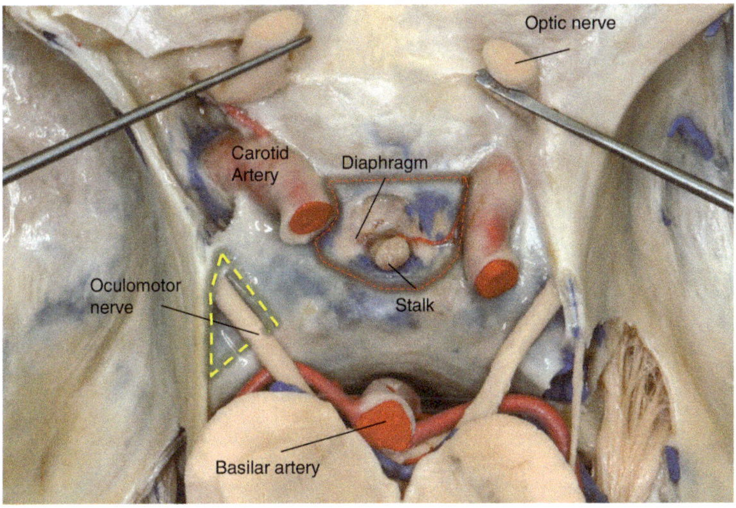

Fig. 15.2 Superior view of the sellar region. The sella (area in the red dashed line) is located between the cavernous sinuses. The diaphragm separates the sella from the suprasellar cisterns. The oculomotor nerves enter the roof of the cavernous sinus, where there is a narrow cistern around the nerve. The oculomotor triangle (yellow dashed line), the triangular patch of dura through which the oculomotor nerve enters the dura in the cavernous sinus roof, is positioned between the anterior and posterior clinoid processes and the petrous apex. The roof of the cavernous sinus extends forward below the anterior clinoid process

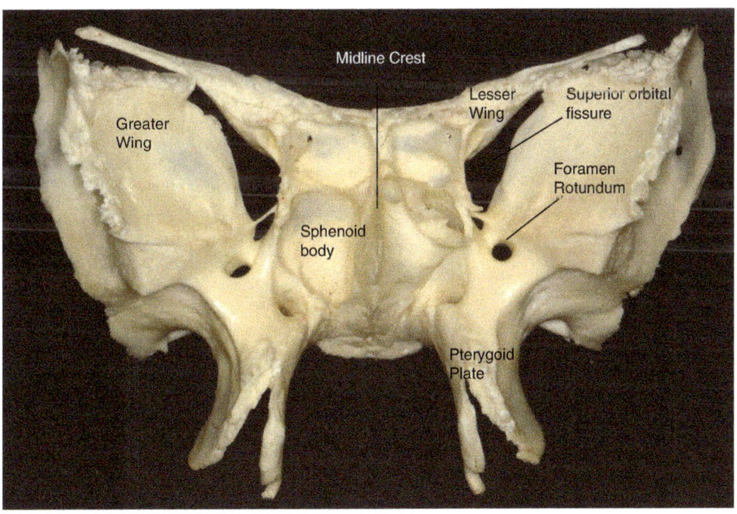

Fig. 15.3 Anterior view of the sphenoid bone. It has a central portion called the body; two lesser wings, which spread outward from the superolateral part of the body; two greater wings, which spread upward from the lower part of the body; and two pterygoid processes, with their medial and lateral pterygoid plates directed downward. The body of the sphenoid bone is cubical and contains the sphenoid sinus. The superior orbital fissure, through which the oculomotor, trochlear, abducens, and ophthalmic nerves pass, is formed on its inferior and lateral margins by the greater wing and on its superior margin by the lesser wing

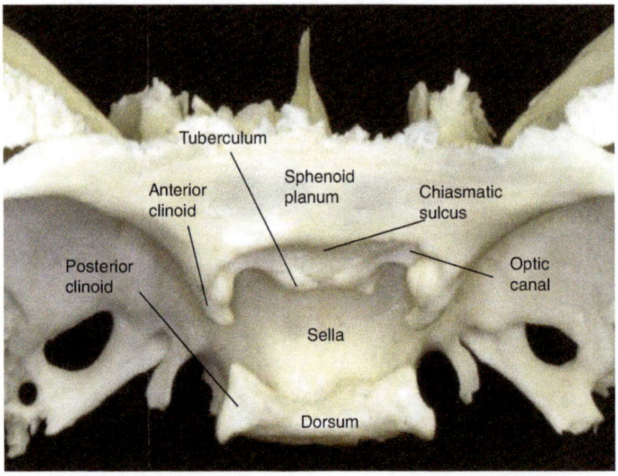

Fig. 15.4 Osseous relationships of the sellar region. Medially, the lesser wings are connected to the sphenoid body through the anterior root, and they form the roof of the optic canal and are continuous with the sphenoid planum. The chiasmatic sulcus is located posterior to the planum. On each side of the chiasmatic sulcus are the endocranial openings of the optic canals. Posteriorly, the chiasmatic sulcus is separated from the sellar cavity by the tuberculum sellae. The posterior limit of the sella includes the dorsum and posterior clinoid processes, which are the medial boundaries between the middle and posterior fossa. The frontal lobes and olfactory tracts rest against the smooth upper surface of the lesser wing and the planum sphenoidale

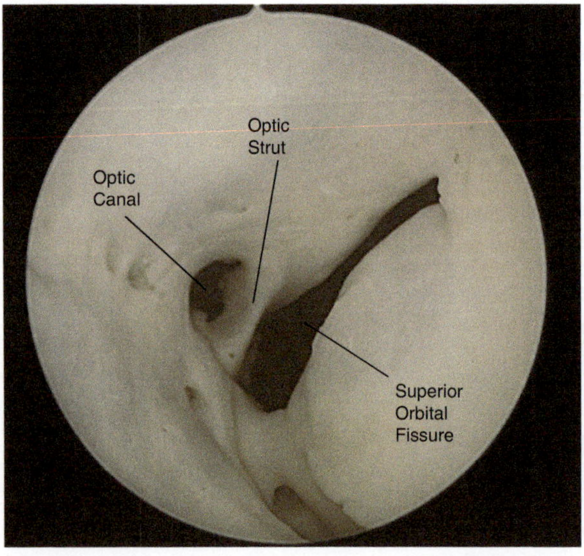

Fig. 15.5 Anterior aspect of the left optic canal (endoscopic view). The optic canal, which transmits the optic nerve and ophthalmic artery, opens in the superomedial corner of the orbital apex. The optic canal is located in the junction of the lesser wing and the sphenoid body. It is separated from the superior orbital fissure by the optic strut, a bony bridge that extends from the anterior clinoid to the sphenoid body

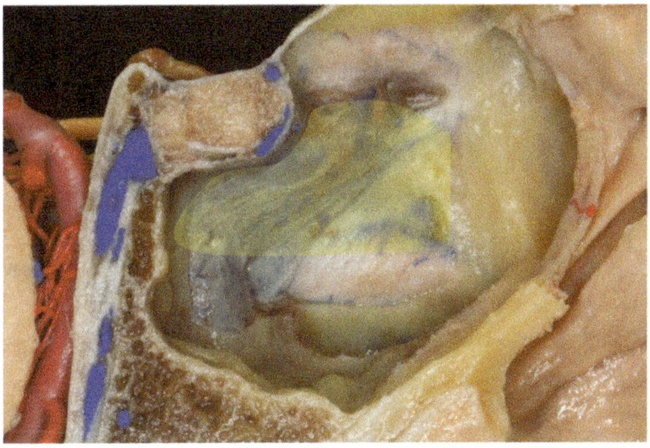

Fig. 15.6 The lateral wall of the sphenoid sinus and the medial wall of the cavernous sinus (yellow shade). The cavernous sinus is located on each side of the sphenoid sinus, sella, and pituitary gland. It extends from the superior orbital fissure (in front) to the petrous apex (behind) and surrounds the horizontal portion of the carotid artery. The medial wall of the paired cavernous sinuses forms the lateral boundary of the sella

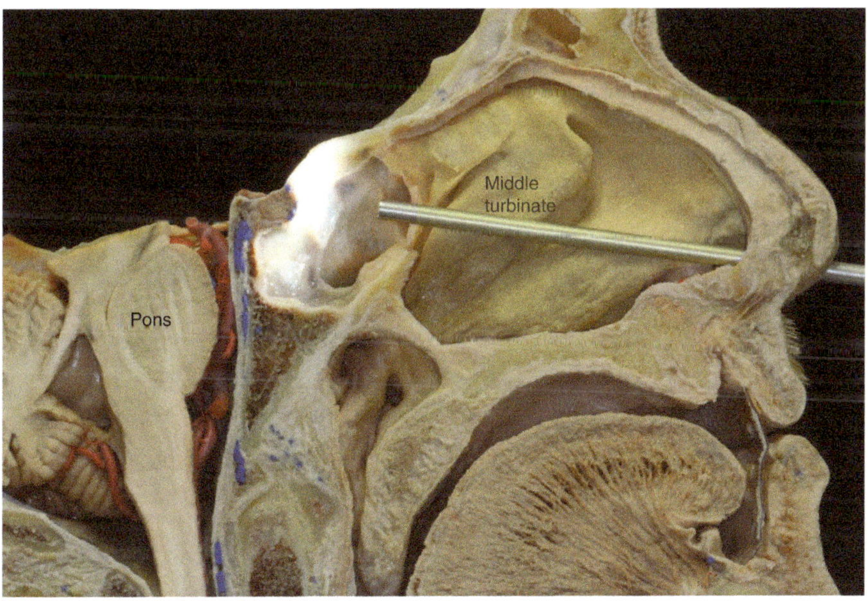

Fig. 15.7 Midsagittal cut showing the endoscopic trajectory through the nasal cavity (yellow shade) and sphenoid sinus. The endoscopic transnasal access is the preferred approach to pituitary tumors. The pneumatized sphenoid sinus gives a clear view of the sellar region. The ethmoidal cells have been removed

sellar type is the most common, and the air cavity extends as far posteriorly as the clivus bellow the sella. A satisfactory pneumatization allows the surgeon to recognize important anatomical landmarks while performing the endoscopic transnasal approach to the sphenoid sinus. It is possible to visualize the carotid prominences, sella, optic nerves, and opticocarotid recess (Fig. 15.9). The recess is particularly important in surgery for defining the localization of the optic nerves and the segments of the carotid artery inside the sphenoid sinus. The lower margin of the recess marks the position of the proximal dural ring, and the upper margin marks the

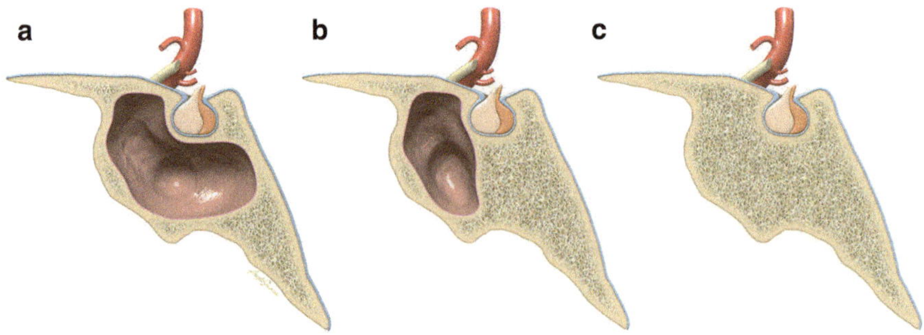

Fig. 15.8 There are three types of sphenoid sinus in the adult, namely, sellar (**a**), presellar (**b**), and conchal (**c**), depending on the extent to which the sphenoid bone is pneumatized. In the conchal type, the area below the sella is a solid block of bone without an air cavity. In the presellar type, the air cavity does not penetrate beyond a vertical plane parallel to the anterior sellar wall. The sellar type of the sphenoid sinus is the most common, where the air cavity extends to the body of the sphenoid below the sella and far posteriorly as the clivus. (Drawings by Angelo Shuman-medical illustrator)

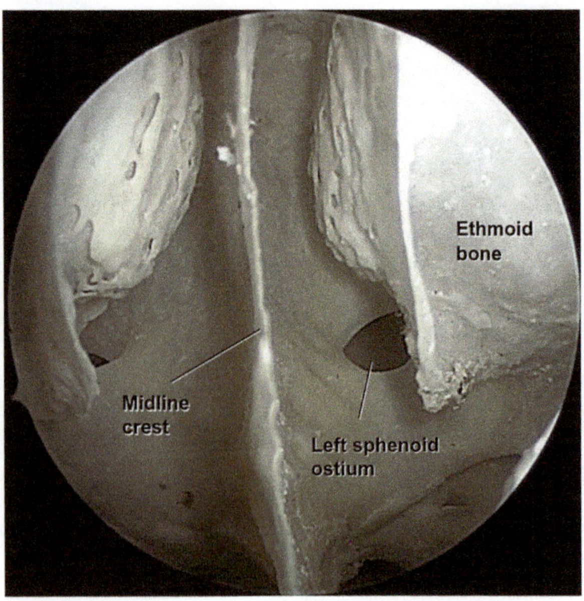

Fig. 15.9 Anterior wall of the sphenoid sinus. The midline crest is a reliable landmark in surgical navigation, and it is the attachment of the posterior part of the osseous nasal septum. The sphenoid ostia is located lateral to the crest and is the route to the sphenoid sinus. It is located above the posterior insertion of the middle turbinate

distal dural ring. The carotid portion between the dural rings is known as the clinoidal carotid. The lateral recess of the sphenoid sinus extends as far as the second division of the maxillary division of the trigeminal nerve (Figs. 15.10, 15.11, and 15.12). The intra-sinus septa are common and can vary in number and position and must be identified prior to surgery with imaging examinations. While drilling the

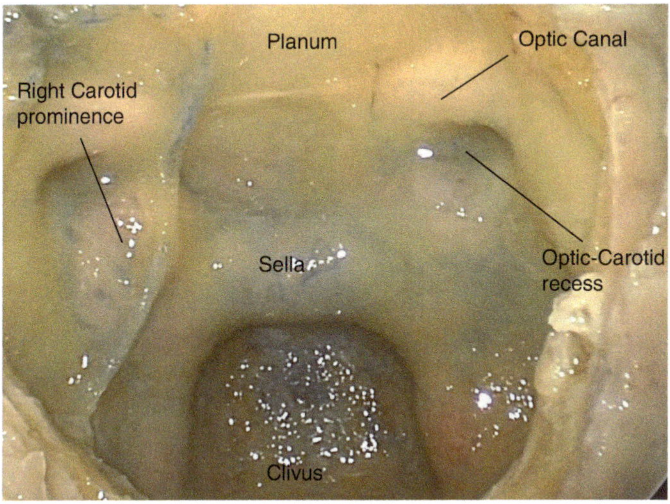

Fig. 15.10 Anterior view of the back wall of the sphenoid sinus showing the relationships of the structures that can be exposed using the transsphenoidal approach. The structures in the exposure include the anterior sellar wall and the prominences over the carotid arteries and optic canals. The tuberculum sellae and planum sphenoidale are located above the anterior sellar wall. The opticocarotid recess extends laterally between the carotid artery and optic canal

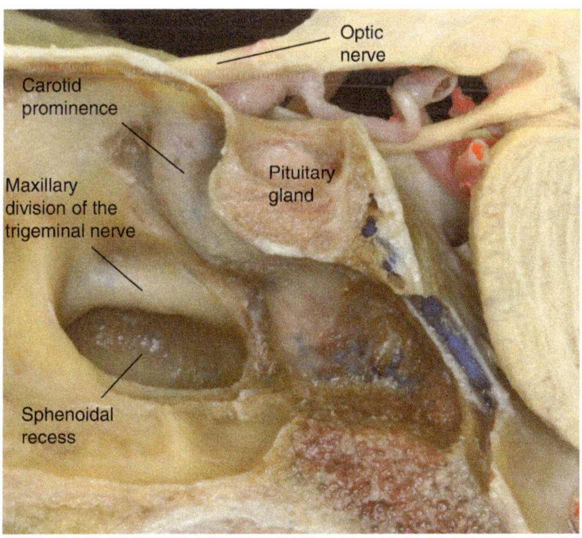

Fig. 15.11 Midsagittal section of the sphenoid sinus. In well-pneumatized sphenoid sinuses, there are several recesses that help delineate adjacent structures on the wall of the sinus or even provide routes to specific areas. In this specimen, there is a prominent recess below the maxillary division of the trigeminal nerve

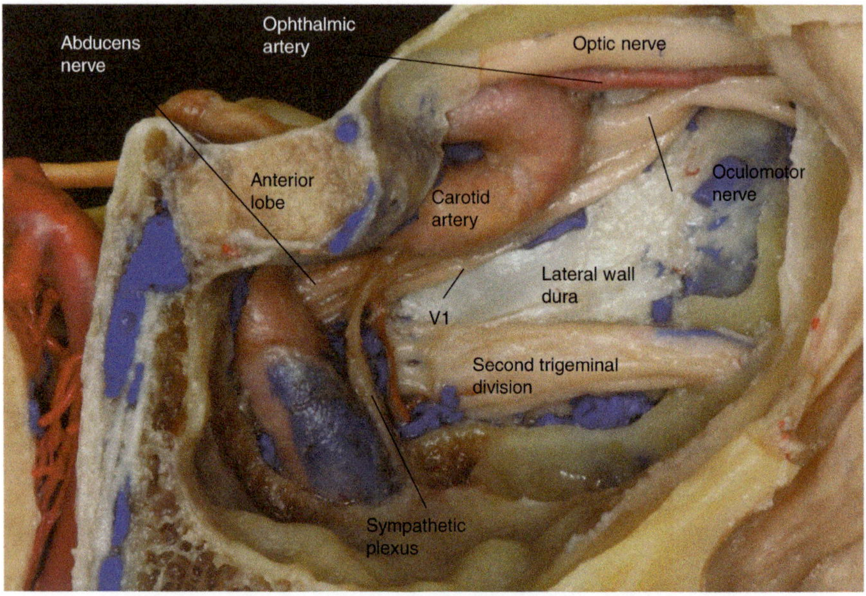

Fig. 15.12 Midsagittal section of the sella extending through the anterior and posterior lobes and sphenoid sinus. The intercavernous carotid produces prominences in the lateral wall of the sphenoid sinus below and anterior to the gland. The basilar sinus, located on the back of the dorsum, is the largest connection across the midline between the posterior edges of the paired cavernous sinuses. The anterior cavernous sinus also connects the paired cavernous sinuses but can be absent in some cases. When present is a site of venous bleeding in transellar approaches. The nerves related to the cavernous sinus are the oculomotor, trochlear, ophthalmic, and abducens nerves and sympathetic plexus around the intracavernous carotid artery. The ophthalmic nerve (first trigeminal division – V1) courses below the trochlear nerve en route to the superior orbital fissure. The maxillary nerve (the second trigeminal division) courses below and does not bulge to the lateral wall of the cavernous sinus. The cavernous sinus ends just above the superior margin of the maxillary nerve, where the medial and lateral walls of the cavernous sinus join in a keel-like formation. The abducens nerve and the sympathetic plexus around the intracavernous carotid artery are the only nerves that have a purely intracavernous course. The ophthalmic artery runs forward along with the optic nerve in the optic canal

lateral wall of the sphenoid sinus, one can reach the medial wall of the cavernous sinus. These important structures must be identified: optic nerve, carotid artery, first and second divisions of the trigeminal nerve, abducens nerve, and oculomotor nerve. The optic canal comprises the optic nerve and the ophthalmic artery. The oculomotor nerve can be seen in an almost parallel trajectory to the floor. The abducens nerve is located between the lateral side of the carotid artery and the medial side of the first trigeminal division. Following the nerve proximally, one can identify the petrolingual ligament at its lower edge, where the nerve crosses the lateral part of the carotid artery and marks its entrance in the cavernous sinus (Fig. 15.13). A venous plexus consisting of an extension of the cavernous sinus can be found surrounding the carotid artery and goes as far as encountering its contralateral pair

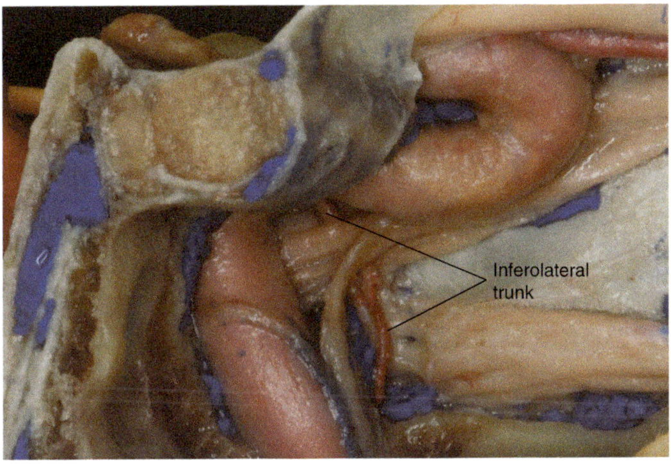

Fig. 15.13 Enlarged view of the sella and medial wall of the cavernous sinus. The intracavernous carotid artery has two main branches. First, the meningohypophyseal trunk arises from the posterior bend of the cavernous paraclival internal carotid artery. Second, the inferolateral trunk, also called the artery of the inferior cavernous sinus, arises from the middle one-third of the inferior or lateral surface of the horizontal segment. It nearly always passes above the abducens nerve and then downward between the abducens and first division of the trigeminal nerve for supplying the dura of the inferolateral wall of the cavernous sinus

through dural canals located on the sellar floor. The first division of the trigeminal nerve is seen as a broad prominence on the lateral wall. The second division of the trigeminal nerve is located at the lower margin of the opening; it can bulge to the lateral wall, which is more evident when a complete pneumatization and consequently enlarged lateral recesses are present. The upper margin marks the inferior limit of the cavernous sinus. The ostium is located in the frontal part of the rostrum, close to the sinus roof, and connects the sinus with the nasal cavity. The sphenoid osseous crest attaches to the vomer and is a reliable mark of the midline of the nose (Fig. 15.14). The roof of the sphenoid sinus extends from the anterior wall to the proximal parts of the optic canal, and it is formed by the planum, chiasmatic sulcus, and tuberculum sellae. Below the tuberculum is the sella on the upper part of the posterior wall of the sinus. The posterior part of the sinus is divided into the upper part (sella) and lower part (clivus). The lateral limits of the posterior wall are the carotid prominences.

15.4 Pituitary Gland and Diaphragma Sellae

The pituitary gland has no bone boundary on its lateral and superior margins. The gland is divided into two lobes (Figs. 15.15 and 15.16): The posterior lobe is softer, darker in appearance, and more adherent to the sellar wall. The anterior lobe is

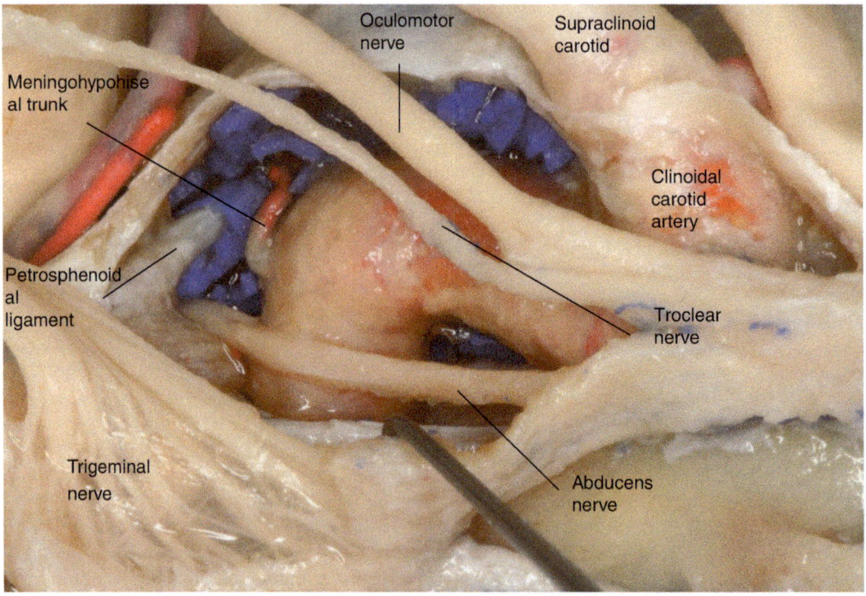

Fig. 15.14 Lateral view of the parasellar region. The branches of the intracavernous portion of the carotid artery that supply the sellar contents is the meningohypophyseal trunk, the largest intracavernous branch, which gives rise to the inferior hypophyseal artery. The meningohypophyseal trunk arises at the level of the dorsum sellae at or just before the apex of the first curve of the carotid, where it turns forward after leaving the carotid canal. The inferior hypophyseal artery arises from the meningohypophyseal trunk and passes medially to the posterior pituitary capsule and lobe and then anastomoses with its mate of the opposite side after supplying the dura of the sellar floor. The oculomotor nerve courses along the lower edge of the anterior clinoid process to enter the superior orbital fissure. The abducens nerve penetrates the cavernous sinus by passing through Dorello's canal, located below the petrosphenoidal ligament (Gruber's ligament). It passes lateral to the posterior vertical segment of the intracavernous carotid artery and courses inside the lateral venous space of the cavernous sinus lateral and inferior to the horizontal segment of the intracavernous carotid and medial to the trigeminal nerve to reach the superior orbital fissure

firmer and can be easily separated from the sellar wall. The pituitary stalk runs from the tuber cinereum in the lower and anterior parts of the third ventricle through the anterior lobe of the pituitary gland. The inferior surface of the pituitary gland has the same shape as the sellar floor; the lateral and upper limits vary according to anatomical variations in the sellar walls and diaphragma. The diaphragma sellae forms the roof of the sella turcica and has a small opening on its center for the pituitary stalk; it is thicker at the periphery and is a natural and anatomical barrier protecting the suprasellar structures during transsphenoidal operation. Patients with large openings in the diaphragma sellae have a higher potential for postoperative cerebrospinal fluid leakage.

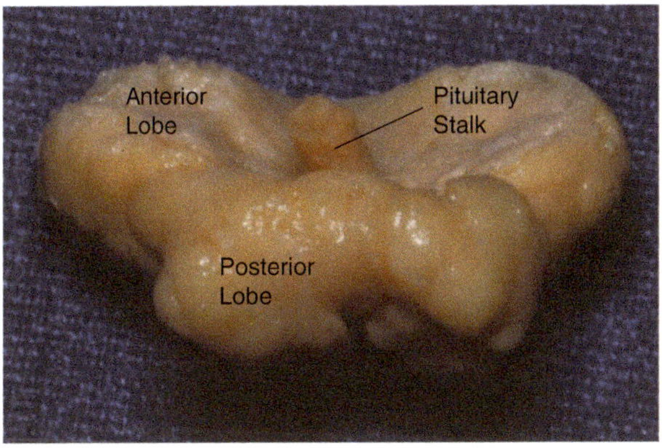

Fig. 15.15 Posterior view of the pituitary gland. The gland is located below the optic nerves and chiasm and between the cavernous segments of the internal carotid arteries. The anterior and posterior lobes form separate distinct nodules. The posterior lobe forms a nodule attached to the posterior edge of the anterior lobe

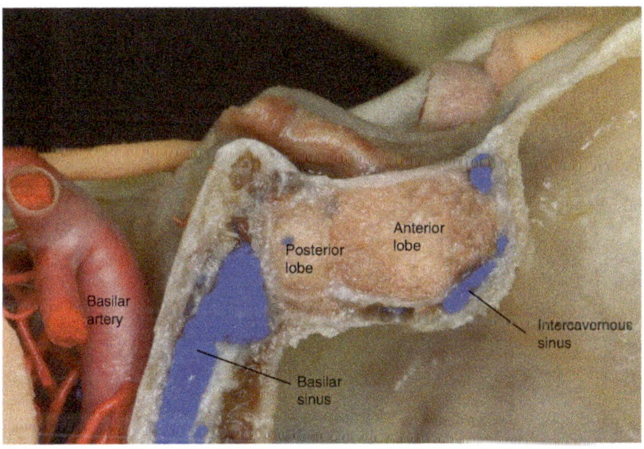

Fig. 15.16 Venous sinuses that interconnect the paired cavernous sinuses can be found around the glands. The intercavernous connections within the sella are named based on their relationship with the pituitary gland; the anterior intercavernous sinuses pass anterior to the hypophysis. These intercavernous connections can occur at any site along the anterior, inferior, or posterior surface of the gland, or all connections between the two sides may be absent. The anterior intercavernous sinus may cover the whole anterior wall of the sella

15.5 Sella and Carotid Artery

According to Rhoton et al., the intracavernous portion of the carotid artery begins lateral to the posterior clinoid process where it exits the foramen lacerum and turns abruptly forward to enter into the cavernous sinus. It then passes forward in a horizontal direction and terminates by passing upward along the medial side to the anterior clinoid process, where it penetrates the roof of the cavernous sinus. The carotid artery is the most medial structure within the cavernous sinus. It rests directly against the lateral surface of the body of the sphenoid bone, producing a prominence, and its course is marked by a groove in the bone, the carotid sulcus, which defines the course of the intracavernous portion of the carotid artery. The carotid prominence can be divided into three parts: the retrosellar, infrasellar, and presellar segments. The retrosellar segment (first part) is located in the posterolateral part of the sinus. The infrasellar segment (second part) is located below the sellar floor. The presellar segment (third part) is located beside the anterior sellar wall. The cavernous carotid is relatively fixed by the bony ring formed by the anterior and middle clinoid processes and the carotid sulcus; however, despite this, large extensions of pituitary tumor may produce lateral displacement of the artery. The meningohypophyseal trunk, the largest intracavernous branch, of the intracavernous portion of the carotid artery supplies the sellar contents, which give rise to the inferior hypophyseal artery, and McConnell's capsular artery arises directly from the internal carotid artery. The meningohypophyseal trunk arises at the level of the dorsum sellae at or just before the apex of the first curve of the carotid, where it turns forward after leaving the carotid canal. The inferior hypophyseal artery arises from the meningohypophyseal trunk and passes medially to reach the lateral surface of the posterior pituitary lobe and capsule and anastomoses with its mate of the opposite side after supplying the dura of the sellar floor (Figs. 15.17 and 15.18).

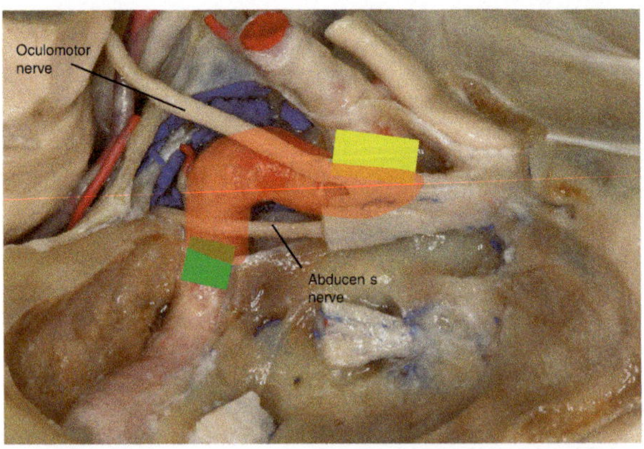

Fig. 15.17 The intracavernous segment of the carotid artery (red shade) begins at the intracranial end of the carotid canal superior to the foramen lacerum and lateral to the posterior clinoid process, where the petrous segment of the internal carotid artery enters the cavernous sinus. The petrous segment of the internal carotid artery passes between the cartilaginous foramen lacerum below and the petrolingual ligament (green line) above to become intracavernous. The petrolingual ligament extends from the lingual process of the sphenoid bone to the petrous apex. The intracavernous segment passes upward and forward along the carotid sulcus posterior to the optic strut and medial to the anterior clinoid process and exits the cavernous sinus by piercing the dura and extending medially from the upper surface of the anterior clinoid process (yellow line)

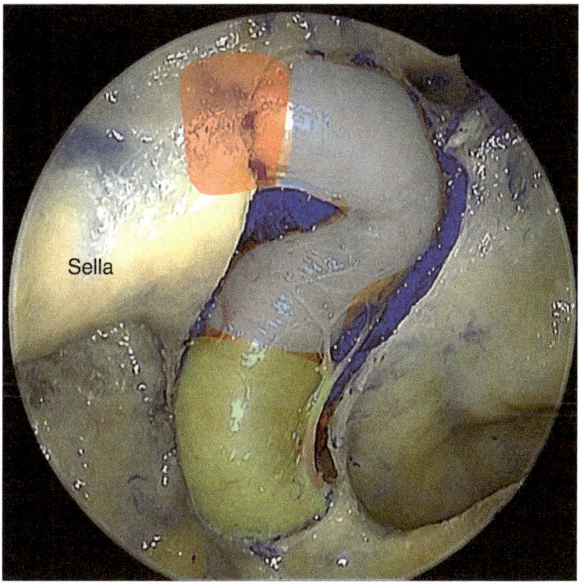

Fig. 15.18 Endoscopic view of the left carotid artery inside the sphenoid sinus after bone removal. The internal carotid artery rests directly against the lateral surface of the body of the sphenoid bone, and its course is marked by a groove in the bone, the carotid sulcus, which defines the course of the intracavernous portion of the carotid artery. As the sphenoid sinus expands and its walls resorb, the carotid sulcus produces a prominence within the sinus wall below the floor and along the anterior margin of the sella. The carotid prominence can be divided into three parts: the retrosellar (red shade), infrasellar (green shade), and presellar (blue shade) segments. The first part, the retrosellar segment, is located in the posterolateral part of the sinus. This segment of the prominence is present only in well-pneumatized sellar-type sinuses in which the air cavity extends laterally to the area below the dorsum. The second part, the infrasellar segment, is located below the sellar floor. The third part, the presellar segment, is located anterolateral to the anterior sellar wall

15.6 Suprasellar Relationships

The suprasellar region corresponds roughly to the anterior incisural space, and it is an area of special interest when dealing with pituitary tumors that extend above the sella. From the front of the midbrain, it extends upward around the optic chiasm, lamina terminalis, and carotid arteries, being its anterolateral limit, the sphenoidal portion of the Sylvian fissure and the posterolateral limit being the uncus, which hangs over the free edge of the tentorium over the oculomotor nerve. The optic and oculomotor nerves and the posterior part of the olfactory tract pass through the suprasellar region. The olfactory tract runs posteriorly and splits above the anterior clinoid process to form the medial and lateral olfactory striae that marks the anterior margin of the anterior perforated substance. The optic nerves enter the optic canals medial to the anterior clinoid processes and are covered by a reflected leaf of dura, the falciform process, which extends medially across the top of the optic nerve. Suprasellar tumors can cause compression of the optic nerve against the sharp edge of the falciform ligament and cause visual loss. The optic chiasm is situated at the junction of the anterior wall and the floor of the third ventricle. The anterior cerebral and communicating arteries, lamina terminalis, and third ventricle are located above the chiasm. The tuber cinereum and infundibulum are posterior, the internal carotid arteries are lateral, and the diaphragm sellae and pituitary gland are inferior to the optic chiasm (Figs. 15.19 and 15.20). A complete understanding of the relationship among the carotid artery, optic nerve, and anterior clinoid process is fundamental to all surgical approaches to the sellar region. The carotid artery and optic nerve are medial to the anterior clinoid process. The artery exits the cavernous sinus beneath and slightly lateral to the optic nerve. The optic nerve takes a posteromedial course toward the chiasm, and the carotid artery, a posterolateral course toward its bifurcation inferior to the anterior perforated substance and into the anterior and middle cerebral arteries in the region. The oculomotor nerve arises in the interpeduncular fossa and has a medial-to-lateral, posterior-to-anterior course. It passes above the

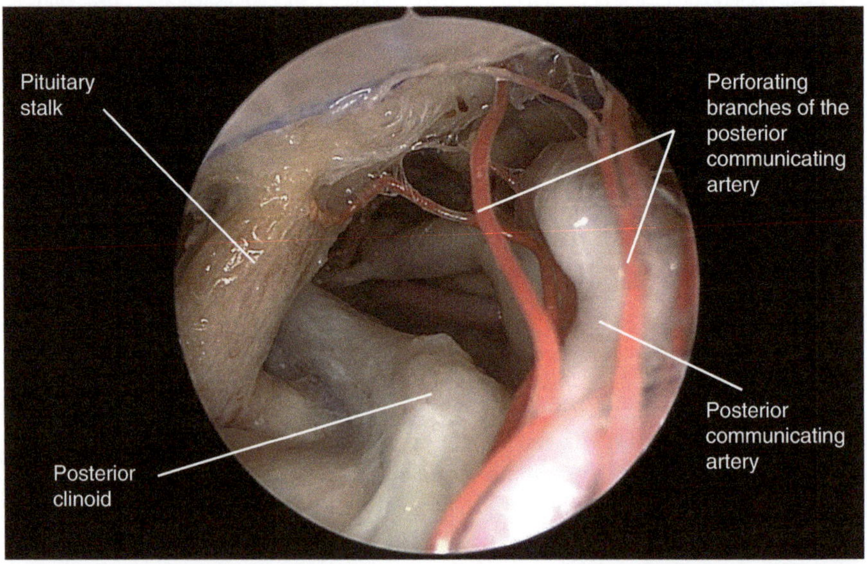

Fig. 15.19 Endoscopic view of the left infrachiasmatic region. The posterior communicating artery arises from the posterior wall of the internal carotid artery and courses posteromedially below the optic tract and the floor of the third ventricle to join the posterior cerebral artery. Its branches penetrate the floor between the optic chiasm and the cerebral peduncle and reach the thalamus, hypothalamus, subthalamus, and internal capsule. It is directed posteromedially above the oculomotor nerve toward the interpeduncular fossa

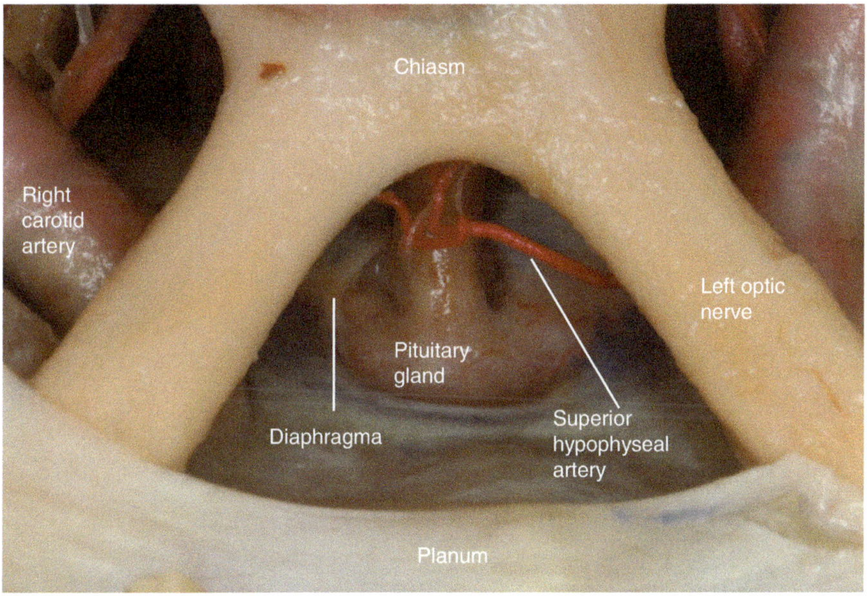

Fig. 15.20 Anterior enlarged view of the sellar region. The superior hypophyseal arteries arise from the medial side of the supraclinoid portion of the internal carotid artery and pass medially to the pituitary stalk and optic chiasm. The diaphragma is thinner around the infundibulum and somewhat thicker at the periphery

superior cerebellar artery and below the posterior cerebral artery. As it moves forward, it has a close relationship with the apex of the uncus, which is located lateral and above the oculomotor nerve. It is also the lateral attachment of the Liliequist membrane. The oculomotor nerve ends at the oculomotor cistern as it enters the cavernous sinus. The posterior part of the circle of Willis and the apex of the basilar artery are located in the anterior incisural space below the floor of the third ventricle; the anterior part of the circle of Willis and the anterior cerebral and anterior communicating arteries are intimately related to the anterior wall of the third ventricle; both the anterior and posterior cerebral arteries send branches into the roof of the third ventricle; perforating branches arise from the internal carotid, anterior choroidal, anterior and posterior cerebral, and anterior and posterior communicating arteries, reaching the walls of the third ventricle and anterior incisural space (Fig. 15.21). The carotid artery arises at the skull base along the medial surface of the anterior clinoid process, and before its bifurcation, it gives off the superior hypophyseal artery, which runs medially below the floor of the third ventricle and the chiasm to reach the tuber cinereum. The ophthalmic artery is usually the first intracranial branch of the internal carotid artery that runs forward along the optic nerve into the optic canal. The posterior communicating artery, a branch of the internal carotid artery, arises from the posterior wall of the internal carotid artery and

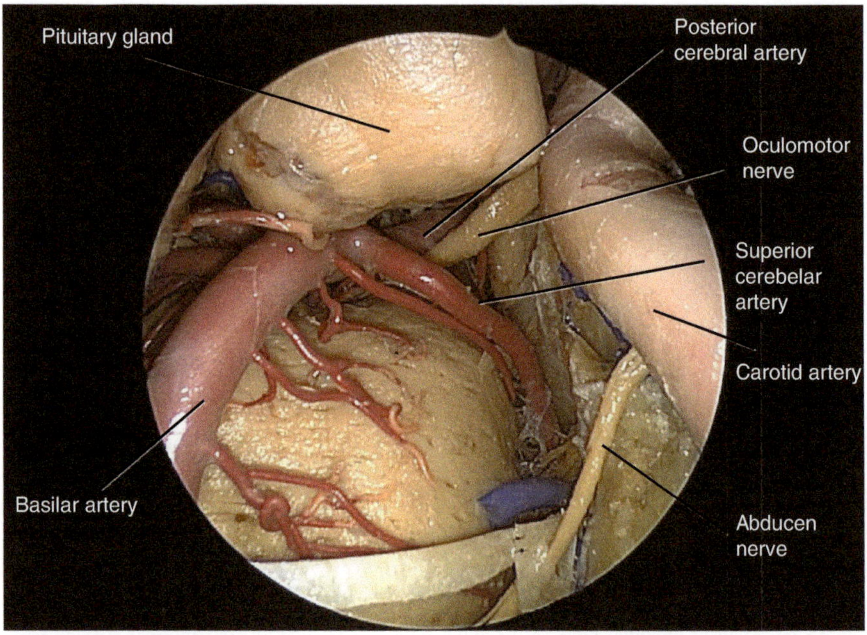

Fig. 15.21 Endoscopic view from below to above of the anterior incisural space. The clivus was drilled. The oculomotor nerve arises from the interpeduncular fossa and passes above the superior cerebellar artery and below the posterior cerebral artery. The abducens nerve has an anterior superior trajectory and usually pierces the dura of the clivus as a single bundle. It arises at the lower margin of the pons

courses posteromedially below the optic tracts and the floor of the third ventricle to join the posterior cerebral artery. Its branches penetrate the floor between the optic chiasm and cerebral peduncle. The premammillary artery is the largest branch that arises from the posterior communicating artery. Between the mammillary body and optic tract, the premammillary artery enters the floor of the third ventricle in front of or beside the mammillary body. The anterior choroidal artery arises from the carotid artery distal to the origin of the posterior communicating artery. It is a single artery that bifurcates as it runs backward below the optic tract between the uncus and the cerebral peduncle ending in the anterior perforated substance and the choroid plexus in the temporal horn. The anterior cerebral artery arises from the internal carotid artery below the anterior perforated substance and courses anteromedially above the optic nerve and chiasm. It is joined by the contralateral anterior cerebral artery ascending in front of the lamina terminalis through the anterior communicating artery (Figs. 15.22, 15.23, and 15.24). The recurrent branch of the anterior cerebral artery arises in the region of the anterior communicating artery, courses laterally above the bifurcation of the internal carotid artery, and enters the anterior perforated substance. The bifurcation of the basilar artery into the posterior cerebral arteries is located in the posterior part of the suprasellar region below the posterior half of the floor of the third ventricle. The thalamoperforating arteries arise from the proximal part of the posterior cerebral artery close to the basilar bifurcation and enter the brain through the floor of the third ventricle.

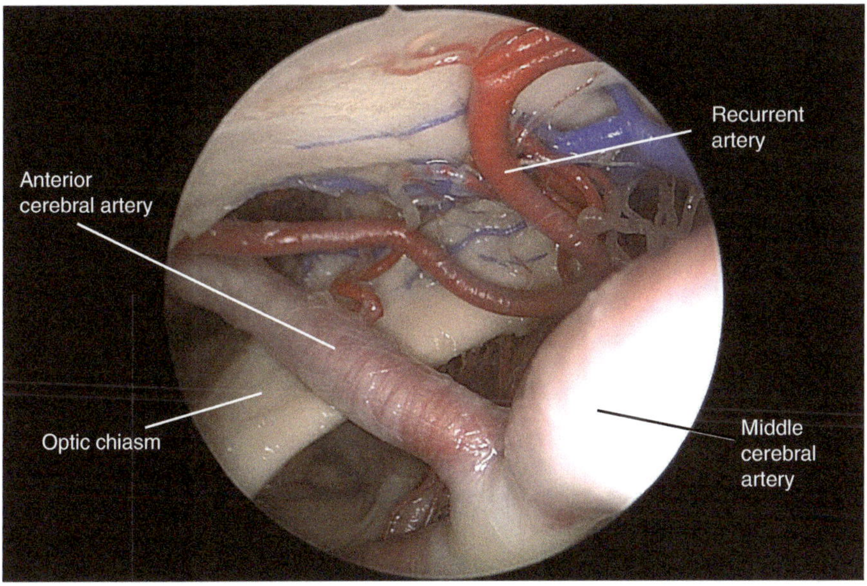

Fig. 15.22 The anterior cerebral artery arises from the internal carotid artery below the anterior perforated substance and courses anteromedially above the optic nerve and chiasm to reach the interhemispheric fissure. The recurrent branch of the anterior cerebral artery arises from the anterior cerebral artery in the region of the anterior communicating artery and courses laterally above the bifurcation of the internal carotid artery and enters the anterior perforated substance

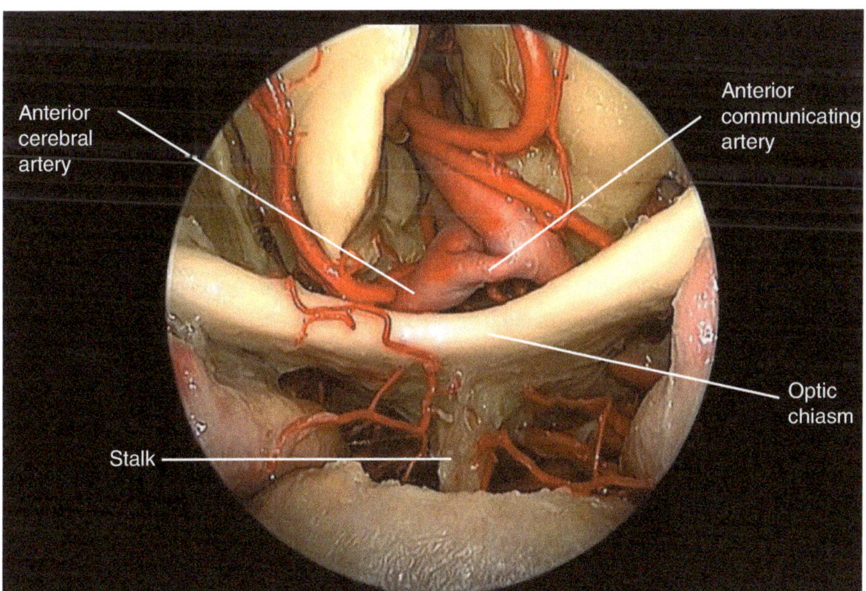

Fig. 15.23 Anterior endoscopic view of the anterior communicating complex. The junction of the anterior communicating artery with the right and left A1 segments is usually above the chiasm in front of the lamina terminalis rather than above the optic nerves

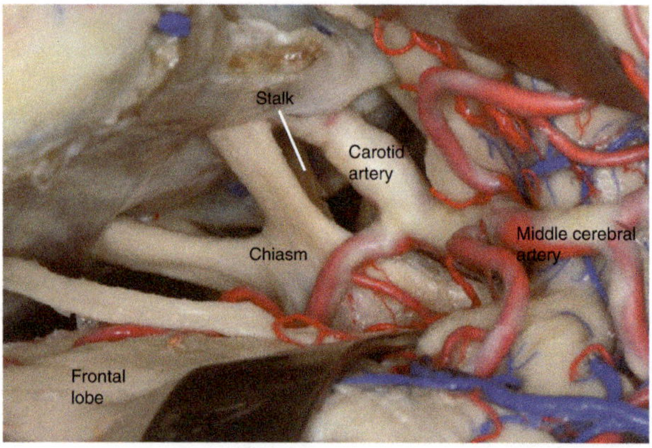

Fig. 15.24 View of the sellar region from a surgical perspective through a pterional approach. The Sylvian fissure was splitted, and the frontal and temporal lobes were retracted. The anterior cerebral artery gives off its recurrent branch earlier than usual, which runs medially above the lamina terminalis

15.7 Final Considerations

The sphenoid sinus has a critical role and is the starting point for ventral skull base surgeries, providing access and anatomical landmarks to relevant neurovascular structures located behind its walls. The endoscopic-assisted technique is widely used to approach pituitary tumors through the sphenoid sinus (Fig. 15.25). It is a multidisciplinary team approach that should be performed by otolaryngologists and neurosurgeons together. An extensive and detailed study of the patient's anatomical features is mandatory when choosing the transsphenoidal approach. The sphenoid ostium should be located and widened inferiorly and medially to give access to the sphenoid sinus (it is located roughly 1.5 cm above the choana). With tactile feedback and direct visualization, the sellar floor can be identified and opened, which is demarcated by the planum sphenoidale superiorly, the rostral clivus inferiorly, and the carotid prominences laterally (this anatomical landmarks are essential for localization and should be identified promptly). In the presence of intersinus septa, extreme caution should be taken when removing these septations, as they often lead to the carotid artery and optic canal. The proximity of the carotid prominences to the midline is important in pituitary surgery. The transverse separation between the carotid prominences on each side should be measured at the level of the tuberculum sellae, anterior sellar wall, sellar floor, dorsum sellae, and clivus. The shortest distance between both carotid bulges into the sphenoid sinus is usually located at the level of the tuberculum sellae. Care must be taken when removing the sellar floor to avoid harming the intercavernous sinus since it can lead to profuse bleeding. The dural opening may vary depending on the surgeon's preference and should always be wide enough to access the tumor (Figs. 15.26, 15.27, and 15.28). The removal of

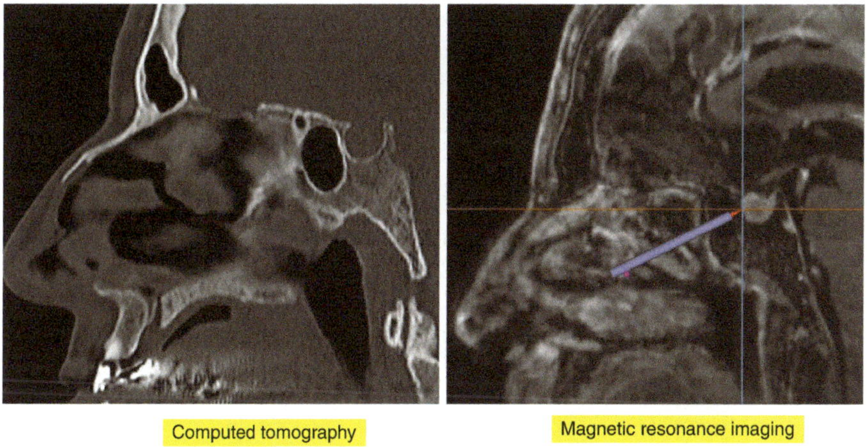

Fig. 15.25 Computed tomography scan (left) and magnetic resonance imaging (right) of a patient with a small pituitary tumor. The patient had a presellar sphenoid sinus. In these cases, as seen in the right picture, neuronavigation is needed because the surgeon cannot identify the anatomical landmarks once entering the sphenoid sinus

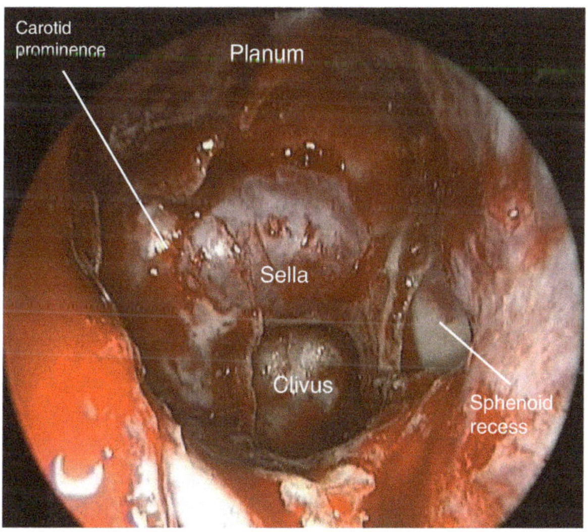

Fig. 15.26 Intraoperative imaging of the sphenoid sinus in an approach to the sellar region. The bone from the sellar floor has been removed to expose the dura. The carotid prominences are lateral to the sella. The intrasinusal septa usually lead to the carotid arteries, and in this case, they are lateral to the clivus. The sinus is well pneumatized, with the recess in the left side extending below the second trigeminal division. With tactile feedback and direct visualization, the sellar floor can be identified and opened, which is being demarcated by the planum sphenoidale superiorly, the clivus inferiorly, and the carotid prominences laterally

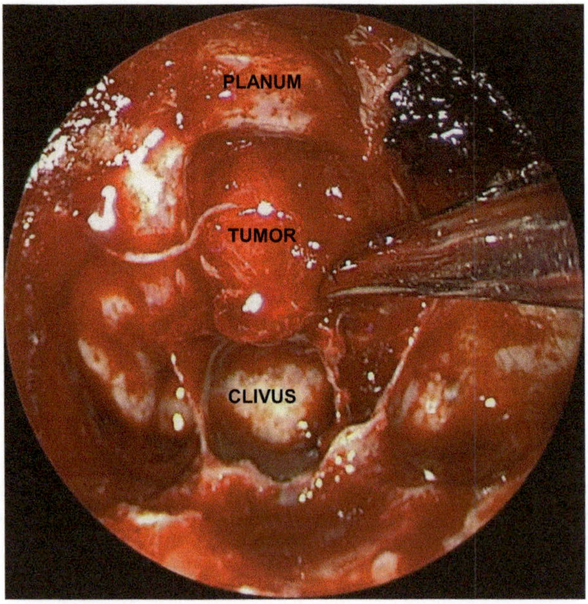

Fig. 15.27 The dural opening may vary depending on the surgeon's preference (it should be always wide enough to access the tumor); usually the opening extends from carotid to carotid and from planum to clivus. The removal of sellar tumor should start on the sellar floor and then move to the lateral parts and, lastly, to the suprasellar extension. Extra capsular dissection is always performed when it is possible (as seen on the picture). It is extremely important to be gentle when working on the lateral aspect of the sella as the medial layer of the cavernous sinus can be thin, and the risk of carotid injury is considerable. The tumor should be removed with a combination of blunt ring curettes and pituitary forceps

the sellar tumor should start on the sellar floor and then move along the lateral parts and, lastly, to the suprasellar extension (Figs. 15.29 and 15.30). It is extremely important to be gentle when working on the lateral aspect of the sella, as the medial layer of the cavernous sinus can be thin, and the risk of carotid injury is considerable. The tumor should be removed with a combination of blunt ring curettes and pituitary forceps. Once the tumor is removed, minor bleeding from the sella should be controlled by packing the sella with neuro-hemostatic agents and a 5-minute waiting, after which an endoscopic inspection of the sella should be performed for ensuring that there is no residual lesion. In the event of a cerebrospinal fluid leak, the skull base defect should be plugged with abdominal fat and sealed with fibrin matrix. The mucosal pediculate flap from the septum is then placed over the defect.

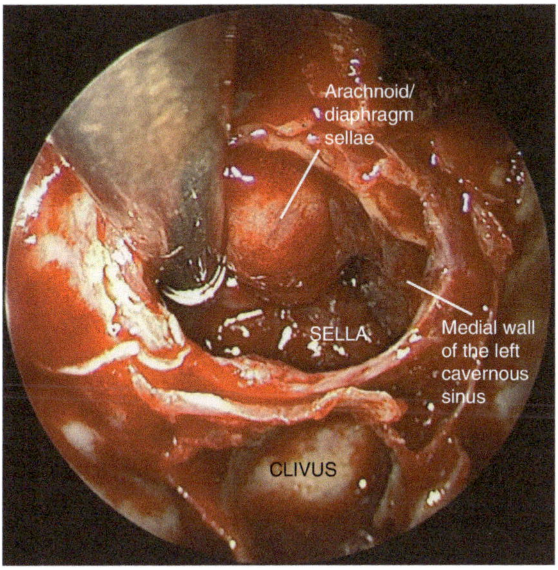

Fig. 15.28 Once the tumor is removed, minor bleeding from the sella should be controlled by packing the sella with neuro-hemostatics and a 5-minute waiting. After this step, an endoscopic inspection of the sella should be performed to ensure that there is no residual lesion. In the picture we can see the complete removal of the tumor, allowing the visualization of the arachnoid falling into the sella through the diaphragm sellae and the medial wall of the cavernous sinus

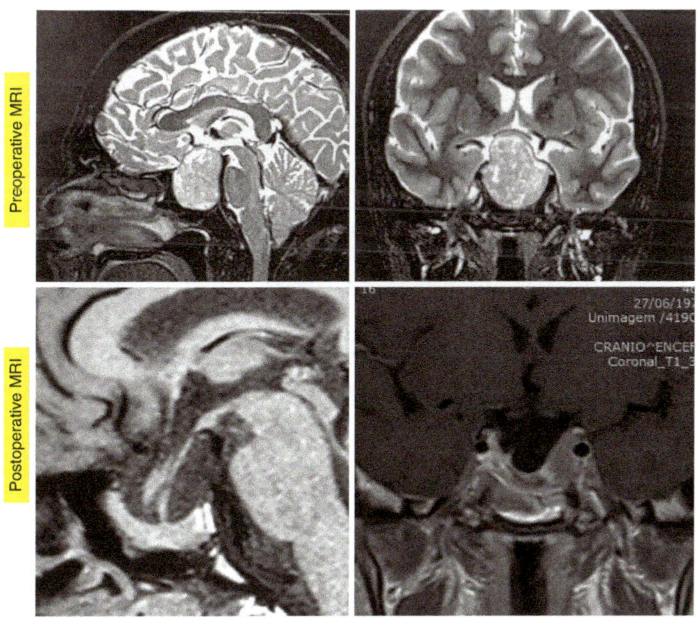

Fig. 15.29 Pre- and postoperative magnetic resonance imaging (MRI) of a patient with a large pituitary tumor. The upper row shows a large pituitary tumor causing compression to the optic chiasm, resulting in headache and consequently bitemporal hemianopsia. The lower row is the postoperative MRI 3 months after the surgery showing a satisfactory extracapsular resection of the tumor

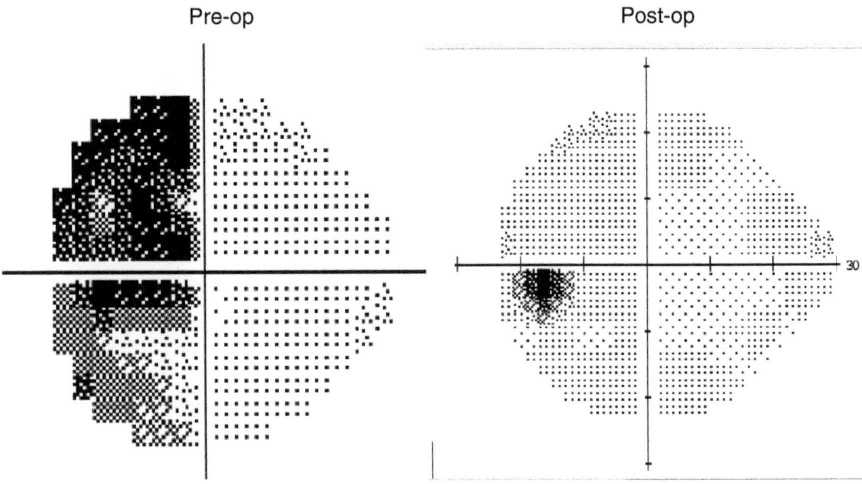

Fig. 15.30 Pre- and postoperative electronic visual test of the previous patient's left eye

Suggested Bibliography

Chaddad-Neto F, da Costa MDS, Santos B, Caramanti RL, Costa BL, Doria-Netto HL, Figueiredo EG. Reproducibility of a new classification of the anterior clinoid process of the sphenoid bone. Surg Neurol Int. 2020 Sep 12;11:281.

da Costa MDS, de Oliveira Santos BF, de Araujo Paz D, Rodrigues TP, Abdala N, Centeno RS, Cavalheiro S, Lawton MT, Chaddad-Neto F. Anatomical Variations of the Anterior Clinoid Process: A Study of 597 Skull Base Computerized Tomography Scans. Oper Neurosurg (Hagerstown). 2016 Sep 1;12(3):289–97.

Rhoton AJ. General and micro-operative techniques. In: Youmans JR, editor. Neurological surgery, vol. 2. 4th ed. Philadelphia: WB Saunders; 1996. p. 724–66.

Rhoton AL Jr, Harris FS, Renn WH. Microsurgical anatomy of the sellar region and cavernous sinus. Clin Neurosurg. 1977;24:54–85. https://doi.org/10.1093/neurosurgery/24.cn_suppl_1.54.

Rhoton AL Jr, Hardy DG, Chambers SM. Microsurgical anatomy and dissection of the sphenoid bone, cavernous sinus and sellar region. Surg Neurol. 1979;12:63–104.

Rhoton AL Jr. Tentorial incisura. Neurosurgery. 2000;47:S131–53. https://doi.org/10.1097/00006123-200009001-00015.

Rhoton AL Jr. The supratentorial arteries. Neurosurgery. 2002a;51:S53–S120.

Rhoton AL Jr. The sellar region. Neurosurgery. 2002b;51:S335–S74.

Rhoton AL Jr. Rhoton's cranial anatomy and surgical approaches. Philadelphia: Lippincott Williams & Wilkins; 2003.

Chapter 16
Surgical Anatomy of the Parasellar Region

16.1 Sphenoid Bone

The sphenoid bone is located at the center of the skull base. The neural relationships of the sphenoid bone are among the most complex of all the bones in the body. The olfactory tract, the straight gyrus, and the posterior part of the frontal lobe rest on the smooth surface of the lesser wing; the temporal lobe rests on the inner surface of the great wing; the pons and the midbrain are located behind the back of the sella; the optic chiasm rests posteriorly to the chiasmatic sulcus; and cranial nerves II to VI are closely associated with the sphenoid bone, leaving the skull through the optic canal, the superior orbital fissure, the round foramen, and the oval foramen, all located in the sphenoid bone (Fig. 16.1).

The sphenoid bone maintains important venous and arterial relationships: the internal carotid artery forms groove on each side of the bone; the basilar artery rests against the posterior surface of the sphenoid bone; the Willis polygon is located above its central portion; and the middle cerebral artery runs parallel to the crest of the sphenoid lesser wing. The cavernous sinuses are located in the sphenoid bone, and the intercavernous venous connections delineate the walls of the back of the sella and the pituitary stalk (Fig. 16.2).

The sphenoid bone contains a central portion called the sphenoid body: two lesser wings extending outside the superolateral portion of the body; two greater wings which project upward from the lower part of the body; and two pterygoid processes with their medial and lateral plates, directed downward. The sphenoid body has a cuboid shape and contains the sphenoid sinus. The superior orbital

Supplementary Information The online version of this chapter (https://doi.org/10.1007/978-3-030-82747-2_16) contains supplementary material, which is available to authorized users.

F. Chaddad-Neto, M. D. Silva da Costa, *Microneuroanatomy and Surgery*, https://doi.org/10.1007/978-3-030-82747-2_16

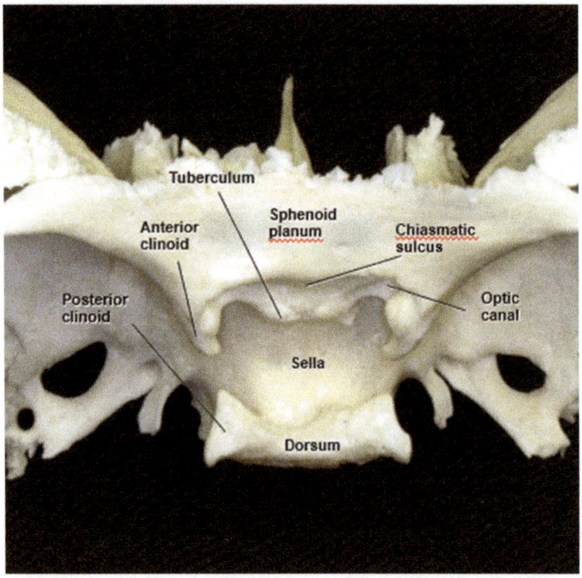

Fig. 16.1 Osseous relationships of the sphenoid bone. Medially, the lesser wings are connected to the sphenoid body through the anterior root, and they form the roof of the optic canal and are continuous with the planum sphenoidale. The chiasmatic sulcus is located posterior to the planum. On each side of the chiasmatic sulcus are the endocranial opening of the optic canals. Posteriorly, the chiasmatic sulcus is separated from the sellar cavity by the tuberculum sellae. The posterior limit of the sella is composed by the dorsum and posterior clinoid processes which are the medial boundaries between the middle and posterior fossa. The frontal lobes and the olfactory tracts rest against the smooth upper surface of the lesser wing and the planum sphenoidale

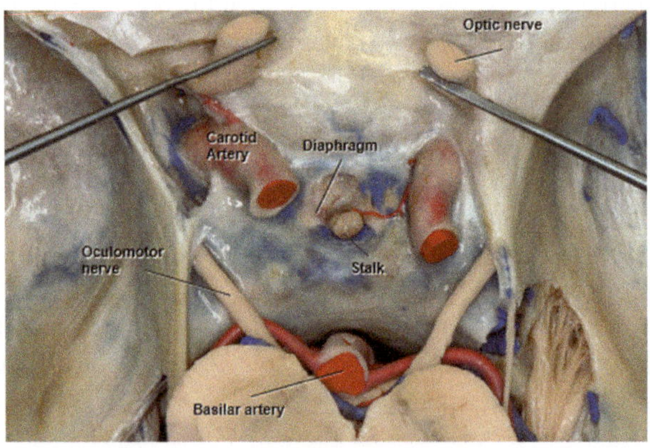

Fig. 16.2 Superior view of the sellar region. The sella is located between the cavernous sinuses. The diaphragm separates the sella from the suprasellar cisterns. The oculomotor nerves enter the roof of the cavernous sinus where there is a narrow cistern around the nerve. The oculomotor triangle, the triangular patch of dura mater through which the oculomotor nerve enters the dura mater in the roof of the cavernous sinus, is positioned between the anterior and posterior clinoid processes and the petrous apex. The roof of the cavernous sinus extends forward under the anterior clinoid process

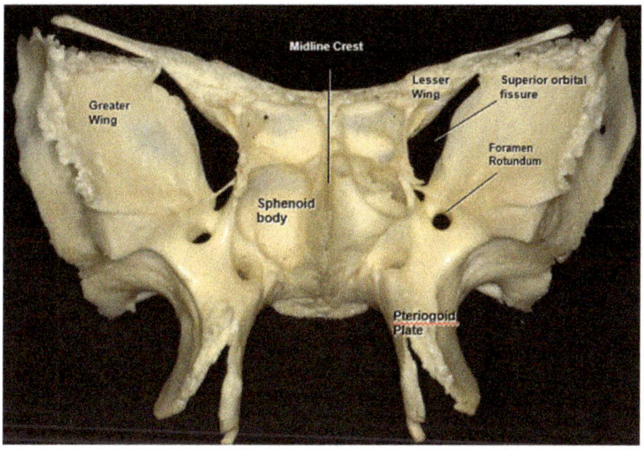

Fig. 16.3 Anterior view of the sphenoid bone. It has a central portion called the body: two lesser wings which spread outward from the superolateral part of the body; two greater wings which spread upward from the lower part of the body; and two pterygoid processes with their medial and lateral pterygoid plates directed downward from the body. The body of the sphenoid bone is more or less cubical and contains the sphenoid sinus. The superior orbital fissure, through which the oculomotor, trochlear, abducens, and ophthalmic nerves pass, is formed on its inferior and lateral margins by the greater wing and on its superior margin by the lesser wing

fissure, through which the oculomotor, trochlear, abducens, and ophthalmic nerves pass, is formed at its lower and lateral edges by the greater wing and at its upper edge by the lesser wing (Fig. 16.3).

The optic channels are located above and are separated from the superior-medial edge of the superior orbital fissure by the optic strut, a bony bridge that extends from the lower edge of the base of the anterior clinoid process to the body of the sphenoid bone.

16.2 Cavernous Sinus

The cavernous sinus is defined as the dural envelope that contains the cavernous segment of the internal carotid artery, in addition to being a region of venous confluence. In general, the cavernous sinus is shaped like a boat, with its narrow keel next to the upper orbital fissure and its wide curvature (posterior wall) located laterally to the back of the sella, above the petrous apex. It has four boundaries: a ceiling and three walls—lateral, medial, and posterior. The roof of the sinus faces upward, while its lower edge, formed by the junction of the medial and lateral walls, below the intracavernous segment of the internal carotid artery, gives the sinus a triangular appearance in axial sections. It is internally covered by a reticular membrane, and the roof, side wall, and floor receive dural reinforcement. The medial wall is composed only of a reticular membrane that is continuous with the diaphragm of the sella.

The roof of the cavernous sinus is formed by the dura mater that lines the lower margin of the anterior clinoid process and by the portion of the dura mater called the oculomotor triangle, located between the anterior and posterior clinoid processes and the petrous apex, by which the oculomotor nerve penetrates the roof of the cavernous sinus and runs through the lateral wall (Fig. 16.4).

The cavernous sinus floor is the body of the sphenoid bone. Under the anterior clinoid process, in the anterior portion of the cavernous sinus roof, the oculomotor nerve is protected by an extra membrane, the distal dural ring, separately from what occurs in the cisternal portion of the nerve. It is easier to enter the cavernous sinus through the back of the ceiling. However, we can also enter the sinus cavity through the anterior portion of the roof, an anterior clinoidectomy, or the lateral wall via a pretemporal craniotomy and "peeling" of the middle fossa, which is nothing more than the separation between the dura mater and the reticular membrane. The main reference is the anterior clinoid process, and the anterior clinoidectomy should not be performed before the "peeling"; otherwise, there is a high risk of injury to the oculomotor nerve. The trochlear nerve enters the cavernous sinus through the postero-inferolateral wall of the roof and walks through the lateral wall above the oculomotor nerve. In the lateral wall of the cavernous sinus, the space between the oculomotor and the trochlear nerves, with a triangular shape and the apex facing the interior of the sinus, is known as the supratrochlear triangle; the space between the

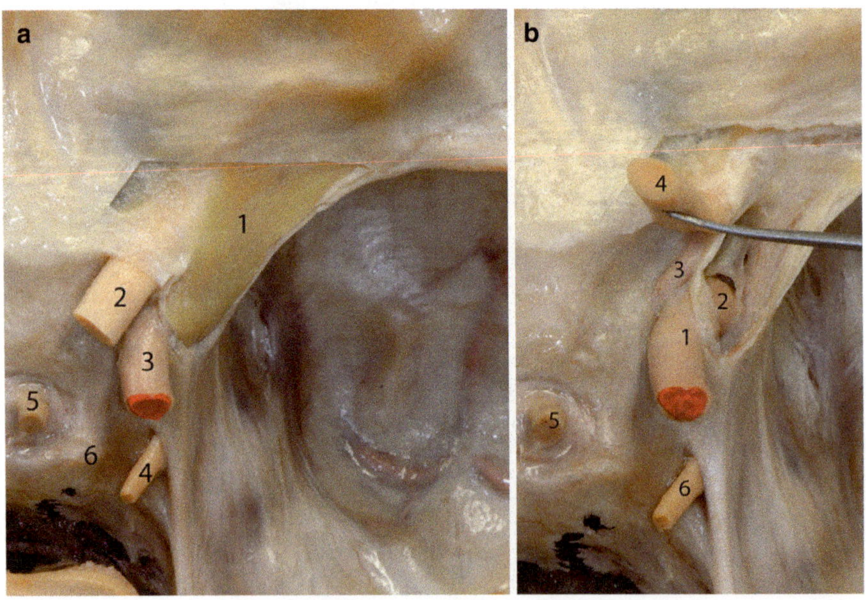

Fig. 16.4 (**a**) Cadaveric dissection demonstrating the roof of the cavernous sinus before de anterior clinoidectomy. (1) Right anterior clinoid process. (2) Right optic nerve. (3) Right supraclinoid segment of ICA. (4) Right third nerve. (5) Pituitary stalk. (6) Posterior clinoid process. (**b**) Roof of the cavernous sinus after anterior clinoidectomy and the clinoidal segment of ICA between the proximal and distal dural ring. (1) Supraclinoid segment of right ICA. (2) Clinoidal segment of right ICA. (3) Right ophthalmic artery. (4) Right optic nerve. (5) Pituitary stalk. (6) Third nerve at the oculomotor

trochlear nerve and the ophthalmic branch of the trigeminal nerve forms the infratrochlear triangle or Parkinson's triangle. The opening of the lateral wall of the cavernous sinus must be transversal to accompany the nerves. After entering the sinus, we must go medially toward the pituitary fossa because laterally, the carotid is located the abducens nerve, which is the only intrinsic nerve of the cavernous sinus.

The medial border of the oculomotor triangle is formed by the interclinoid dural fold, which extends from the anterior clinoid process to the posterior clinoid process; the lateral border is formed by the anterior petroclinoid fold, which extends from the anterior clinoid process to the petrous apex; and the posterior border is formed by the posterior petroclinoid fold, which extends from the posterior clinoid process to the petrous apex. The cavernous sinus is not related to the optic nerve but to the superior orbital fissure, which is the anterior wall of the cavernous sinus and drains into the superior orbital vein. The posterior limit, on the other hand, is the inferior petrous sinus and the basilar venous plexus, the Dorello canal, and the trigeminal hiatus. The roof of the Dorello canal is delimited by the only ligament of the skull, which is the petroclinoid ligament that extends from the apex of the petrous (lateral to the oculomotor nerve) to the back of the sellae. Next to the petrolingual ligament, which runs from the apex of the petrous to the sphenoid bone lingula and fixes the carotid artery, they are known as the petrosphenoidal ligaments (Fig. 16.5).

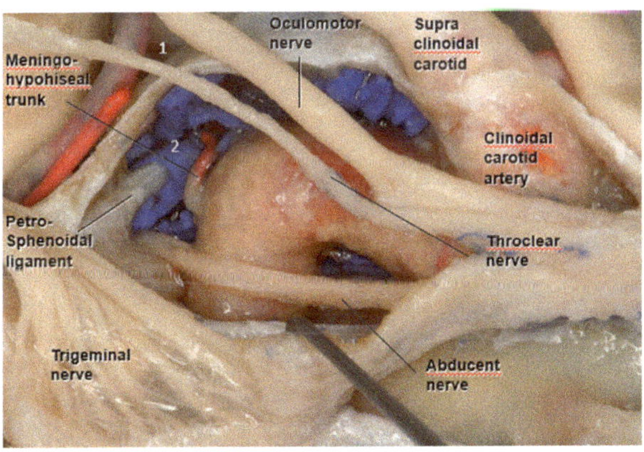

Fig. 16.5 Lateral view of the parasellar region. A branch of the intracavernous portion of the carotid artery that supplies the sellar contents is the meningohypophyseal trunk, the largest intracavernous branch, which gives rise to the inferior hypophyseal artery. The meningohypophyseal trunk arises at the level of the dorsum sellae at or just before the apex of the first curve of the carotid, where it turns forward after leaving the carotid canal. The inferior hypophyseal artery arises from the meningohypophyseal trunk and passes medially to the posterior pituitary capsule and lobe, and anastomoses with its mate of the opposite side after supplying the dura of the sellar floor. The oculomotor nerve courses along the lower edge of the anterior clinoid process to enter the superior orbital fissure. The abducens nerve penetrates the cavernous sinus by passing through Dorello's canal, located below the petrosphenoid ligament (Gruber's ligament). It passes lateral to the posterior vertical segment of the intracavernous carotid artery and courses inside the lateral venous space of the cavernous sinus lateral and inferior to the horizontal segment of the intracavernous carotid and medial to the trigeminal nerve to reach the superior orbital fissure. (1) Supratrochlear triangle; (2) infratrochlear triangle

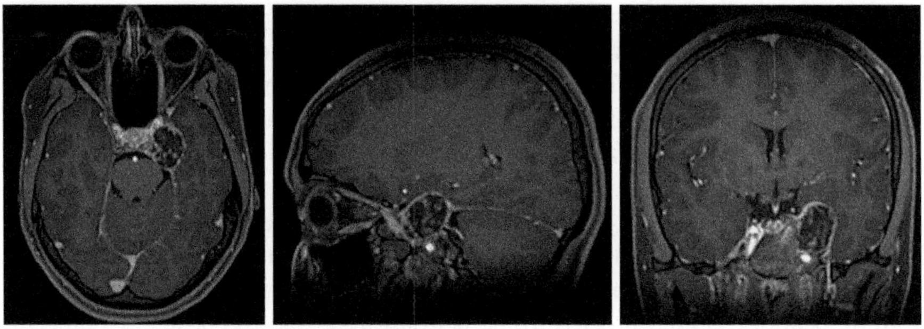

Fig. 16.6 Brain MRI T1 post gadolinium showing a left fifth nerve schwannoma in axial, sagittal, and coronal view, located in the cavernous sinus

Pathologies lateral either to the oculomotor nerve or to the lateral wall of the cavernous sinus are best addressed by pretemporal craniotomy, via the temporopolar approach. When we intend to address lesions between the oculomotor nerves, at the level of the interpeduncular fossa, posterior brain, and basilar top, the best craniotomy is performed via the pretemporal, transcavernous route, by removing the posterior clinoid process.

Tumors of the trigeminal nerve have a cisternal portion and a cavernous portion, because they insinuate into the cavernous sinus through the trigeminal hiatus (Fig. 16.6). The cavernous sinus ends at the upper edge of the maxillary branch of the trigeminal nerve (*another case example is also demonstrated in the* Video 16.1).

16.3 Trigeminal Nerve

The trigeminal nerve has two roots and three branches: the motor root, which accompanies the mandibular branch of the trigeminal and innervates the chewing muscles, and the sensory root, whose three branches are the ophthalmic, which goes to the upper orbital fissure through the lateral wall of the cavernous sinus (different from the oculomotor and trochlear nerves that enter the roof and run through the sidewall); the jaw, which goes toward the round foramen; and the mandibular, which goes to the foramen ovale (medial to the spinous foramen). Proximal to the trigeminal ganglion, we find the lacerum segment of the internal carotid artery surmounted by the greater superficial petrous nerve; below the ganglion, we find the petrous segment; and from the upper margin of the maxillary branch, we find the cavernous segment of internal carotid artery.

The trigeminal nerve is the main anatomical reference in the approaches to the skull base, mainly the midbrain region, anterior pontine, and anterolateral pontine. Anterior petrosectomy, which removes the petrous apex, greatly expands the angle of surgical vision to the posterior fossa. Medial to the trigeminal nerve, we find the structures related to the roof of the cavernous sinus, which are intradural. Laterally to the trigeminal nerve, we find the structures related to the lateral wall of the cavernous sinus, which are extradural. The anterior petrosectomy expands the view of the structures lateral to the trigeminal nerve. The "peeling" of the middle fossa begins by identifying the anterior clinoid process, which is the roof of the upper orbital fissure, as well as the meningo-orbital fold, through which we will start to enter the lateral wall of the cavernous sinus. After the removal of the dura mater, we find the reticular membrane that covers the trigeminal nerve and continues with the carotid-oculomotor membrane that forms the true roof of the cavernous sinus and the proximal and distal dural rings. Opening the lateral wall of the cavernous sinus, we can visualize the supratrochlear triangle, between the oculomotor and the trochlear nerves, and the infratrochlear or Parkinson's triangle, between the trochlear nerves and the ophthalmic branch of the trigeminal, through which we best visualize the cavernous carotid artery. The two triangles have an apex facing the upper orbital fissure. The exact point of entry is determined by the end of the apex of the anterior clinoid process; posteriorly, the Parkinson's triangle expands, allowing safe manipulation. The anteromedial triangles, between the ophthalmic and maxillary and anterolateral branches and between the maxillary and mandibular branches, have apex facing posterior and lead to the sphenoid sinus. The anteromedial triangle is also part of the cavernous sinus. The ophthalmic branch continues until the upper orbital fissure; the maxillary branch, up to the pterygopalatine fossa; and the mandibular branch, up to the infratemporal fossa. Another important point of reference is the middle meningeal artery. The drilling of the floor of the middle fossa, to enlarge the space between the basal surface of the temporal lobe and the middle fossa, can be done safely outside the middle meningeal artery because this anterior portion houses only the infratemporal fossa. The middle meningeal artery is the first ascending branch of the maxillary artery and enters the skull through the spinous foramen, which is posterolateral to the oval foramen. Between the spinous foramen, the lateral arched eminence, the medially greater superficial petrous nerve, and the mandibular branch at the base form the posterolateral or Glasscock triangle, where we can access the horizontal portion of the carotid artery. On the other hand, between the posterior edge of the Gasser ganglion anteriorly, the greater superficial petrous nerve laterally, and the petrous edge with the superior petrous sinus medially, we find the posteromedial or Kawase triangle, which gives access to the petrous apex. Special attention to the cochlea, located anteromedial to the geniculate ganglion and to the facial canal and immediately after the posterior "looping" of the internal carotid (Figs. 16.7, 16.8, and 16.9).

Medially, the cavernous segment of the internal carotid artery emits the meningohypophyseal trunk that originates from the dorsal meningeal artery and goes through the Dorello canal. The anterior lobe of the pituitary gland is irrigated by the upper pituitary artery, a branch of the medial wall of the ophthalmic segment of the internal carotid artery. The upper pituitary also irrigates the optic chiasm. On the other hand, the posterior pituitary lobe is irrigated by the inferior pituitary artery (which enters the posterior flexion of the carotid artery). The meningohypophyseal

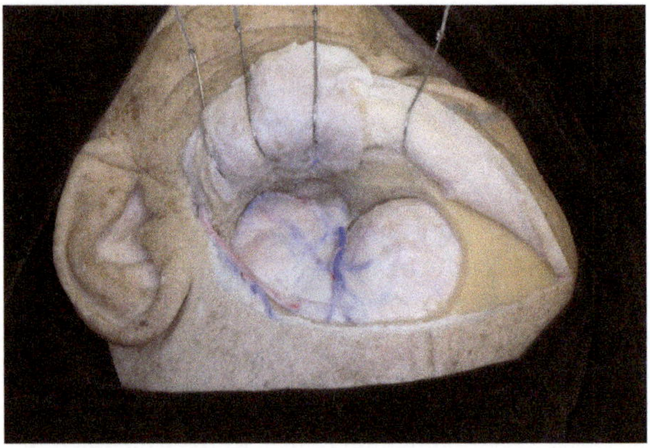

Fig. 16.7 Cadaveric dissection showing the left pretemporal craniotomy to achieving the lateral wall of the cavernous sinus, before middle fossa peeling

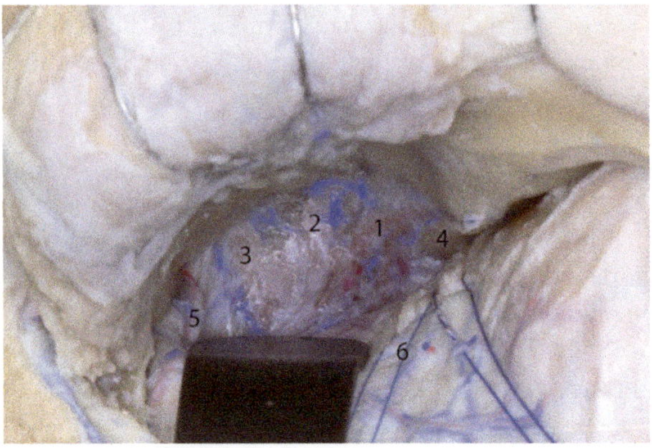

Fig. 16.8 Sequence of dissection to reach the lateral wall of the cavernous sinus. Left pretemporal craniotomy after peeling of the middle fossa, demonstrating the structures within the cavernous sinus, and other nearby anatomical landmarks. (1) V1 segment of the fifth nerve. (2) V2 segment of the fifth nerve. (3) V3 segment of the fifth nerve. (4) Left anterior clinoid process. (5) Left middle meningeal artery arising from the spinous foramen. (6) Left temporal dura mater

trunk originates in addition to the dorsal meningeal artery, Bernasconi and Cassinari's tentorial artery, and inferior hypophyseal artery (Fig. 16.10). Functioning pituitary tumors tend to be friable and invade the reticular membrane and the cavernous sinus, extending laterally. In contrast, non-functioning pituitary tumors tend to be firmer and may extend cranially through the sellar diaphragm, acquiring an hourglass shape (extradural and intradural portions). They are better approached via pterional craniotomy because they are on the same axis as the carotid. However, they can be approached by orbitozygomatic craniotomy if they extend superiorly to the floor of the third ventricle.

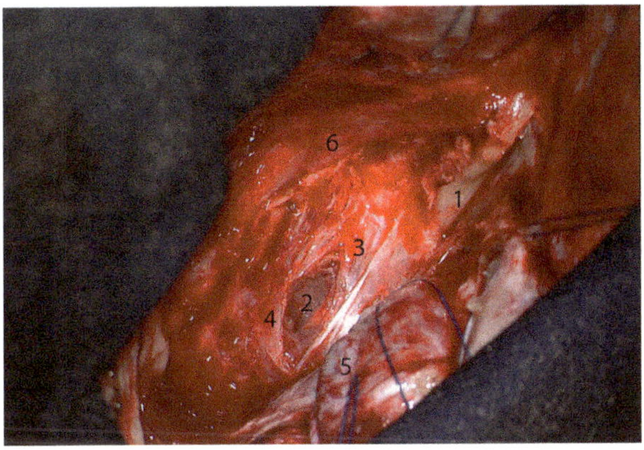

Fig. 16.9 Surgical view of the case exposed on Fig. 16.8. Left pretemporal craniotomy, after peeling of the middle fossa and opening the lateral wall of the cavernous sinus. (1) Left anterior clinoid process. (2) Tumor of the cavernous sinus, after opening the lateral wall between V2 and V3. (3) V2 segment of the left fifth nerve. (4) V3 segment of the left fifth nerve. (5) Left temporal pole retracted. (6) Left greater wing of the sphenoid bone

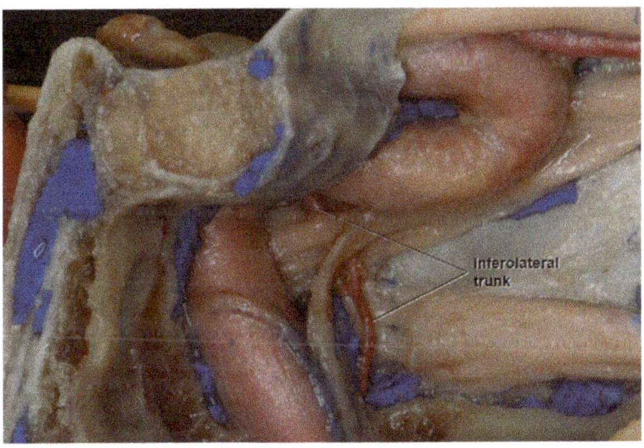

Fig. 16.10 Enlarged view of the sellae and medial wall of the cavernous sinus. The intracavernous carotid artery has two main branches. The first, the meningohypophyseal trunk, arises from the posterior bend. The second, the inferolateral trunk, also called the artery of the inferior cavernous sinus, arises from the horizontal segment. The inferolateral trunk, also called the artery of the inferior cavernous sinus, usually arises from the middle one-third of the inferior or lateral surface of the horizontal segment. It nearly always passes above the abducens nerve and then goes downward between the abducens and the first division of the trigeminal nerve to supply the dura of the inferolateral wall of the cavernous sinus

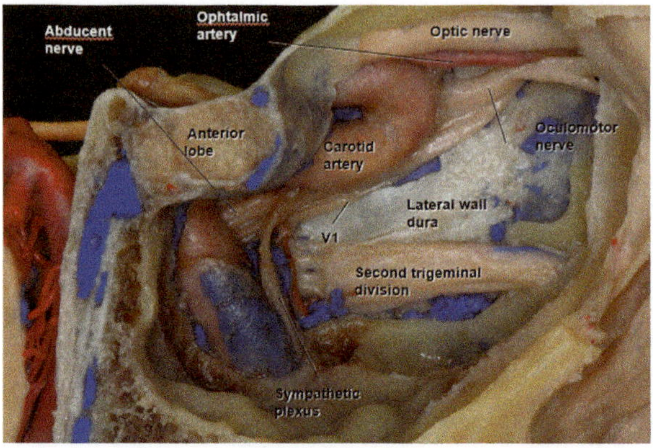

Fig. 16.11 Midsagittal section of the sella extending through the anterior and posterior lobes and sphenoid sinus. The intercavernous carotid produces prominences in the lateral wall of the sphenoid sinus below and anterior to the gland. The basilar sinus, located on the back of the dorsum, is the largest connection across the midline between the posterior edge of the paired cavernous sinuses. The anterior circulation sinus also connects the paired cavernous sinus but can be absent in some cases. The nerves associated with the cavernous sinus are the oculomotor, trochlear, ophthalmic, and abducens nerves and the sympathetic plexus around the intracavernous carotid artery. The ophthalmic nerve (first trigeminal division – V1) courses below the trochlear nerve on its way to the superior orbital fissure. The maxillary nerve (the second trigeminal division) courses below and does not belong to the lateral wall of the cavernous sinus. The cavernous sinus ends just above the superior margin of the maxillary nerve, where the medial and lateral walls of the cavernous sinus join in a keel-like formation. The abducens nerve and the sympathetic plexus around the intracavernous carotid artery are the only nerves that have a purely intracavernous course. The ophthalmic artery runs forward along with the optic nerve in the optic canal

16.4 Ophthalmic Artery

The ophthalmic artery (Fig. 16.11) originates in the medial portion of the anterior wall of the carotid artery and passes through the inferolateral portion of the optic nerve in most cases; less frequently, in the inferocentral portion; and in a few cases, in the inferomedial portion. Medial to the clinoid segment of the carotid artery, we have the sphenoid sinus, so CSF fistula is a complication after the clinoidectomy. The clinoid segment of the carotid artery is found between the proximal and distal dural rings and is extracavernous and extradural, whereas the supraclinoid carotid is extracavernous and intradural.

16.5 Anterior Clinoid Process

The anterior clinoid process is projected posteriorly from the lesser wing of the sphenoid bone, above the anterior portion of the roof of the cavernous sinus. The base of the anterior clinoid process has three fixation points with the adjacent portion of the sphenoid bone: laterally, the medial border of the sphenoid crest, formed by the lesser sphenoid wing, and, medially, it attaches to the anterior and posterior roots of the lesser wing. The anterior root of the lesser wing extends medially from the base of the anterior clinoid process to the body of the sphenoid, forming the roof of the optic canal, also called the sphenoid plane. The posterior root of the lesser wing of the sphenoid bone, called the optic strut, extends medially below the optic nerve, reaching the body of the sphenoid, forming the floor of the optic canal and the roof of the upper orbital fissure. The limits of the optical channel are the lateral wall, anterior clinoid; medial wall, sphenoid body; roof, the sphenoid plane; and floor, optical strut (Fig. 16.12).

The base of the anterior clinoid process forms the lateral margin of the optical channel. The segment of the internal carotid that runs along the medial surface of the anterior clinoid process, and that can be exposed by removing it, is called the clinoid segment. The anterior clinoid process corresponds to the point of attachment of the anteromedial portion of the cerebellar tent and the anterior petroclinoid and interclinoid dural folds. Another dural fold, called the sickle cell ligament, extends from the base of the anterior clinoid process, crosses the roof of the optic canal, and reaches the sphenoid plane.

The anterior clinoid process has a dense surface of cortical bone and a fragile diploe of medullary bone that is eventually crossed by small venous channels that join the cavernous sinus and diploic veins of the orbit's roof.

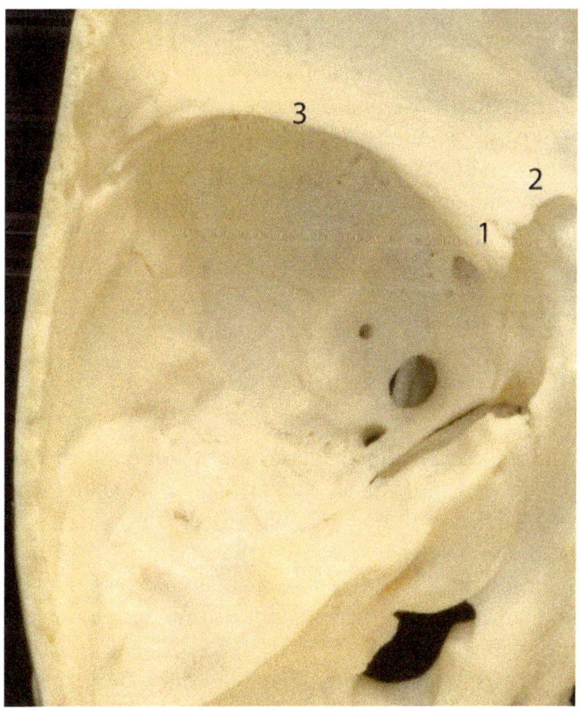

Fig. 16.12 Superior view of the left sphenoid bone, highlighting the points of fixation of the anterior clinoid process visible in this angle of view. (1) Anterior clinoid process. (2) Planum sphenoidale. (3) Lesser sphenoid wing

16.6 Optic Strut

The optic strut corresponds to a small bony bridge that extends from the inferior medial surface of the base of the clinoid process anterior to the sphenoid body, immediately in front of the carotid artery. The optic strut, from its junction with the anterior clinoid process, gently tilts down as it approaches the sphenoid body. The optic strut separates the optic channel from the superior orbital fissure. The upper surface of the optic strut, which tilts slightly downward and forward from its intracranial end, forms the floor of the optical channel. The lower surface of the optic strut forms the medial portion of the roof of the superior orbital fissure and the anterior portion of the roof of the cavernous sinus. The superior orbital fissure forms the anterior wall of the cavernous sinus.

The optic strut is located at the junction of the orbit apex, anteriorly, with the superior orbital fissure and with the optical canal posteriorly. The front edge of the optic strut corresponds to a narrow ridge located at the junction of its upper and lower surfaces. The posterior face of the optic strut, turned slightly downward, is formed in order to accommodate the anterior surface of the anterior curve of the intracavernous carotid, which juxtaposes itself to the posterior surface of the optic strut as it ascends medially to the anterior clinoid process. This is a region of change in the flow of the cavernous segment of the internal carotid artery and is a frequent site of aneurysms. The posterior face of the optic strut widens as it tilts medially. The point at which the optic strut joins the body of the sphenoid is marked on the surface of the sphenoid bone juxtaposed to the sphenoid sinus by a recess called the optic carotid recess, which extends laterally from the superolateral portion of the sphenoid sinus, between the prominences existing in the sinus wall, which covers the carotid groove and the optic canal. This recess can extend to the depth of the optic strut, which causes it to be partially or completely pneumatized by a lateral extension of the sphenoid sinus. This pneumatization can extend through the optic strut to the anterior clinoid process. The venous channels that connect the cavernous sinus with the diploid veins of the orbital roof and the anterior clinoid process can extend into or through the optic strut (Fig. 16.13).

16.7 Anterior Clinoidectomy

Within the explored context of the anatomy of the sphenoid bone, the anterior clinoid process has strategic importance with regard to the surgical approach, since its removal is a critical step in the process of treating paraclinoid lesions, that is, diseases that are closely related to the anterior clinoid process, with the medial portion of the lesser wing of the sphenoid bone and the roof of the cavernous sinus. Examples of these diseases are aneurysms of the ophthalmic segment of the internal carotid artery, meningiomas of the medial third of the lesser wing of the sphenoid bone, cavernous sinus meningiomas (chordomas, chondromas, chondrosarcomas), low bifurcation aneurysms of the basilar artery, and giant pituitary adenomas.

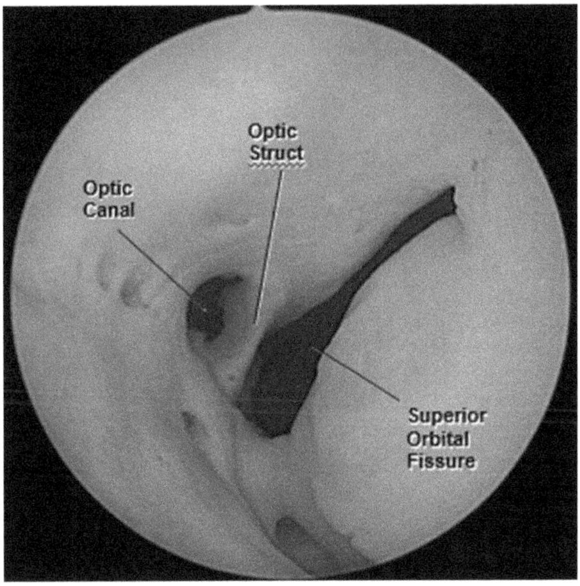

Fig. 16.13 Anterior aspect of the left optic canal, endoscopic view. The optic canal, which transmits the optic nerve and ophthalmic artery, opens in the superomedial corner of the orbital apex. The optic canal is located in the junction of the lesser wing and the sphenoid body. It is separated from the superior orbital fissure by the optic strut, a bony bridge that extends from the anterior clinoid process to the sphenoid body

Technically, there are two basic strategies described for anterior clinoidectomy in which the final objective is to disconnect the anterior clinoid process at its three points of attachment: lesser sphenoid wing, planum sphenoidale, and optic strut. One is performed before opening the dura mater, therefore extradural, and the other is performed after opening the dura mater, therefore intradural.

Intradural clinoidectomy starts after durotomy and, most of the time, after dissection and opening of the lateral fissure of the brain (Sylvian fissure) and other cisterns at the base. The starting point is the identification of the anterior clinoid process laterally to the optic nerve and the meningoorbitary fold in the superolateral portion of the superior orbital fissure and superolaterally to the internal carotid artery; then, thermocoagulation (with bipolar forceps) of the dura mater that surrounds the sphenoid plane, the base of the anterior clinoid process and the lateral surface of the anterior clinoid process is performed, so that thermo-coagulated region durotomy is performed and, then, the exposure of the anterior clinoid process. It is important before the durotomy to check if there is any bone defect in the anterior clinoid process, avoiding inadvertent damage to the optic nerve.

Disconnection of the anterior clinoid process is done by drilling its attachment points with small diamond drills coupled to high-speed rotors. This step requires extreme skill, given the importance of the structures surrounding the anterior clinoid process, as explained in the text above. Special attention should be given to protecting the entire length of the drill to prevent injury to early branches of the middle cerebral artery. Disconnection begins with the drilling of the sphenoid plane, parallel to the optic nerve, and goes to the lesser sphenoid wing, transversal to the optic nerve and then to the base of the anterior clinoid process, parallel to the clinoid, until the exposure of the optic strut after opening of the distal dural ring and exposure of the ophthalmic artery, when the drilling plane changes direction, occupying an oblique and medial position, according to the layout of the optic strut. After disconnecting the optic strut, the anterior clinoid process can be removed en bloc. It may often be necessary to disconnect the attachment points of the carotid artery to improve mobility and exposure, which are the petrolingual ligament and the posterior communicating artery in addition to the distal dural ring, just as there may be a need to mobilize the oculomotor nerve laterally through the opening of the anterior and posterior and interclinoid petroclinoid ligaments. In the intradural technique, during the anterior clinoidectomy process, the surrounding anatomy, as well as the lesion in question, will be under the direct visual control of the surgeon (Figs. 16.14 and 16.15) (*another case example is also demonstrated in the* Video 16.2).

Extradural clinoidectomy, initially described by Dolenc in 1985, received several modifications, but it basically starts right after the craniotomy and is based on the exposure of the anterior clinoid process with extradural detachment of the dura that surrounds the sphenoid plane, anterior clinoid process, lesser wing of the sphenoid bone, and medium fossa. Then, the anterior clinoid process is disconnected in the same way as described in intradural clinoidectomy; however, in this technique, the structures neighboring the previous clinoid process and the lesion in question are not under direct visual control and are protected by the dura mater.

There is a debate among authors as to the best strategy, so the decision as to when to perform one technique or another may vary according to the disease in question and the experience of neurosurgeon.

The microsurgical procedure of anterior clinoidectomy is the key for the treatment of pathologies that are located in the parasellar region; however, it involves a

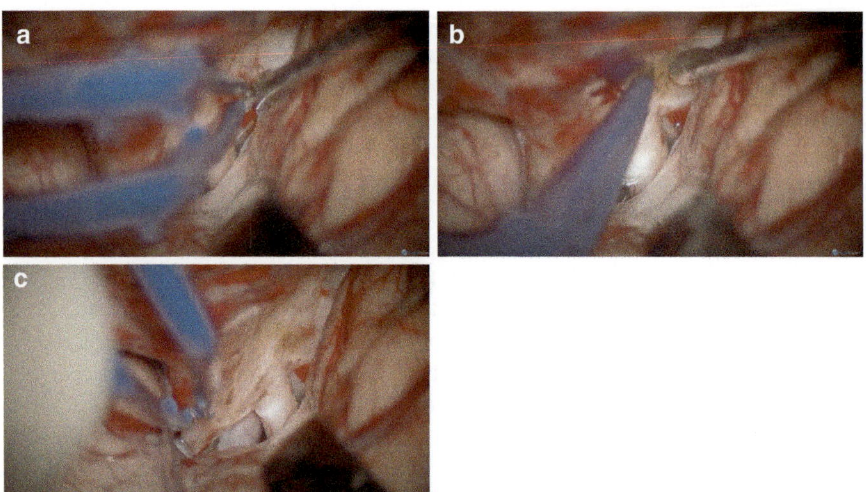

Fig. 16.14 Surgical view showing the sequence of dural coagulation followed by a cutting in the dura mater over the anterior clinoid process at the skull base to improve exposure for anterior clinoidectomy. First, the coagulation is directed parallel to the optic nerve (**a**), thereafter transversal to the optic nerve (**b**), and, finally, at the lateral border of the anterior clinoid process (**c**)

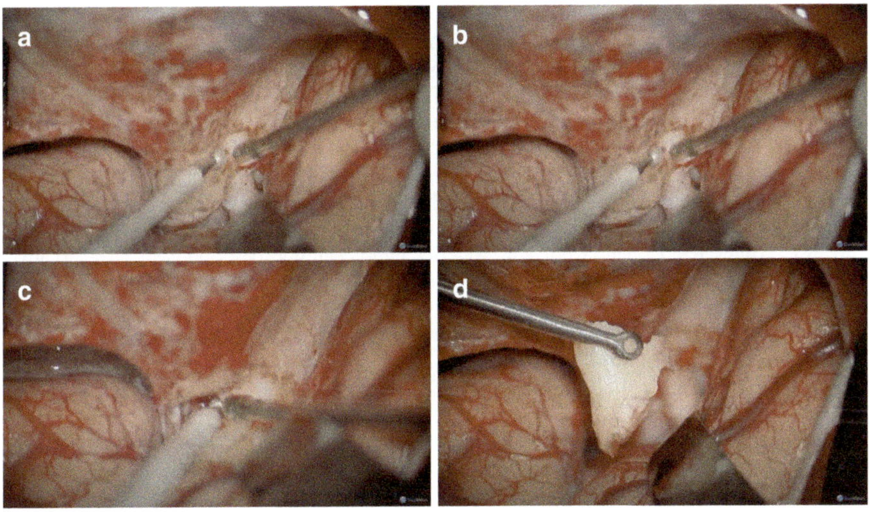

Fig. 16.15 Sequence of surgical images, showing the steps involved in performing an anterior clinoidectomy. First, the drill is directed parallel and transverse to the ipsilateral optic nerve to disconnect the planum sphenoidale (**a**). In the second step, the drill is transverse to the lesser sphenoid wing to disconnect this structure (**b**), and, finally, the drill is directed to the optic strut to complete the resection of the anterior clinoid process which (**c**) is extracted in block (**d**)

risk of developing visual disturbances, oculomotor nerve paresis/plegia, hemorrhage from the opening of the cavernous sinus, injury of the internal carotid artery as well as of the ophthalmic artery, opening of the paranasal sinuses, pneumocranium, cerebrospinal fluid fistula associated or not with meningitis, and death. In the literature, for example, there are high rates of morbidity, mortality, and presence of CSF fistula associated or not with meningitis. Even so, anterior clinoidectomy is an imperative procedure for treatment, even with the impact that the complications listed above may have on the clinical outcome of patients with paraclinoid diseases. The wide and safe access to pathologies in this region also involves an in-depth knowledge of the anatomy of the middle fossa floor and the complex relationships between neurovascular structures and the paraclinoid region.

Suggested Bibliography

Alejandro SA, Carrasco-Hernández JP, da Costa MDS, Ferreira DS, Lima JVF, de Amorim BL, Paz-Archila JA, Chaddad-Neto F. Anterior clinoidectomy: Intradural step-by-step en bloc removal technique. World Neurosurg. 2021;146:217–31.

Caramanti RL, da Costa MDS, Hernández YAU, Alves Filho CAF, Wainberg R, Palmiero H, Saick RP, Chaddad-Neto F. Intradural anterior clinoidectomy for ophthalmic artery aneurysm clipping: 3-dimensional operative video. Oper Neurosurg (Hagerstown). 2018;14(6):708.

Chaddad-Neto F, da Costa MDS, Santos B, Caramanti RL, Costa BL, Doria-Netto HL, Figueiredo EG. Reproducibility of a new classification of the anterior clinoid process of the sphenoid bone. Surg Neurol Int. 2020;11:281.

da Costa MDS, de Oliveira Santos BF, de Araujo PD, Rodrigues TP, Abdala N, Centeno RS, Cavalheiro S, Lawton MT, Chaddad-Neto F. Anatomical variations of the anterior clinoid process: a study of 597 skull base computerized tomography scans. Oper Neurosurg (Hagerstown). 2016;12(3):289–97. https://doi.org/10.1227/NEU.0000000000001138.

Dolenc VV. A combined epi- and subdural direct approach to carotid-ophthalmic artery aneurysms. J Neurosurg. 1985;62(5):667–72.

Giannotta SL. Ophthalmic segment aneurysm surgery. Neurosurgery. 2002;50(3):558–62.

Javalkar V, Banerjee AD, Nanda A. Paraclinoid carotid aneurysms. J Clin Neurosci. 2011;18(1):13–22.

Kulwin C, Tubbs RS, Cohen-Gadol AA. Anterior clinoidectomy: description of an alternative hybrid method and a review of the current techniques with an emphasis on complication avoidance. Surg Neurol Int. 2011;2:140.

Lehmberg J, Krieg SM, Meyer B. Anterior clinoidectomy. Acta Neurochir. 2014;156(2):415–9; discussion 9.

Rhoton AL Jr. The sellar region. Neurosurgery. 2002a;51(4 Suppl):S335–74.

Rhoton AL Jr. The cavernous sinus, the cavernous venous plexus, and the carotid collar. Neurosurgery. 2002b;51(4 Suppl):S375–410.

Takahashi JA, Kawarazaki A, Hashimoto N. Intradural en-bloc removal of the anterior clinoid process. Acta Neurochir. 2004;146(5):505–9.

Chapter 17
Surgical Anatomy of the Foramen Magnum

The structures related to the region of the foramen magnum (Fig. 17.1) are as follows: the lower cranial nerves (i.e., the glossopharyngeal, vagus, accessory, and hypoglossal nerves), the vertebral artery (V3 and V4 segments), the posterior inferior cerebellar artery, the suboccipital surface of the cerebellum, the cerebellomedullary fissure, the inferior half of the fourth ventricle, the atlas, and the axis.

There are two main surgical approaches to the foramen magnum, the median suboccipital approach and the far lateral approach, which has several variations.

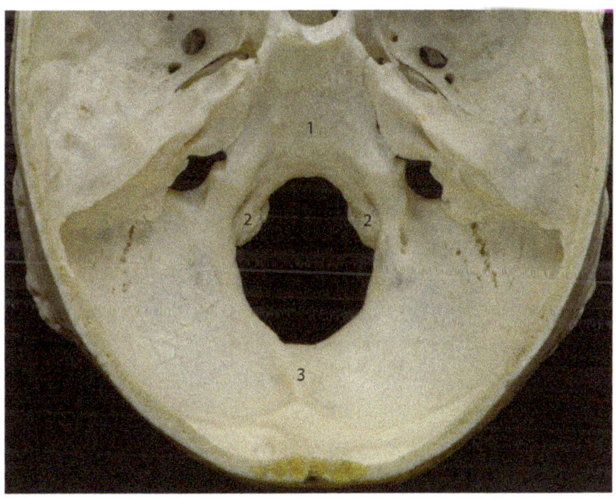

Fig. 17.1 Superior view of the skull base showing the three parts of the occipital bone: (1) the basilar part, (2) the condylar part, (3) the squamous part

Supplementary Information The online version of this chapter (https://doi.org/10.1007/978-3-030-82747-2_17) contains supplementary material, which is available to authorized users.

F. Chaddad-Neto, M. D. Silva da Costa, *Microneuroanatomy and Surgery*,
https://doi.org/10.1007/978-3-030-82747-2_17

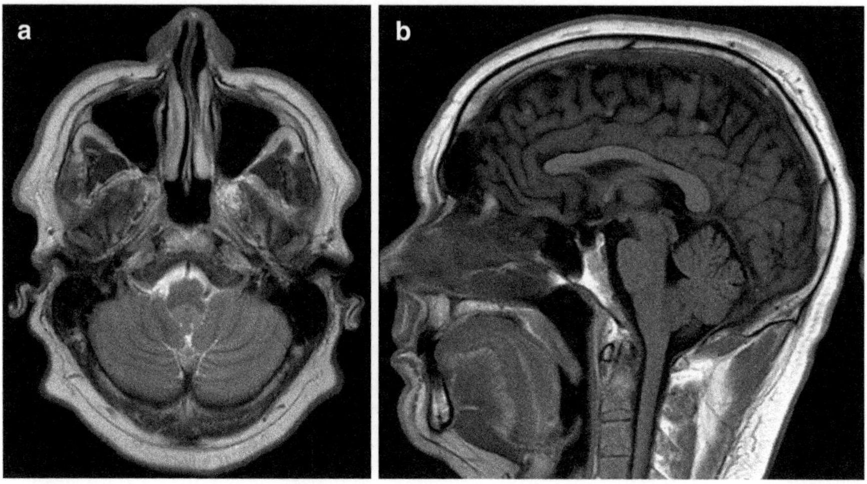

Fig. 17.2 (**a**) Brain T2-weighted axial MRI showing a fourth ventricle lesion. (**b**) Brain T1-weighted sagittal MRI showing the tumor in relation to the bottom half and the median aperture of the fourth ventricle

17.1 Surgical Case 1

A 45-year-old man with progressive headache that worsened when lying down.

Based on the MRI above, where is the lesion located? What is the preferable approach to treat this lesion? What are the main surface and deep landmarks? Which artery is related to the lesion? Which segment of the artery?

The lesion shown in Fig. 17.2 is situated in the bottom half of the fourth ventricle, which for surgical purposes is considered to be in the region of the foramen magnum. Median suboccipital craniotomy is the best approach to reach lesions related to the medium line because it offers a view from the posterior surface of the medulla and the suboccipital surface of the cerebellum. Further, by using the telovelar route, the fourth ventricle can be reached. Positioning is a key step to this procedure. In order to promote cerebellar relaxation after dissection of the cisterna magna, the patient is placed in a semi-seated position during the far lateral and suboccipital craniotomy. Cervical flexion is crucial to allow the surgeon access to the tentorium as well as the structures of the foramen magnum.

In terms of surface landmarks, the muscles of the posterior aspect of the occipital and upper cervical vertebrae are the key points for delimitation of the craniotomy. The superior nuchal line delimits the inferior border of the transverse sinus and can be easily palpated or found posteriorly to the zygomatic arch (Figs. 17.3, 17.4, and 17.5).

The occipital artery (Figs. 17.4 and 17.5), which courses between the semispinalis capitis and splenius capitis muscles, is the surface landmarks. The point where this artery pierces the splenius capitis is deep in relation to the lateral border of the foramen magnum; therefore, it should serve as the lateral border of the craniotomy. The vertebral artery courses on the superior face of the posterior arch of the atlas, making the dissection safer on the inferior face of the posterior arch.

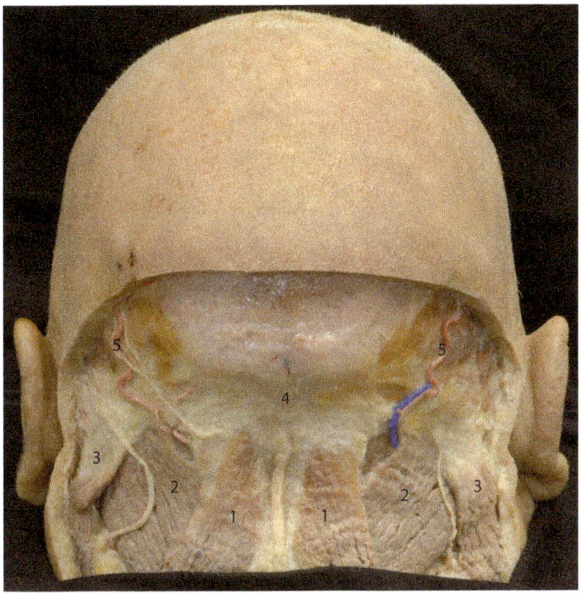

Fig. 17.3 Cadaveric dissection of the superficial posterior craniocervical junction muscles. (1) Trapezius muscle, (2) splenius capitis muscle, (3) sternocleidomastoid muscle, (4) external occipital protuberance, (5) occipital artery

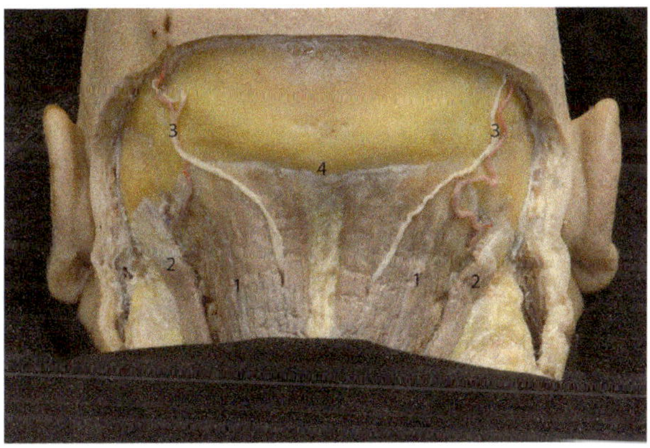

Fig. 17.4 Cadaveric dissection of a deeper muscle layer after dissection of the trapezius and the splenius capitis muscle. (1) Semispinalis capitis muscle, (2) sternocleidomastoid muscle, (3) occipital artery and greater occipital nerve, (4) external occipital protuberance

The artery related to this case is the posterior inferior cerebellar artery (PICA). The lesions inside the fourth ventricle are irrigated by the choroidal branches of the PICA from both the tonsillomedullary and the telovelotonsillar segments (Figs. 17.6, 17.7, and 17.8).

17.2 Surgical Case 2

A 65-year-old woman with sudden headache, vomiting, and stiff neck.

Based on the digital subtraction angiography (DSA) above, where is the lesion located? What is the preferred treatment approach? What parameters form the basis of this choice in approach? What is the angle of view required for this approach and what are its disadvantages?

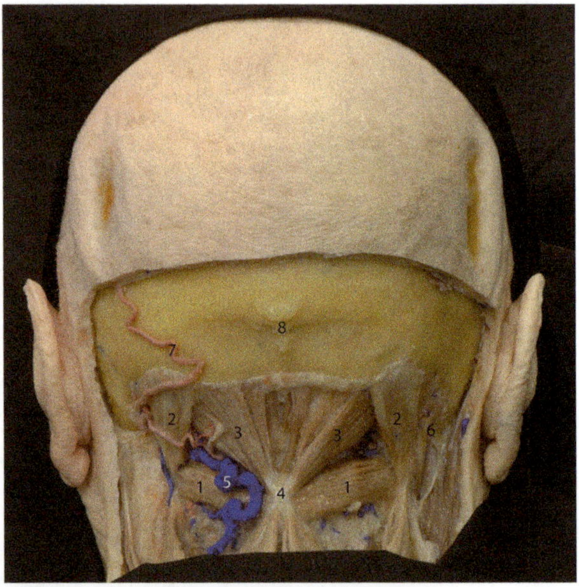

Fig. 17.5 After dissection of the semispinalis capitis muscle, the suboccipital triangle is exposed. (1) Inferior oblique muscle, (2) superior oblique muscle, (3) major rectus capitis posterior muscle, (4) C2 spinous process, (5) venous plexus of the vertebral artery, (6) posterior belly of the digastric muscle, (7) occipital artery, (8) external occipital protuberance

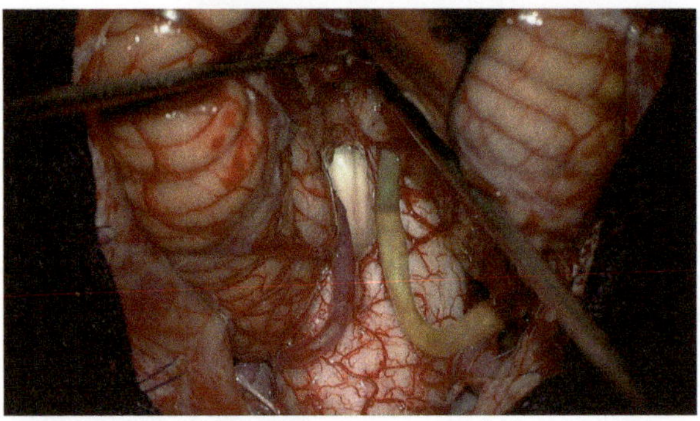

Fig. 17.6 Surgical view of a suboccipital craniotomy by the telovelar route. Here, the image shows both the right and left PICA. The right PICA is highlighted in yellow and green, which show the tonsillomedullary segment and the telovelotonsillar segment, respectively. This surgical view shows the choroidal branch occurring between the telovelotonsillar segment and the cortical segment

The aneurysm in Fig. 17.9 is located at the origin of the PICA, which is related to the foramen magnum zone. Based on this information, in which segment of the PICA has the aneurysm occurred? Which portion of the foramen magnum is thus affected? In which portion of the foramen magnum has the aneurysm occurred?

As previously stated, the foramen magnum region is associated with the PICA, the lower cranial nerves, the suboccipital surface of the cerebellum, the cerebellomedullary fissure, and the bottom half of the fourth ventricle.

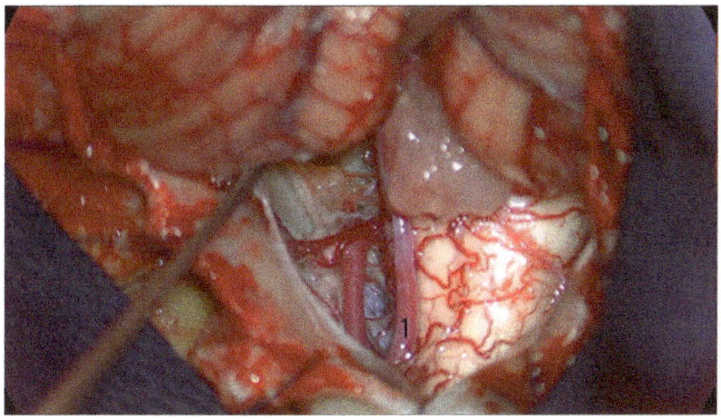

Fig. 17.7 Surgical view of another case following the use of the telovelar route. (1) A lower caudal loop of the PICA

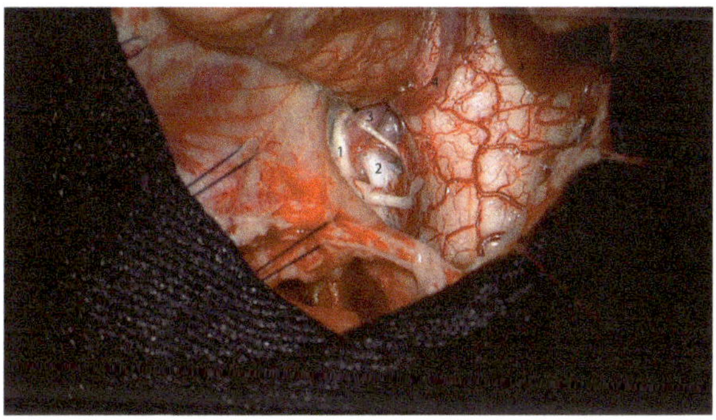

Fig. 17.8 – Surgical view from a median suboccipital craniotomy, showing the structures related to the foramen magnum. (1) Eleventh nerve, (2) denticulate ligament and its insertion in the foramen magnum, (3) V4 segment of the vertebral artery, (4) caudal loop of the PICA

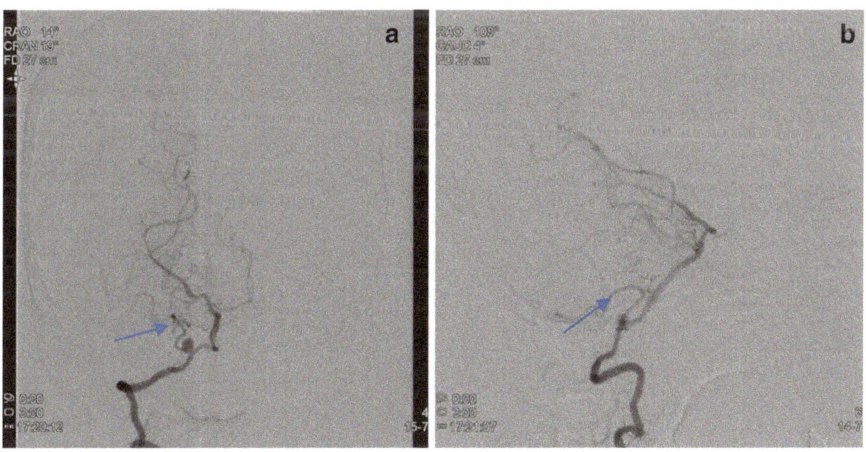

Fig. 17.9 (**a**) Digital subtraction angiography in the frontal projection showing a PICA aneurysm located at the origin of the artery. (**b**) Lateral projection demonstrating the close relation with the anterior portion of the foramen magnum

The PICA is the most variable artery in the posterior fossa and is related to the cerebellomedullary fissure, the inferior half of the fourth ventricle, and the suboccipital surface of the cerebellum. It can be extradural and has variations in its segments. Bifurcation of its cortical branches gives rise to arteries supplying the vermis (via the medial trunk) and the hemisphere (via the lateral trunk).

The proximal branches of all arteries of the posterior fossa are in close proximity to the brainstem. The aneurysm shown in Fig. 17.9 at the origin of the PICA is in the first segment, called the anterior medullary segment. The arteries of the posterior fossa follow a path anterior, lateral, and posterior to the brainstem before the choroidal and cortical branches supply the tela choroidea and the cerebellum, respectively (Fig. 17.10).

If the aneurysm is found in the anterior medullary segment of the PICA and is closely related to the anterior surface of the medulla, how is this lesion reached?

The anterior surface of the medulla and anterior part of the foramen magnum can be reached by a far lateral craniotomy because this approach gives a lateral to anterior view of the foramen magnum region, and the angle of work is thus adequate to achieve lesions between the lower cranial nerves, vertebral arteries, and anteromedial surface of the medulla. This approach can only be employed after studying the angiography in order to build a three-dimensional view of the aneurysm and its surroundings.

The variations of the classic far lateral approach are the transcondylar approach, the paracondylar approach, and the supracondylar approach. In terms of vascular neurosurgery, we believe that the transcondylar approach and the classic far lateral approach are adequate strategies to reach the majority of PICA aneurysms.

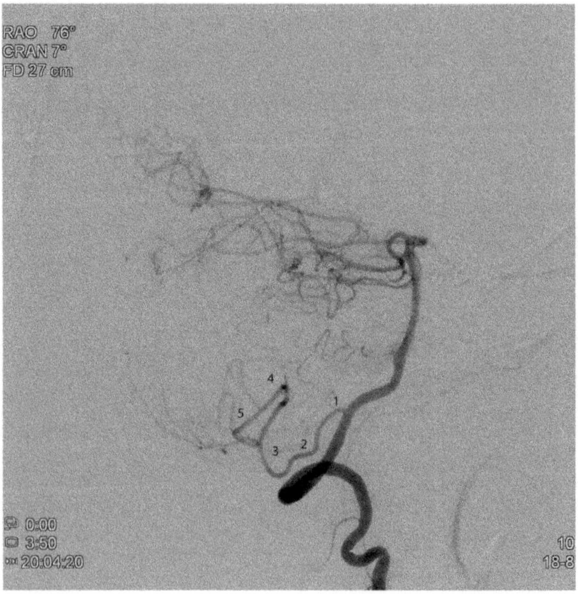

Fig. 17.10 Lateral DSA of the vertebral artery showing the segments of the PICA and the path anterior, lateral, and finally posterior to the brainstem. (1) Anterior medullary segment, (2) lateral medullary segment, (3) tonsillomedullary segment, (4) telovelotonsillar segment including the choroidal branches, and (5) cortical branches, including the medial and lateral trunks

The relationship between the aneurysm and the lower cranial nerves will direct the choice in surgical approach. If the PICA aneurysm is located between the hypoglossal nerves, the far lateral transcondylar approach is preferable. Based on our experience, aneurysms located between the 9th, 10th, and 11th nerves can be reached by either a transcondylar approach or the classic far lateral approach. Finally, aneurysms located lateral to the 9th, 10th, and 11th nerves are reachable by the classic far lateral approach (Fig. 17.11).

The sternocleidomastoid and trapezius muscles are the most superficial muscles, followed by the splenius capitis muscle. After dissecting and cutting these, the sternocleidomastoid is detached and placed laterally. The next muscle, the longissimus capitis, is reached by reflecting medially the trapezius and splenius capitis muscles. The semispinalis capitis is next in the sequence of the muscle layer, and this is the final muscle before the suboccipital triangle.

The vertebral artery becomes intradural between the first and second denticulate ligaments and is then positioned behind the first denticulate ligament. The posterior (sensitive) roots of the cervical nerves are located posterior to the ligaments, while the anterior (motor) roots of the cervical nerves are located anterior to the ligaments. Most of the denticulate ligament's cranial inserts are on the medial border of the foramen magnum. Thus, the posterior fossa is located above the first denticulate ligament, while the cervical vertebral canal is located below it.

The rootlets of the hypoglossal nerve are anterior to the olive, and the glossopharyngeal, vagus, and accessory nerves emerge posterior to the olive. The glossopharyngeal nerve generally has a unique root, making it easier to identify. The vagus nerve has many rootlets, and the accessory nerve has medullary and ascending cervical rootlets. The V3 segment of the vertebral artery passes posterior to the atlantooccipital joint and is followed by the ventral ramus of the C1 nerve.

The far lateral approach has a lateral to anterior view, and the angle goes from inferior to superior. This makes it difficult to reach the lesion at the junction of the vertebral arteries as well as to surgically treat pathologies of the pontomedullary sulcus. For this reason, these lesions are better approached with a presigmoid craniotomy.

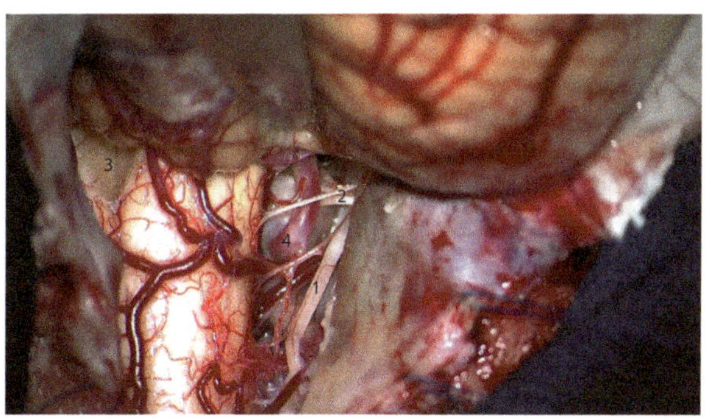

Fig. 17.11 Surgical view of a far lateral approach showing several labeled structures. (1) Cervical rootlets of the accessory nerve, (2) complex of the 9th and 10th nerves, (3) floor of the fourth ventricle, (4) PICA, including the aneurysm related to the lower cranial nerves

17.3 Surgical Case 3

A 64-year-old woman with a severe headache, described as the worst of her life, 5 days prior.

Based only on the DSA in Fig. 17.12, where is the aneurysm located? In which artery? What is the best way to approach treatment?

Figure 17.12 shows another case of PICA aneurysm, but in a different location from the one shown in Fig. 17.9. The DSA in the lateral view shows an aneurysm located at the distal portion of the PICA. Thus, the median suboccipital approach (Fig. 17.13) is best because it offers a posterior to anterior view in the midline, which is required because the distal segments of the PICA are tonsillomedullary and telovelotonsillar, posterior to the medulla.

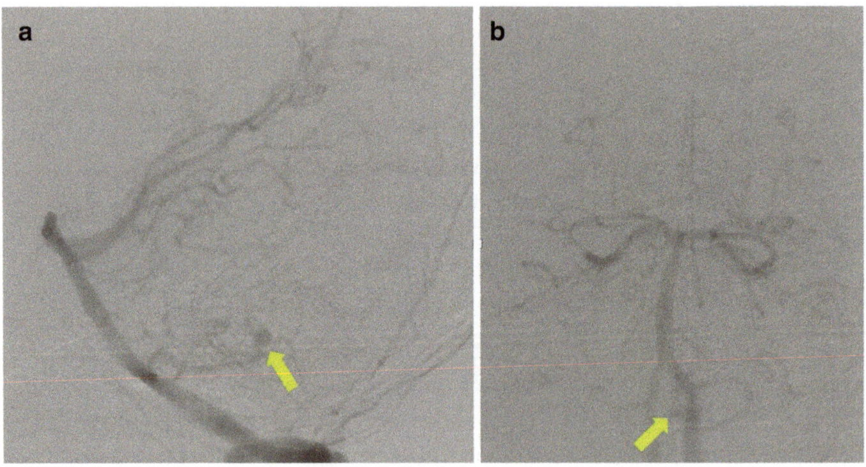

Fig. 17.12 (**a**) Lateral vertebral DSA demonstrating a ruptured distal PICA aneurysm related to the telovelotonsillar segment. (**b**) Frontal vertebral DSA showing the aneurysm posterior to the brainstem

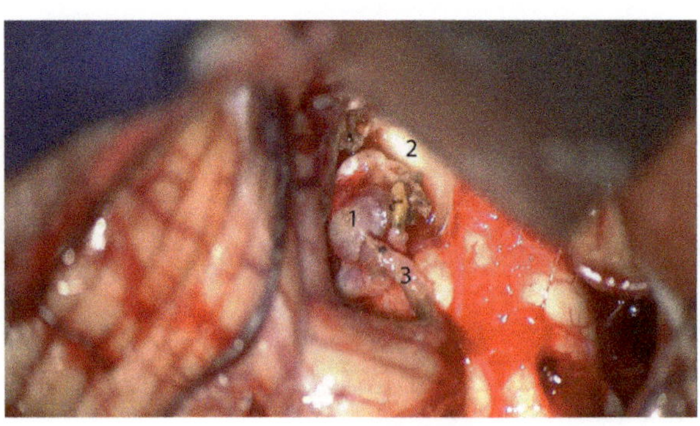

Fig. 17.13 Surgical view of a suboccipital craniotomy showing the aneurysm in Fig. 17.12 using the telovelar route. (1) The PICA aneurysm, (2) floor of the fourth ventricle, (3) televelotonsillar segment of the PICA

17.4 Conclusion

The foramen magnum is an important structure of the skull base and is the point of transition between the medulla and the cervical spine. It can be affected by several diseases, and knowledge of key anatomical points can help clinicians choose better surgical approaches, which in turn can lead to more favorable outcomes.

Further, knowledge of PICA position and the positions of its segments can help the surgeon find better approaches and better predict surgical difficulties. This can improve treatment of vascular, tumoral, and skull base diseases.

Suggested Bibliography

Chaddad Neto F, Doria-Netto HL, Campos Filho JM, et al. The far-lateral craniotomy: tips and tricks. Arq Neuropsiquiatr. 2014;72(9):699–705. https://doi.org/10.1590/0004-282X20140130.

de Oliveira E, Tedeschi E, Rhoton A, Peace D. Microsurgical anatomy of the posterior circulation: vertebral and basilar arteries. In: Carter L, Spetzler R, editors. Neurovascular surgery. New York: McGraw Hill, Inc.; 1995. p. 25–34.

Osborn AG. Osborn's brain: imaging, pathology, and anatomy. Salt Lake City: Amirsys Publishing, Inc.; 2012.

Rhoton AL. The posterior cranial fossa: microsurgical anatomy and surgical approaches. Neurosurgery. 2000;47(suppl_3):S5–6. https://doi.org/10.1097/00006123-200009001-00005.

Yasargil M. Microneurosurgery, volume 1: microsurgical anatomy of the basal cisterns and vessels of the brain, diagnostic studies, general operative techniques and pathological considerations of the intracranial aneurysms. New York: George Thieme Verlag; 1984.

Chapter 18
Surgical Anatomy of the Pineal Region

18.1 Anatomy of the Pineal Region or Posterior Incisural Space

18.1.1 Introduction

The pineal region is named after the pineal gland, also known as the cerebral epiphysis, and corresponds to the posterior incisural space. Neuroanatomical knowledge of this region is fundamental in the field of neurosurgery and particularly with regard to the relationships between the pineal body, posterior cerebral artery, superior cerebellar artery, vein of Galen, basal vein of Rosenthal, internal cerebral vein, straight sinus, bridging vein, the size of the tentorial notch, and the fourth cranial nerves inside the quadrigeminal cistern.

The complexity of this region is a result of the large number of venous structures and the dense arachnoid that makes dissection difficult. It is for this reason that lesions located in this region represent a challenge for the neurosurgeon, both from the point of view of neuroanatomical knowledge and microsurgical technique.

In the present chapter, we focus on the neuroanatomy of the region and its relationship to current surgical techniques for pathologies in the pineal region.

Supplementary Information The online version of this chapter (https://doi.org/10.1007/978-3-030-82747-2_18) contains supplementary material, which is available to authorized users.

F. Chaddad-Neto, M. D. Silva da Costa, *Microneuroanatomy and Surgery*,
https://doi.org/10.1007/978-3-030-82747-2_18

18.1.2 Neural Relationships

The posterior incisural space lies posterior to the midbrain and corresponds to the pineal region (Fig. 18.1). The pineal region has a roof, floor, and anterior and lateral walls and extends posteriorly to the level of the tentorial apex.

Anterior Wall The quadrigeminal plate is located at the center of the anterior wall. The anterior wall rostral to the colliculi is formed by the pineal body. The habenular commissure forms the upper half, and the posterior commissure forms the lower half of the attachment of the pineal body to the posterior part of the third ventricle. The part of the anterior wall below the colliculi is formed in the midline by the lingula of the vermis and laterally by the superior cerebellar peduncles as they ascend beside the lingula (Fig. 18.2).

When visualizing the posterior fossa, the most important reference point is the fourth cranial nerve (the trochlear nerve); above this nerve lies the quadrigeminal lamina and below it lies the superior medullary veil along with the interpeduncular sulcus, which separates the superior and middle cerebellar peduncles.

Lateral Wall The lateral wall is divided into anterior and posterior components, and each lateral wall is formed by the pulvinar, crus of the fornix, and the medial surface of the cerebral hemisphere. The anterior part of the lateral wall is formed by the part of the pulvinar located immediately lateral to the pineal body. The lateral wall posterior to the pulvinar is formed by the segment of the crus of the fornix that wraps around the posterior margin of the pulvinar (Fig. 18.3). The posterior part of the lateral wall is formed by the cortical areas located below the splenium on the medial surface of the hemisphere. These cortical areas include the posterior part of the parahippocampal and dentate gyri (Figs. 18.4 and 18.5).

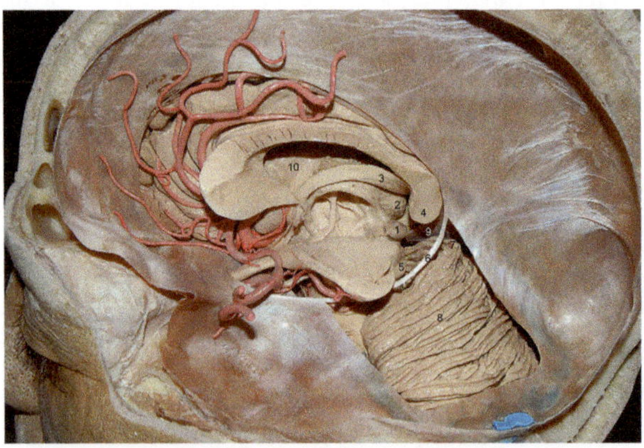

Fig. 18.1 Lateral view of a head after the removal of the left cerebral hemisphere and the tentorium. The posterior incisural space is exposed. (1) Pineal body; (2) pulvinar of the thalamus; (3) crus of the fornix; (4) splenium of corpus callosum; (5) quadrigeminal plate; (6) tentorial edge; (7) culmen; (8) tentorial surface; (9) tentorium; (10) frontal horn of the lateral ventricle

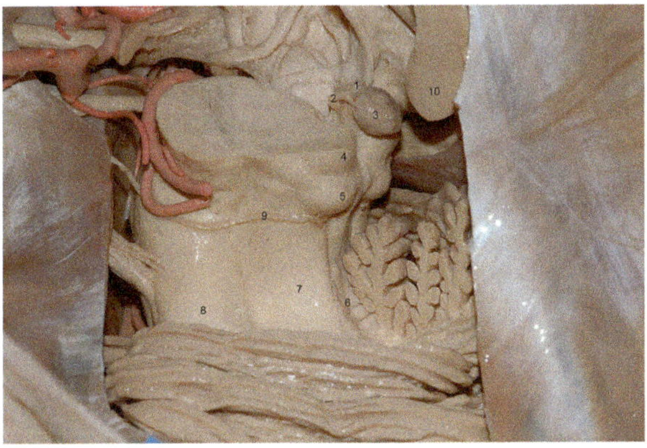

Fig. 18.2 Posterolateral view of the brainstem. The quadrangular lobule of the cerebellum was removed to expose the anterior wall of the pineal region. (1) Habenular commissure; (2) posterior commissure; (3) pineal body; (4) superior colliculus; (5) inferior colliculus; (6) lingula; (7) superior cerebellar peduncle (brachium conjunctivum); (8) middle cerebellar peduncle (brachium pontis); (9) trochlear nerve; (10) splenium of the corpus callosum

Fig. 18.3 Anatomical specimen dissected using the Klingler's technique. The tentorium was removed to expose the basal surface of the left cerebral hemisphere. The anterior part of the lateral wall is formed by the part of the pulvinar located immediately lateral to the pineal body. (1) Splenium of the corpus callosum; (2) pineal body; (3) parahippocampal gyrus; (4) crus of the fornix; (5) pulvinar of the thalamus; (6) dentate gyrus; (7) body of the hippocampus; (8) optic radiation; (9) tentorium

Roof The splenium of the corpus callosum is a brain structure that is an important point of reference in neurosurgery. The roof of the posterior incisural space is formed by the lower surface of the splenium, the terminal part of the crura of the fornices, and the hippocampal commissure. Each crus arises as a continuation of the fimbria, passing around the posterior margin of the pulvinar before blending into the lower margin of the splenium. The hippocampal commissure is an oblique band of

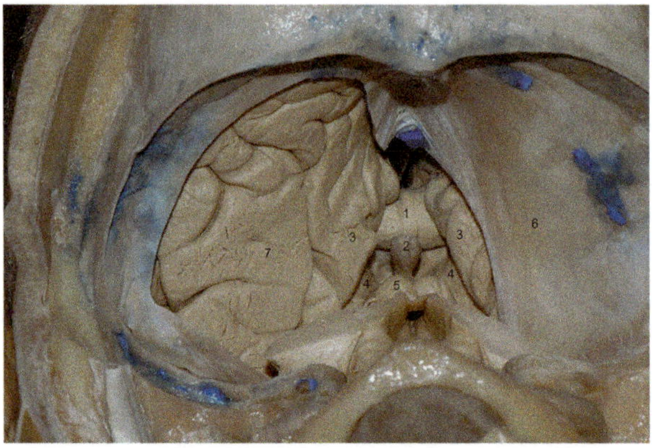

Fig. 18.4 View of the basal surface of the left cerebral hemisphere. The posterior part of the lateral wall is formed by the cortical areas located below the splenium on the medial surface of the hemisphere. These cortical areas include the posterior part of the parahippocampal and dentate gyri. (1) Splenium of the corpus callosum; (2) pineal body; (3) parahippocampal gyrus; (4) pulvinar of the thalamus; (5) inferior colliculus; (6) tentorium; (7) fusiform gyrus

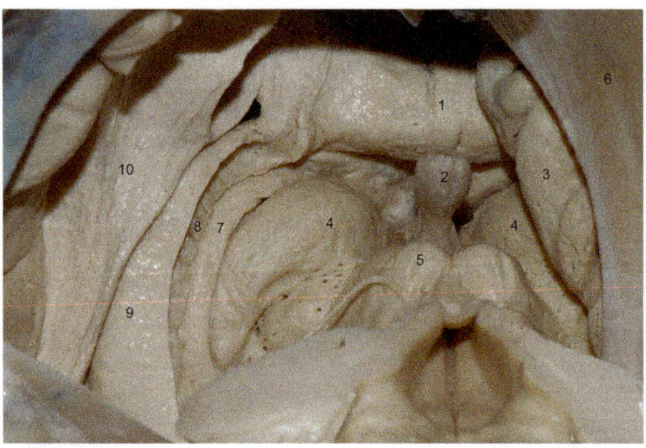

Fig. 18.5 Anatomical specimen dissected using the Klingler's technique. The neural relationships lateral to the pineal region are exposed. (1) Splenium of the corpus callosum; (2) pineal body; (3) parahippocampal gyrus; (4) pulvinar of the thalamus; (5) inferior colliculus; (6) tentorium; (7) fimbria; (8) dentate gyrus; (9) body of the hippocampus; (10) optic radiation

fibers that courses below the splenium between the medial margins of the crura (Figs. 18.6 and 18.7).

Floor The floor of the posterior incisural space is formed by the anterosuperior part of the cerebellum and consists of the culmen of the vermis in the midline and the quadrangular lobules of the hemispheres laterally. The posterior incisural space extends inferiorly into the cerebellomesencephalic fissure (Figs. 18.8 and 18.9). (Fig. 18.8)

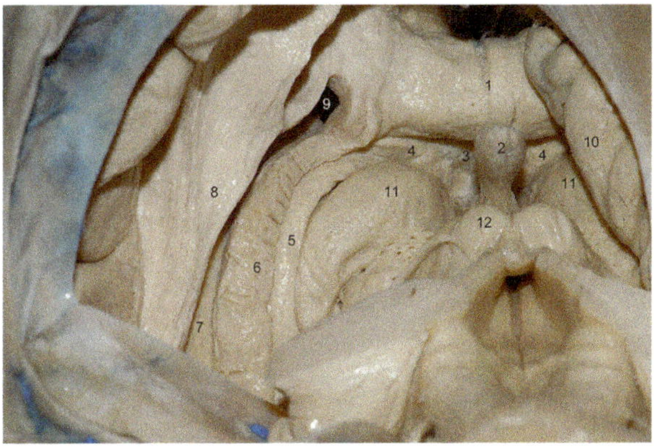

Fig. 18.6 Neural relationships of the lateral pineal region. The white matter in the depth of the parahippocampal gyrus and fusiform gyrus has been exposed, in addition to the temporal horn of the lateral ventricle. (1) Splenium of the corpus callosum; (2) pineal body; (3) hippocampal commissure; (4) crus of the fornix; (5) fimbria; (6) dentate gyrus; (7) temporal horn of the lateral ventricle; (8) optic radiation; (9) atrium of the lateral ventricle; (10) parahippocampal gyrus; (11) pulvinar of the thalamus; (12) inferior colliculus

Fig. 18.7 Basal surface of the brain. The two thalami have been removed to expose the roof of the lateral ventricles and the fornix along with its commissure. The roof of the posterior incisural space is formed by the lower surface of the splenium, the terminal part of the crura of the fornices, and the hippocampal commissure. (1) Mamillary body; (2) column of the fornix; (3) body of the fornix; (4) crus of the fornix; (5) hippocampal commissure; (6) fimbria; (7) Ammon's horn; (8) splenium of the corpus callosum; (9) posterior bundle of the optic radiation; (10) tapetum

The trunks and branches of the posterior cerebral artery and the superior cerebellar artery enter through the anterior part of the floor of this region to transit inferolaterally to the membrane that surrounds the vein of Galen and its tributaries. Note that the arrangement of the vascular structures within the cistern includes the superomedial veins and the inferolateral arteries.

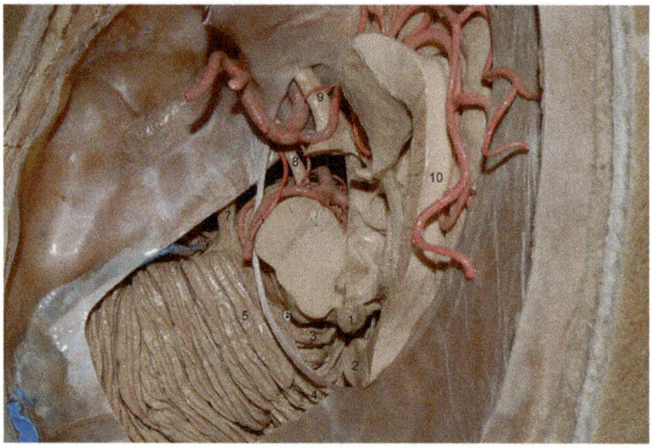

Fig. 18.8 Superior view of a head after removal of the left cerebral hemisphere and tentorium. The tentorial incisura is located between the tentorial edges and is the only site of communication between the supra- and infratentorial spaces. The tentorial apex is located at the junction of the vein of Galen and the straight sinus. The floor of the posterior incisural space is formed by the anterosuperior part of the cerebellum and consists of the culmen of the vermis in the midline and the quadrangular lobules of the hemispheres laterally. (1) Pineal body; (2) splenium of the corpus callosum; (3) central lobule; (4) culmen; (5) quadrangular lobule; (6) cerebellomesencephalic fissure; (7) trigeminal nerve; (8) oculomotor nerve; (9) optic nerve; (10) corpus callosum

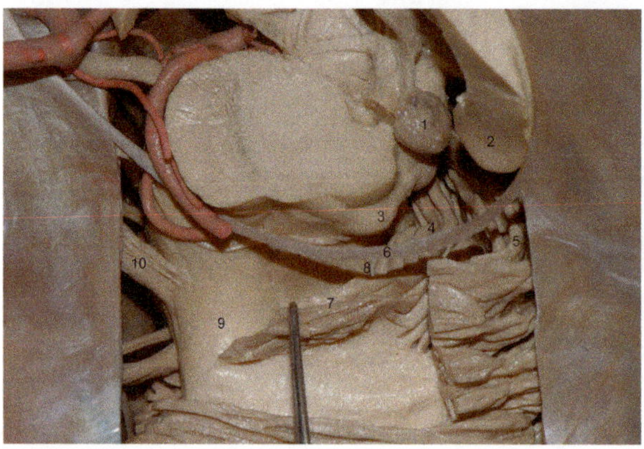

Fig. 18.9 Superolateral view of the same anatomical specimen. The quadrangular lobule of the cerebellum was removed to expose the floor of the posterior incisural space. (1) Pineal body; (2) splenium of the corpus callosum; (3) inferior colliculus; (4) central lobule; (5) culmen; (6) tentorial edge; (7) wing of the central lobule; (8) cerebellomesencephalic fissure; (9) middle cerebellar peduncle (brachium pontis); (10) trigeminal nerve

18.1.3 Cisternal Relationships

Quadrigeminal Cistern (Cisterna Venae Magnae Galeni) The quadrigeminal cistern, situated posterior to the quadrigeminal plate, is the major cistern in the posterior incisural space. The anterior limits of the cistern are the dorsal mesencephalon,

Fig. 18.10 Posteroinferior view of the brainstem and the pineal region. The cerebellum and tentorium on the left side have been removed. The neural relationships between the pineal gland, the thalamus, and the quadrigeminal plate are shown. The region of the quadrigeminal cistern is shaded in blue. (1) Splenium of corpus callosum; (2) pineal body; (3) parahippocampal gyrus; (4) pulvinar of the thalamus; (5) inferior colliculus; (6) superior colliculus; (7) medial geniculate body; (8) lateral geniculate body; (9) superior cerebellar peduncle (brachium conjunctivum); (10) crus cerebri; (11) Tentorium

the quadrigeminal plate, and the pineal gland. Posteriorly, arachnoid is attached to the tentorium and extends from the splenium of the corpus callosum inferiorly to the lingula of the cerebellar vermis, above the anterior medullary velum of the fourth ventricle (Fig. 18.10). The quadrigeminal cistern communicates above with the posterior pericallosal cistern, inferiorly into the cerebellomesencephalic fissure, inferolaterally into the posterior part of the ambient cistern located between the midbrain and the parahippocampal gyrus, and laterally into the retrothalamic areas medial to where the crus of the fornix wraps the posterior part of the pulvinar. The quadrigeminal cistern contains the medial posterior choroidal arteries, the great vein of Galen, the terminal portions of its tributaries, and the internal cerebral, basal, pericallosal, and occipital veins. The origins of the posterior pericallosal arteries and the continuation of the posterior cerebral arteries are also contained within this cistern.

18.1.4 Ventricular Relationships

The posterior portion of the third ventricle and the cerebral aqueduct are anterior, and the atrium and occipital horns of the lateral ventricles are lateral to the posterior incisural space. The aqueduct passes ventral to the anterior wall of the posterior incisural space. The atrium is separated from the posterior incisural space by the crus of the fornix as it passes posterior to the pulvinar and also by the cortical gyri located in the lateral wall of the posterior incisural space (Fig. 18.11).

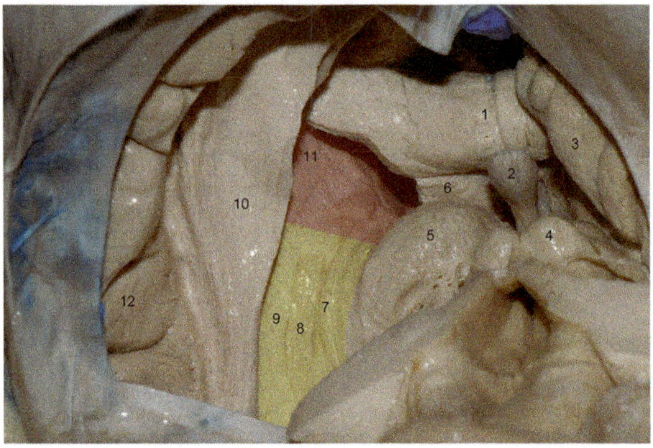

Fig. 18.11 Basal surface view of the left cerebral hemisphere. The relationship of the pineal region with the ventricular cavities has been exposed. In addition, the white matter found in the roof of the atrium and temporal horn of the lateral ventricle has been exposed. The atrium is separated from the posterior incisural space by the crus of the fornix as it passes posterior to the pulvinar and by the cortical gyri located in the lateral wall of the posterior incisural space. The atrium of lateral ventricle is shaded in red, and the temporal horn of the lateral ventricle is shaded in yellow. (1) Splenium of corpus callosum; (2) pineal body; (3) parahippocampal gyrus; (4) inferior colliculus; (5) pulvinar of the thalamus; (6) crus of the fornix; (7) stria terminalis; (8) tail of caudate nucleus; (9) central bundle of the optic radiation; (10) anterior bundle of the optic radiation; (11) tapetum; (12) fusiform gyrus

18.1.5 Arterial Relationships

The trunks and branches of the posterior cerebral artery (PCA) and superior cerebellar artery (SCA) enter the posterior incisural space from an anterior perspective.

The *posterior cerebral artery* can been described using several different classifications. We used the classification that divides the PCA into four segments: P1, which extends from the origin of the PCA to its junction with the posterior communicating artery; P2, which is divided into an anterior and posterior segment, with the P2A segment beginning at the posterior communicating artery and ending at the most lateral aspect of the cerebral peduncle and the P2P segment extending from the most lateral aspect of the cerebral peduncle to the posterior edge of the lateral surface of the midbrain; P3, which extends from the posterior edge of the lateral surface of the midbrain to the origin of the parieto-occipital sulcus along the calcarine fissure; and P4, which corresponds to the parts of the PCA that run along or inside both the parieto-occipital sulcus and the distal part of the calcarine fissure (Fig. 18.12).

The third segment, called P3, courses posteriorly from the posterior edge of the lateral surface of the midbrain and ends at the point of the origin of the parieto-occipital sulcus along the calcarine fissure, hence lying along or inside the proximal part of the calcarine fissure. Also called the quadrigeminal segment, the P3 segment

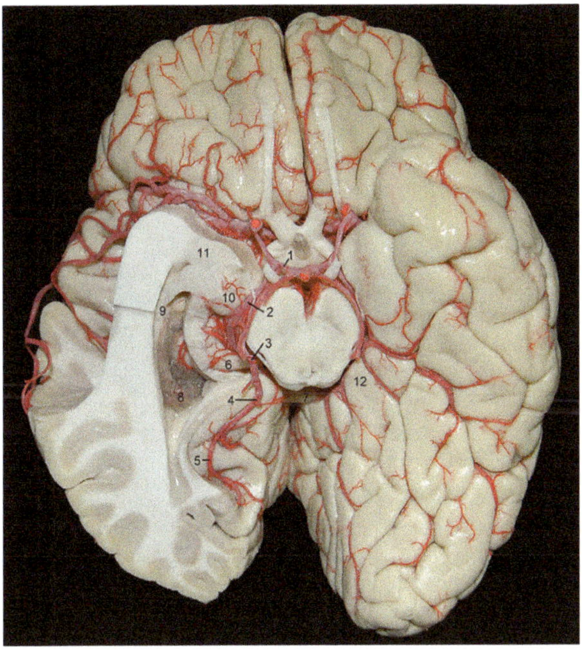

Fig. 18.12 Basal surface of the brain: the right temporal pole and opercula have been dissected to expose the Sylvian fissure. The right parahippocampal gyrus has been removed from the dentate gyrus and fimbria, and the temporal horn has been exposed. The hippocampal arteries arise either from the anterior choroidal artery or posterior cerebral artery and course through the hippocampal notch, hippocampal sulcus, or fimbrodentate sulcus to supply the hippocampal formation. Unlike the hippocampal arteries, the anterior choroidal and lateral posterior choroidal arteries course through the choroidal fissure (between the fornix and the thalamus) to supply the choroid plexus. (1) Posterior cerebral artery, P1 segment; (2) posterior cerebral artery, P2A segment; (3) posterior cerebral artery, P2P segment; (4) posterior cerebral artery, P3 segment; (5) posterior cerebral artery, P4 segment; (6) pulvinar of the thalamus; (7) splenium of corpus callosum; (8) atrium of lateral ventricle; (9) temporal horn of the lateral ventricle; (10) uncus; (11) amygdala; (12) parahippocampal gyrus

begins between the ambient and quadrigeminal cisterns at the most lateral and posterior aspect of the midbrain, and both P3 segments converge medially, forming the basis of the so-called angiographic collicular point. The PCA divides in to its major terminal branches (the calcarine and parieto-occipital arteries) before reaching the parieto-occipital sulcus (Fig. 18.13).

The *long circumflex arteries*, also called the quadrigeminal arteries, originate from the P1 and P2A segments of the PCA. They usually consist of one or two branches, and, after their origin, these arteries travel through the crural, ambient, and quadrigeminal cisterns to supply the quadrigeminal plate. During their course they give off branches to the cerebral peduncle, medial geniculate body, and pulvinar, ending in the quadrigeminal plate where they form a rich vascular network to supply the superior colliculi; here, they anastomose with branches of the superior cerebellar artery that supply the inferior colliculi.

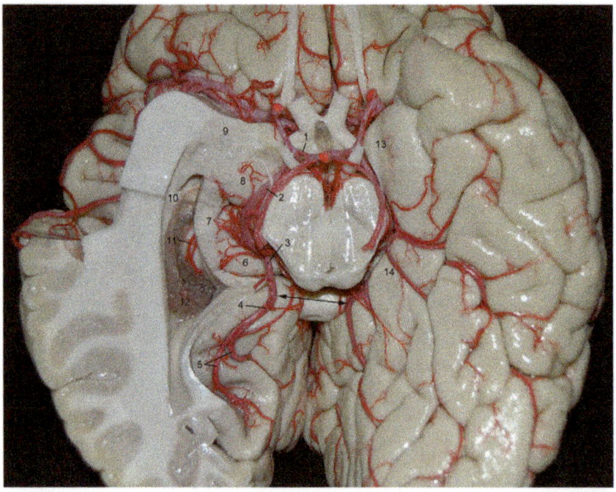

Fig. 18.13 View of the basal surface of the brain with an axial section of the right temporal lobe at the level of the calcarine fissure. The P3 segment of the posterior cerebral artery proceeds posteriorly from the posterior edge of the lateral surface of the midbrain and ends at the point of origin of the parieto-occipital sulcus along the calcarine fissure, hence lying along or inside the proximal part of the calcarine fissure. Also called the quadrigeminal segment, the P3 segment begins between the ambient and quadrigeminal cisterns at the most lateral and posterior aspect of the midbrain, and both P3 segments converge medially, characterizing the so-called angiographic collicular point (shown by arrows). (1) Posterior cerebral artery, P1 segment; (2) posterior cerebral artery, P2A segment; (3) posterior cerebral artery, P2P segment; (4) posterior cerebral artery, P3 segment; (5) posterior cerebral artery, P4 segment; (6) pulvinar of the thalamus; (7) dentate gyrus; (8) uncus; (9) amygdala; (10) temporal horn of the lateral ventricle; (11) posterolateral choroidal arteries; (12) atrium of the lateral ventricle; (13) uncus; (14) parahippocampal gyrus

The *posteromedial choroidal arteries* (PMChA) originate from the proximal half of the P2A segment in 70%, from the P1 segment in 14.3%, from the P2P and P3 segments in 5.7%, and from the P4 segment in 1.4% of humans. The number of PMChA branches varies from one to three. After their origin, the PMChA branches border the lateral wall of the midbrain medial to the PCA and enter the quadrigeminal cistern, changing trajectory to travel anteriorly and laterally to the pineal gland; they subsequently enter the roof of the third ventricle within the velum interpositum cistern and run along the choroid fissure, pass through the interventricular foramen, and end along the choroid plexus of the lateral ventricle (Figs. 18.14 and 18.15). The PMChA supply the cerebral peduncle, mesencephalic tegmentum, lateral and medial geniculate bodies, superior colliculi, pulvinar, pineal gland, and the medial and dorsal surfaces of the thalamus.

The *posterolateral choroidal arteries* (PLChAs) originate from the P2P segment of the PCA in 87.1%, from the P2A segment in 7.1%, from the P2A and P2P segments in 4.3%, and from the P3 segment in 1.4% of humans. They originate from

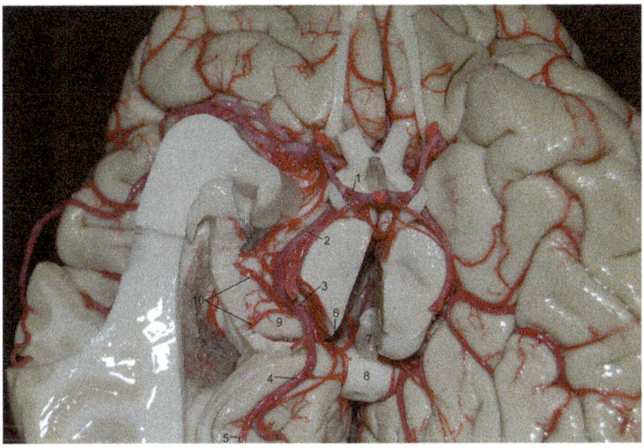

Fig. 18.14 View of the basal surface of the brain. A cut was made in the midbrain to expose the posteromedial choroidal artery. After their origin, they border the lateral wall of the midbrain medially to the posterior cerebral artery and enter the quadrigeminal cistern, changing their trajectory to travel anteriorly and laterally to the pineal gland; they subsequently enter the roof of the third ventricle within the velum interpositum cistern. (1) Posterior cerebral artery, P1 segment; (2) posterior cerebral artery, P2A segment; (3) posterior cerebral artery, P2P segment; (4) posterior cerebral artery, P3 segment; (5) posterior cerebral artery, P4 segment; (6) posteromedial choroidal arteries; (7) pineal body; (8) splenium of the corpus callosum; (9) pulvinar of the thalamus; (10) posterolateral choroidal arteries

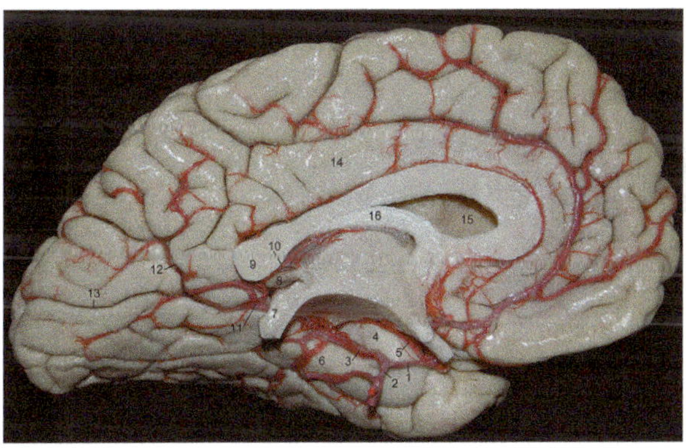

Fig. 18.15 View of the medial surface of the left hemisphere. The brainstem has been dissected, preserving the quadrigeminal plate. The arterial relationships of the pineal region have been exposed. (1) Posterior communicating artery; (2) posterior cerebral artery, P1 segment; (3) posterior cerebral artery, P2A segment; (4) uncus; (5) anterior choroidal artery; (6) parahippocampal gyrus; (7) quadrigeminal plate; (8) pineal body; (9) splenium of the corpus callosum; (10) posteromedial choroidal arteries; (11) posterior cerebral artery, P3 segment; (12) parieto-occipital sulcus; (13) calcarine fissure; (14) cingulate gyrus; (15) frontal horn of the lateral ventricle; (16) fornix

the lateral surface of the PCA and from the superior surface. There are, on average, two PLChAs (Fig. 18.14). The PLChAs course laterally and enter the temporal horn through the choroid fissure to supply the choroid plexus of the lateral ventricle and then continue posteriorly to enter the ventricular atrium where they anastomose in the glomus with branches of the anterior choroidal artery and PMChA (Fig. 18.16). The PLChAs supply the cerebral peduncle, posterior commissure, crura and body of the fornix, lateral geniculate body, pulvinar, superomedial thalamic nuclei, and the body of the caudate nucleus.

The *superior cerebellar artery* (SCA) originates from the basilar artery and encircles the upper pons/lower mesencephalon parallel to the course of the basal vein of Rosenthal, the PCA, and the free edge of the tentorium and courses toward the superior surface of the cerebellar hemisphere and the superior cerebellar vermis. The SCA is coursing within the cerebellomesencephalic fissure when it reaches the posterior incisural space. The SCA branches, upon exiting the cerebellomesencephalic fissure, are anterior to the free edge but pass below the free edge to supply the tentorial surface of the cerebellum.

The perforating branches of the PCA and SCA and the PMChAs supply the walls of the posterior incisural space. The PCAs supply the structures above the level of the lower margin of the superior colliculi, and the SCAs supply the structures below the upper margin of the inferior colliculus.

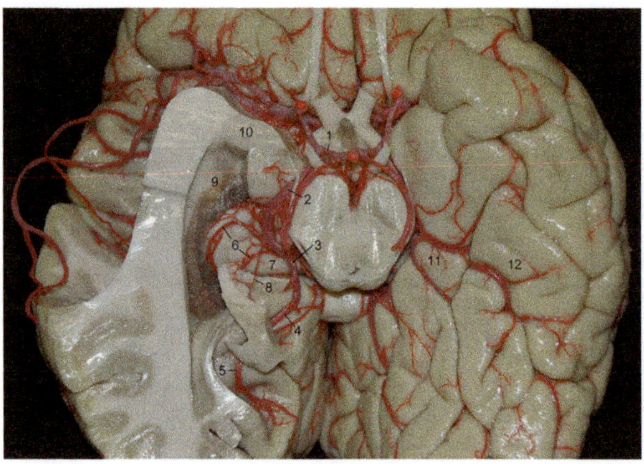

Fig. 18.16 Basal view. The right temporal pole and opercula have been dissected to expose the Sylvian fissure. The head of the hippocampus, fimbria, and dentate gyri has been dissected to expose the inferior choroidal point and the posterior cerebral artery as well as its ventricular branches. The amygdala is anterior and superior to the hippocampal head and is the anterior and superior limit of the temporal horn. The temporal lobe connects with the frontal lobe anterolaterally along the limen insulae and anteromedially with the globus pallidus through the superior aspect of the amygdala. (1) Posterior cerebral artery, P1 segment; (2) posterior cerebral artery, P2A segment; (3) posterior cerebral artery, P2P segment; (4) posterior cerebral artery, P3 segment; (5) posterior cerebral artery, P4 segment; (6) posterolateral choroidal arteries; (7) pulvinar of the thalamus; (8) hippocampal artery; (9) frontal horn of the lateral ventricle; (10) amygdala; (11) parahippocampal gyrus; (12) fusiform gyrus

18.1.6 Venous Relationships

The posterior incisural space contains the most complex venous relationships in the cranium, because the internal cerebral and basal veins and many of their tributaries converge on the vein of Galen within this area (Fig. 18.17).

The *internal cerebral veins* originate immediately posterior to the foramen of Monro and course posteriorly within the velum interpositum by the union of the septal, thalamostriate, and choroidal veins. Paired internal cerebral veins run posteriorly within the tela choroidea, adjacent to the midline, and unite with the subependymal veins, the basal vein of Rosenthal, and/or the internal occipital vein to form the vein of Galen. The tributaries of the internal cerebral vein from the lateral and third ventricles include the anterior septal, anterior caudate, posterior septal, posterior caudate, thalamostriate, thalamocaudate, anterior thalamic, anterior superficial thalamic, superior choroidal, superior thalamic, and superior superficial thalamic veins, as well as the veins draining the stria medullaris thalami.

The *vein of Galen* is formed by the union of the paired internal cerebral veins. Tributaries that join the vein of Galen comprise the internal cerebral vein, precentral cerebellar vein, internal occipital vein, basal vein of Rosenthal, posterior pericallosal vein, pineal vein, posterior mesencephalic vein, and the posterior ventricular vein. The vein of Galen then courses superoposteriorly under the splenium of the corpus callosum and joins the inferior sagittal sinus to form the straight sinus (Fig. 18.18).

The *basal vein of Rosenthal* is divided into three segments: the anterior (or striate) segment, the middle (or peduncular) segment, and the posterior (or posterior mesencephalic) segment. The middle segment is further subdivided into anterior

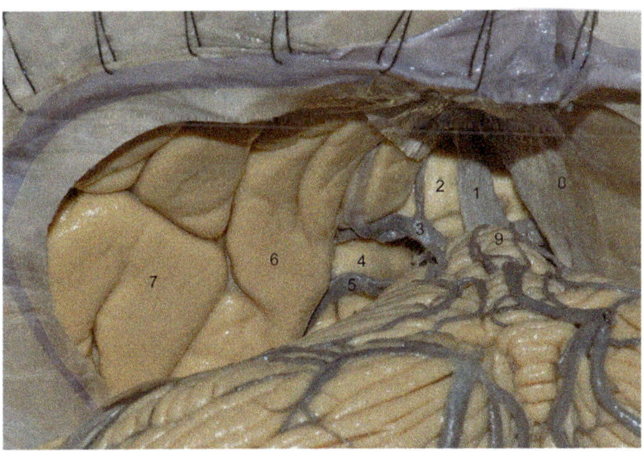

Fig. 18.17 Basal view of the left cerebral hemisphere through a supracerebellar transtentorial approach. The venous complex emptying into the vein of Galen blocks access to the pineal region. This venous complex includes the internal occipital, basal, and internal cerebral veins and the vein of the cerebellomesencephalic fissure. (1) Vein of Galen; (2) splenium of the corpus callosum; (3) internal occipital vein; (4) pulvinar of the thalamus; (5) basal vein of Rosenthal; (6) parahippocampal gyrus; (7) fusiform gyrus; (8) Tentorium; (9) Culmen of cerebellar Vermis

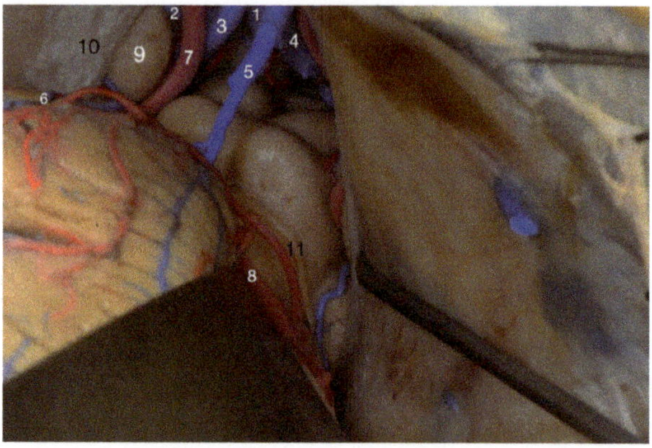

Fig. 18.18 The paramedian infratentorial supracerebellar approach. (1) Vein of Galen, (2) internal occipital vein; (3) basal vein of Rosenthal; (4) internal cerebral vein; (5) precentral cerebellar vein; (6) superior vermian vein; (7) posterior cerebral artery; (8) superior cerebellar artery; (9) parahippocampal gyrus; (10) tentorium; (11) fourth nerve

and posterior portions by the most lateral point of the vein as it turns around the peduncle. The main tributaries of the anterior segment of the basal vein of Rosenthal are the fronto-orbital, olfactory, inferior striate, anterior cerebral, deep middle cerebral, and anterior pericallosal veins. The anterior peduncular segment starts from the site where the peduncular vein joins the basal vein of Rosenthal, and it runs laterally between the upper part of the posteromedial surface of the uncus and the upper part of the crus cerebri, under the optic tract, to reach the most lateral part of the crus cerebri, which corresponds to the most lateral point of the vein as it turns around the peduncle; this is usually where the inferior ventricular vein joins the basal vein. After this point, the posterior peduncular segment turns medially, superiorly, and posteriorly to the plane of the lateral mesencephalic sulcus, behind the crus cerebri, to constitute the posterior mesencephalic segment. The main tributaries of the middle segment are the peduncular or interpeduncular vein and the inferior ventricular, inferior choroidal, hippocampal, and anterior hippocampal veins. The posterior segment starts at the lateral mesencephalic sulcus and then runs medially, superiorly, and posteriorly under the pulvinar of the thalamus to penetrate the quadrigeminal cistern and drain into the vein of Galen. The main tributaries of the posterior segment are the lateral mesencephalic, posterior thalamic, posterior longitudinal hippocampal, medial temporal, and medial occipital veins. The basal vein of Rosenthal originates on the surface of the anterior perforated substance by the union of multiple veins and passes through the crural and ambient cisterns. It courses posteromedially above the uncus to reach the anterior portion of the cerebral peduncle. At the most medial point of the basal vein of Rosenthal, anterior to the peduncle, it turns posterolaterally to reach the lateral most point of the cerebral peduncle and then turns posteromedially around the inferior and posterior aspects of the pulvinar to join the vein of Galen or the internal cerebral vein in the quadrigeminal cistern (Fig. 18.19). Many anatomical variations to the drainage pattern of the basal vein of Rosenthal have been reported, including drainage into the straight sinus,

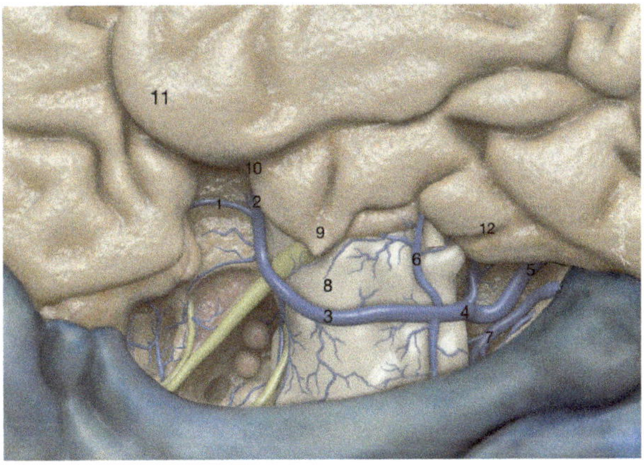

Fig. 18.19 Lateral view of the contents of the crural and ambient cisterns. The basal vein of Rosenthal is divided into three segments: the anterior (or striate) segment originates from the junction of the anterior cerebral, inferior striate, olfactory, fronto-orbital, and deep middle cerebral veins under the anterior perforated substance; it then runs posteriorly and medially, under the optic tract, to the anterior portion of the crus cerebri, depicting the location of the apex of the uncus. (1) Anterior cerebral vein; (2) deep middle cerebral vein; (3) anterior peduncular segment of the basal vein of Rosenthal; (4) posterior mesencephalic segment of the basal vein of Rosenthal; (5) vein of Galen; (6) lateral mesencephalic vein; (7) precentral cerebellar vein; (8) crus cerebri; (9) uncus; (10) rhinal sulcus; (11) temporal pole; (12) parahippocampal gyrus. (Illustration created by Angelo Shuman)

lateral sinus, superior petrosal sinus via the anastomotic lateral mesencephalic vein, and the sphenoparietal sinus.

The internal cerebral veins exit the velum interpositum, and the basal veins of Rosenthal exit the ambient cistern to reach the posterior incisural space, where they join to form the vein of Galen. The vein of Galen passes below the splenium to enter the straight sinus at the tentorial apex. The junction of the vein of Galen and the straight sinus varies in configuration from being almost flat if the tentorial apex is located below the splenium, to forming a sharp angle if the apex is located above the splenium, in which case the vein of Galen turns sharply upward to reach the straight sinus at the apex.

The *internal occipital vein* originates on the inferior and medial surface of the occipital lobe and then courses anteromedially to end in the vein of Galen. Rarely, it also joins the internal cerebral vein and the basal vein of Rosenthal.

The *posterior pericallosal vein* has also been referred to as the "splenial vein," the "posterior cerebral vein," the "posterior marginal vein," the "posterior vein of the corpus callosum," and the "dorsal callosal vein." It originates on the dorsal surface of the corpus callosum and traverses around the splenium parallel to the posterior pericallosal artery to enter the internal cerebral vein or the vein of Galen.

The *precentral cerebellar vein* originates in the precentral cerebellar fissure, usually as two brachial veins uniting into a single common trunk. This vessel courses upward to join the vein of Galen or the posterior portion of the internal cerebral vein and drains into the vein of Galen.

The *straight sinus* is formed by the union of the inferior sagittal sinus and the vein of Galen at the posterior end of the splenium and continues in a posteroinferior direction following the line of junction of the falx cerebri with the tentorium cerebelli to the torcular Herophili.

The large number of venous structures in the pineal region, as well as their variability in course and relationships, makes this region one of the most complex to deal with surgically.

18.2 Surgical Considerations

Lesions in the posterior incisural space include pineal tumors; meningiomas arising at the falcotentorial junction and from the tela choroidea of the velum interpositum and atrium; gliomas of the splenium, pulvinar, quadrigeminal plate, and cerebellum; aneurysms of the vein of Galen; and arteriovenous malformations involving the medial occipital lobe and upper cerebellum.

Pineal region lesions are most commonly approached in one of three ways: by the supracerebellar infratentorial approach of Krause and Stein, by the occipital transtentorial approach of Poppen, or by Sekhar's combined occipital, transtentorial, supracerebellar, and transsinus approach. Dandy's parietal transcallosal approach is rarely used, because of the adverse effects associated with division of the posterior corpus callosum.

Lesions in the posterior incisural space may be approached from above the tentorium along the medial surface of the occipital lobe using an occipital transtentorial approach, through the posterior part of the lateral ventricle using a posterior transventricular approach, through the corpus callosum using a posterior interhemispheric transcallosal approach, or from below the tentorium through the supracerebellar space using a supracerebellar infratentorial approach. The supracerebellar infratentorial and occipital transtentorial approaches, which are most commonly used for pineal region tumors, may be combined with incision of the tentorium lateral to the straight sinus and, less commonly, with division of the tentorium and transverse sinus. A tentorial branch of the PCA or SCA may enter the dura lateral to the straight sinus. Venous sinuses are more commonly encountered in the posterior than in the anterior parts of the tentorium. Part of the tentorium may be removed during the resection of tumors that arise from or invade it.

The *supracerebellar infratentorial approach* was first used by Krause and further developed by Stein. It can be used for lesions in the pineal region located below the vein of Galen and its major tributaries (Figs. 18.20, 18.21, and 18.22) (*another case example is also demonstrated in the* Video 18.1). The approach is best suited to tumors in the midline that grow into the lower half of the posterior incisural space, displacing the quadrigeminal plate and apex of the tentorial cerebellar surface. With the patient in the sitting position, the cerebellum falls away from the tentorium,

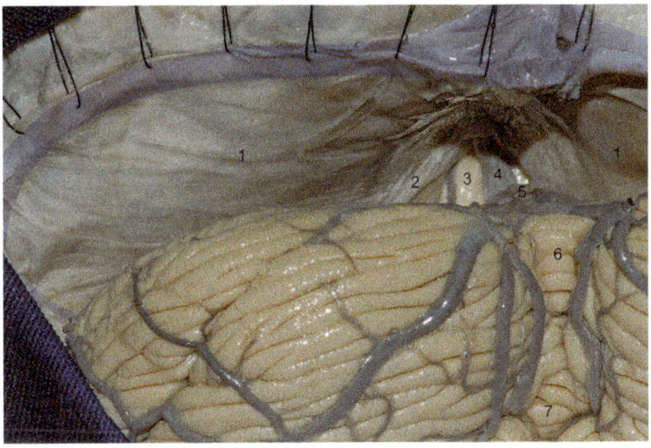

Fig. 18.20 Anatomical specimen demonstrating the supracerebellar infratentorial approach. The approach is best suited to tumors in the midline that grow into the lower half of the posterior incisural space, displacing the quadrigeminal plate and apex of the tentorial cerebellar surface. (1) Tentorium; (2) tentorial edge; (3) splenium of corpus callosum; (4) vein of Galen; (5) culmen; (6) tuber; (7) pyramid

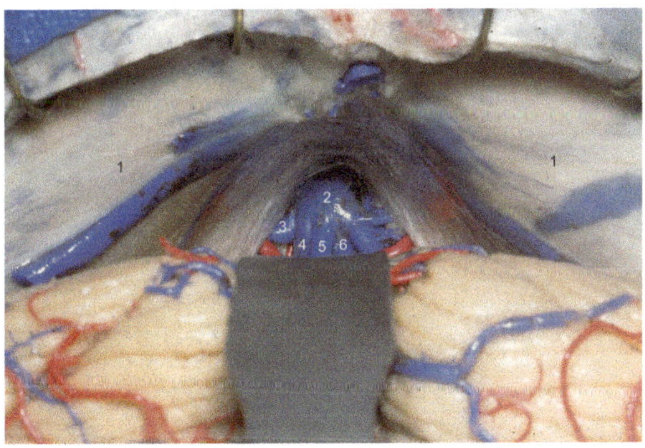

Fig. 18.21 The supracerebellar infratentorial approach. The venous complex of the pineal region is visible. (1) Tentorium; (2) vein of Galen; (3) internal occipital vein; (4) basal vein of Rosenthal; (5) internal cerebral vein; (6) precentral cerebellar vein

facilitated by gravity. Minimal retraction is needed to provide a natural corridor with an unobstructed view of the pineal region and without interference of the venous structures. Furthermore, the sitting position minimizes venous bleeding, which could otherwise obscure the operative field. The risk of air embolism can be minimized by using proper intraoperative monitoring. As an alternative, the supracerebellar infratentorial approach can be performed in the three-quarter prone or Concorde position, which may be desirable for patients younger than 3 years of age or patients who have excessively large ventricles that might be predisposed to ventricular collapse.

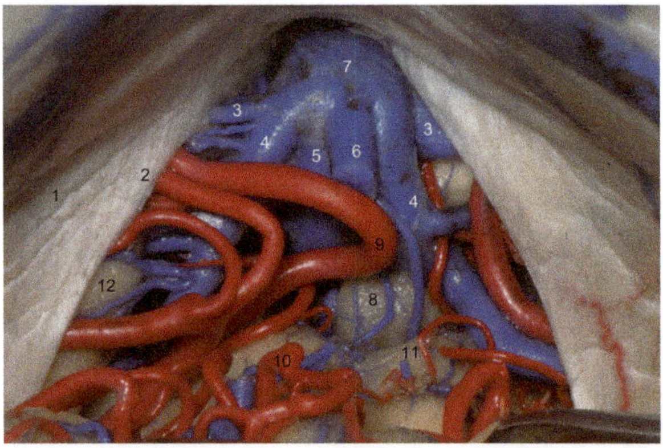

Fig. 18.22 The supracerebellar infratentorial approach. The venous complex emptying into the vein of Galen blocks access to the pineal region. This complex includes the internal occipital, basal, and internal cerebral veins and the vein of the cerebellomesencephalic fissure. A tentorial branch of the superior cerebellar artery traverses the exposure. (1) Tentorium; (2) tentorial edge; (3) internal occipital vein; (4) basal vein of Rosenthal; (5) internal cerebral vein; (6) precentral cerebellar vein; (7) vein of Galen; (8) pineal body; (9) posterior cerebral artery, P3 segment; (10) superior cerebellar artery; (11) superior colliculus; (12) parahippocampal gyrus

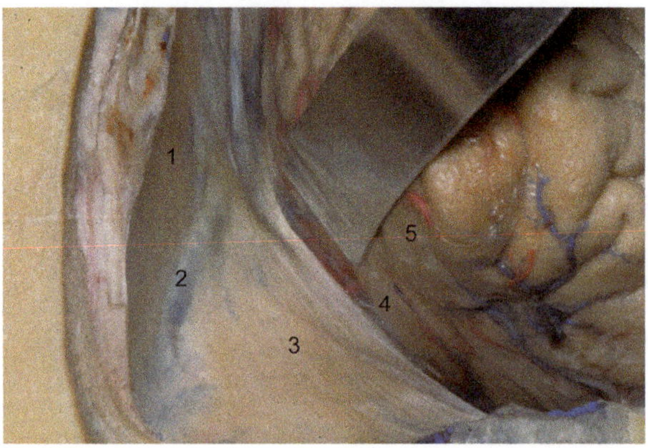

Fig. 18.23 The occipital transtentorial approach is directed along the medial surface of the occipital lobe inferior to the lambdoid suture. The occipital lobe inferior to the lambdoid suture is commonly free of bridging veins to the superior sagittal sinus, making it a less challenging route for the occipital transtentorial approach. There are no large bridging veins between the posterior 6 cm of the occipital lobe and the superior sagittal sinus. The first vein encountered is the internal occipital vein that passes from the anterior part of the medial occipital lobe to the vein of Galen. (1) Falx; (2) straight sinus; (3) tentorium; (4) lingula; (5) cuneus

The *occipital transtentorial approach* of Poppen is preferred for lesions centered at or above the tentorial edge, especially if they are located above the vein of Galen (Figs. 18.23 and 18.24). The latter approach may also provide a better angle of access for some lesions involving the ipsilateral half of the cerebellomesencephalic

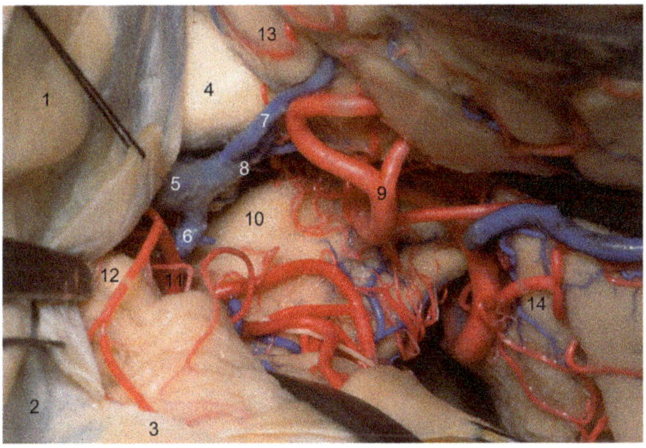

Fig. 18.24 The occipital transtentorial approach. The tentorium has been opened lateral to the straight sinus, and the vein of Galen has been displaced to the left side to expose the pineal gland as well as the superior and inferior colliculi. Elevating the branches of the vein of Galen provides a satisfactory view into the quadrigeminal cistern, with a better view into the cerebellomesencephalic fissure. (1) Falx; (2) straight sinus; (3) tentorium; (4) splenium of corpus callosum; (5) vein of Galen; (6) precentral cerebellar vein; (7) internal occipital vein; (8) basal vein of Rosenthal; (9) posterior cerebral artery, P3 segment; (10) superior colliculus; (11) superior cerebellar artery; (12) culmen; (13) isthmus of the cingulate gyrus; (14) parahippocampal gyrus

fissure and the posterior part of the ambient cistern, although they may be located below the level of the vein of Galen. Indications for direct surgery using the occipital transtentorial approach include pineal region tumors such as teratomas, germ cell tumors, and meningiomas, with or without obstructive hydrocephalus. Vascular lesions, such as varices of the vein of Galen, arteriovenous malformations in the pineal region, and aneurysms of the P3 and P4 segments of the posterior cerebral artery, are other indications for this approach.

The *combined supra- and infratentorial transsinus approach* was developed by Sekhar for the removal of large pineal region tumors and involves the section of the nondominant transverse sinus. The exposure of large pineal tumors with this approach is better than that achieved with the other two approaches, and it substantially minimizes the risk of brain retraction. The deep venous structures are well exposed, allowing the exposure and resection of lateral, superior, and inferior tumor extensions. A semiprone position can be used, which is comfortable for the surgeon. The disadvantages with this approach include the longer exposure time and the need to divide the sinus. It is indicated for large tumors with a diameter greater than 4.5 cm; tumors extending well above and below the planes of the tentorium or tumors arising from the tentorium; tumors well below the plane of cerebellar retraction (more than 2 cm below the superior surface of the cerebellum); tumors encasing the important venous structures of the region; and tumors that are vascular to the extent that, during tumor resection, the surgeon has to come around the tumor without being able to internally debulk the tumor beforehand.

The *posterior transcallosal approach* was first performed by Dandy. This approach requires splitting the corpus callosum and resecting 2–4 cm of the splenium. Occasionally, an occipital lobectomy is also required to facilitate surgical access. Cushing noted the difficulty of removing pineal region tumors and wrote that he himself had never succeeded in exploring a pineal tumor sufficiently to justify an attempt at removing it. This approach should only be used if the lesion appears to arise from the splenium above the vein of Galen and extends into the posterior incisural space.

The *posterior transventricular approach*, first described by Van Wagenen, was another surgical technique developed to facilitate the removal of pineal tumors. He used a 6–7 cm, reversed, L-shaped incision in the cortex, extending from the posterior end of the superior temporal gyrus posterosuperiorly to the superior parietal lobule. This approach provides adequate exposure of the atrium and posterior portion of the body of the lateral ventricle and would be the preferred approach to a tumor involving the posterior incisural space if the tumor extends into the pulvinar or involves the atrium or the glomus of the choroid plexus. The preferred approach to access the ventricle is through the superior parietal lobule, although a cortical incision in the superior temporal gyrus directed to the atrium has also been described.

18.3 Illustrative Case

A 20-year-old female reported symptoms of headache, dizziness, and difficulty looking upward. Parinaud syndrome was observed during the physical examination, and a magnetic resonance imaging study was performed, which demonstrated a lesion in the pineal region (Fig. 18.25).

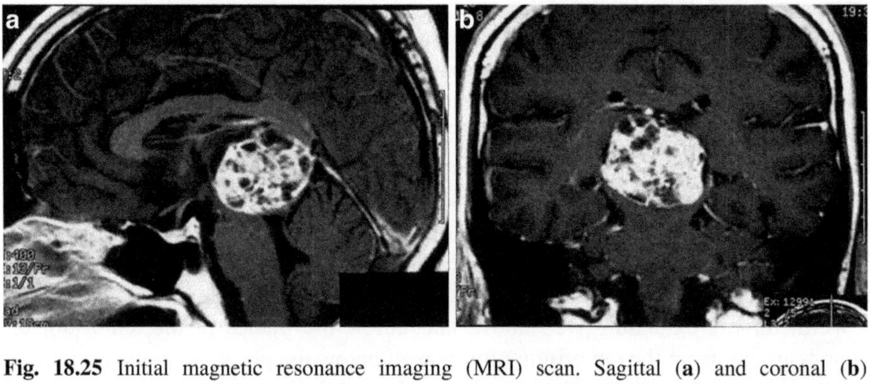

Fig. 18.25 Initial magnetic resonance imaging (MRI) scan. Sagittal (**a**) and coronal (**b**) T1-weighted MRI with contrast showing a heterogeneous lesion, with regular edges and irregular gadolinium enhancement, located in the posterior incisural space at the level of the pineal region

A standard management protocol for tumors of the pineal region was performed.

Based on the previous magnetic resonance images, the following questions arose: On which side is the lesion bigger at the posterior tentorial incisura? Which structures should be considered when choosing the surgical approach?

The tumor is large and situated in the midline, occupying the entire lamina quadrigeminal cistern and extending into the lower half of the posterior incisural space, displacing the quadrigeminal plate and apex of the tentorial cerebellar surface. One of the most important features to consider when deciding on the correct surgical approach is the relation between the vein of Galen, its major tributaries, and the tumor. If the majority of the tumor is located in the lower half of the tentorial notch and is displacing venous structures upward, the *supracerebellar infratentorial approach* in the semisitting position is the best surgical option; this is because it allows a natural corridor facilitated by gravitational forces and an unobstructed view of the pineal region without interference from the venous structures.

The semisitting position was chosen for this patient's surgery (Fig. 18.26). The patient underwent placement of a central venous catheter, transesophageal echocardiogram, and sequential compression stockings.

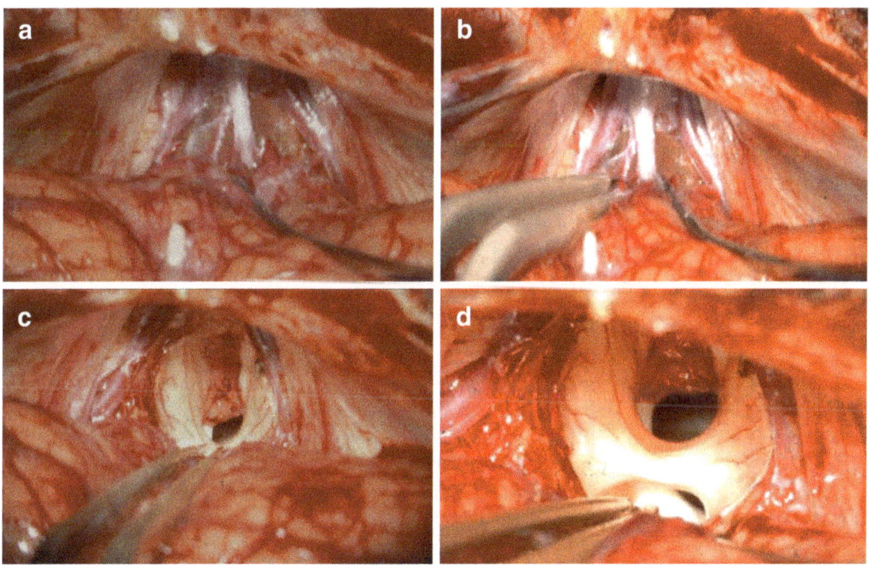

Fig. 18.26 Intraoperative pictures. Minimal retraction was needed to provide a natural corridor, and the sitting position minimized venous bleeding. (**a**, **b**) The tumor was located posterior to the arachnoid membrane, which was thickened, as is often seen with these pathologies. The precentral cerebellar vein is located in the midline and can often be sacrificed. The bilateral veins of Rosenthal are displaced laterally. (**c**, **d**) The resection cavity is observed in the depth of the surgical corridor, with the thalamus, choroid plexus of the third ventricle, bilateral veins of Rosenthal, and the tentorium all visible

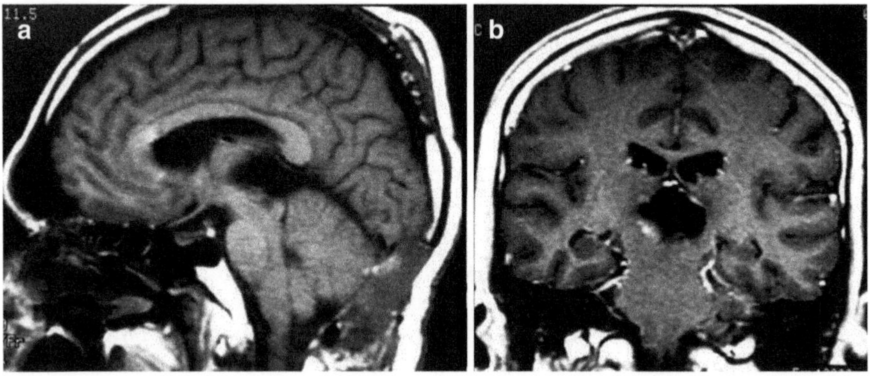

Fig. 18.27 Postoperative MRI. (**a**) Axial T1 image showing postoperative changes at occipital bone, pineal region with absence of tumor lesion. (**b**) Coronal T1 with contrast image showed the resection cavity and postoperative changes

The postoperative MRI showed a complete resection of the tumor (Fig. 18.27). The pathological diagnosis confirmed that the tumor was a teratoma, and the patient was referred to the oncology department for further management.

Suggested Bibliography

Párraga RG, Ribas GC, Andrade SE, de Oliveira E. Microsurgical anatomy of the posterior cerebral artery in three-dimensional images. World Neurosurg. 2011;75:233–57.

Poppen JL. The right occipital approach to a pinealoma. J Neurosurg. 1966;25:706–10.

Rhoton AL Jr. Cranial anatomy and surgical approaches. Schaumburg: Lippincott Williams & Wilkins; 2003.

Sekhar LN, Fessler R. Atlas of neurosurgical techniques. New York: Thieme Medical Publishers; 2016.

Sekhar LN, Goel A. Combined supratentorial and infratentorial approach to large pineal-region meningioma. Surg Neurol. 1992;37:197–201.

Stein BM. The infratentorial supracerebellar approach to pineal lesions. J Neurosurg. 1971;35:197–202.

Yamamoto I, Rhoton AL Jr, Peace DA. Microsurgery of the third ventricle: part 1—microsurgical anatomy. Neurosurgery. 1981;8:334–56.

Yasargil MG, Antic J, Laciga R, Jain KK, Boone SC. Arteriovenous malformations of vein of Galen: microsurgical treatment. Surg Neurol. 1976;3:195–200.

Index

F. Chaddad-Neto, M. D. Silva da Costa, *Microneuroanatomy and Surgery*,
https://doi.org/10.1007/978-3-030-82747-2

D

E

F